Communication Technology Update and Fundamentals

14th Edition

Communication Technology Update and Fundamentals

14th Edition

Editors

August E. Grant
Jennifer H. Meadows

In association with Technology Futures, Inc.

NEW YORK AND LONDON

Editors:

August E. Grant
Jennifer H. Meadows

Technology Futures, Inc.:

Production & Art Editor: Helen Mary V. Marek
Production Assistants: Vickie Flaugher and Carrie Vanston

First published 2014
by Focal Press
70 Blanchard Road, Suite 402, Burlington, MA 01803

Simultaneously published in the UK
by Focal Press
2 Park Square, Milton Park, Abingdon, Oxon OX14 4RN

Focal Press is an imprint of the Taylor & Francis Group, an informa business

ISSN 2373-258X (print)
ISSN 2373-261X (online)

ISBN: 978-0-415-73294-9 (hbk)
ISBN: 978-0-415-73295-6 (pbk)
ISBN: 978-1-315-84878-5 (ebk)

Companion website for Recent Developments: http://www.tfi.com/ctu

Printed and bound in the United States of America by Sheridan Books, Inc. (a Sheridan Group Company).

Table of Contents

Glossary and Updates can be found on the

Communication Technology Update and Fundamentals website

http://www.tfi.com/ctu/

Preface

You are reading an oxymoron. This book purports to present you with the latest information on a wide range of communication technologies as of 2014, but many of you are probably reading these words from an old-fashioned paper book. Reading about new technologies on an old technology may seem incongruous or even ironic at first, but it is also a natural part of understanding communication technologies.

As we explore a wide range of technologies in this book, one of the goals of the authors is to share a variety of perspectives that you can use to understand and use communication technologies. Even though the focus is on new media—because that's where the largest changes are taking place—there is still plenty of room for these new media to co-exist alongside traditional media.

So this book explores a wide range of new media technologies from the latest television and telephone technologies to communicating cars and networked health care. Across these technologies, the focus is on understanding the systems in which these technologies operate, along with the interrelationships among technologies. As you read, keep in mind that these system-level factors apply as much to traditional media as to new media, so understanding the emerging technologies discussed in this book will help you to understand other communication technologies as well.

In comparing this edition to previous editions, the most important changes are in the Table of Contents. For the first time, this edition includes chapters on e-health and automotive telematics, as well as brand new chapters on the Internet, digital signage, and multichannel television.

One thing shared by all of the contributors to this book is a passion for communication technology. In order to keep this book as current as possible we asked the authors to work under extremely tight deadlines. Authors begin working in January 2014, and most chapters were submitted in April 2014 with the final details added in May 2014. Individually, the chapters provide snapshots of the state of the field for individual technologies, but together they present a broad overview of the role that communication technologies play in our everyday lives. The efforts of these authors have produced a remarkable compilation, and we thank them for all for their hard work in preparing this volume.

One fortunate constant in production of this book is the involvement of TFI's Helen Mary Marek, who played a pivotal role in production, moving all 25 chapters from draft to camera-ready in weeks. Helen Mary also provided on-demand graphics production, adding visual elements to help make the content more understandable. Almost everyone on the TFI team contributed to the process, including Larry Vanston, Carrie Vanston, and Vickie Flaugher. Our editorial and marketing team at Focal Press, led by Katy Morrissey, ensured that production and promotion of the book were as smooth as ever.

For us, producing this book every two years is a labor of love that we can't perform without the support of our spouses. Diane Grant and Floyd Meadows have kept us both grounded and motivated, providing the support that enables us to immerse ourselves in producing this book every two years. But when it comes down to the most important reason we create a new version of this book every two years, it is the satisfaction we get from being part of the process of training students in how to understand and apply new media technologies.

Your search for insight into the technologies explored in this book can continue on the companion website for the *Communication Technology Update and Fundamentals:* www.tfi.com/ctu. The complete Glossary for the book is on the site, where it is much easier to find individual entries than in the paper version of the book. We have also moved the vast quantity of statistical data on each of the communication technologies that were formerly printed in Chapter 2 to the site. As always, we will periodically update the Web site to supplement the text with new information and links to a wide variety of information available over the Internet.

As a reader of this book, you are also part of the *Communication Technology Update* community. Each edition of this book has been improved over previous editions with the help and input from readers like you. You are invited to send us updates for the website, ideas for new topics, and other contributions that will inform all members of the community. You are invited to communicate directly with us via e-mail, snail mail, or voice.

Thank you for being part of the CTU community!

Augie Grant and Jennifer Meadows

May 7, 2014

Augie Grant
School of Journalism and Mass Communications
University of South Carolina
Columbia, SC 29208
Phone: 803.777.4464
augie@sc.edu

Jennifer H. Meadows
Department of Communication Design
California State University, Chico
Chico, CA 95929-0504
Phone: 530.898.4775
jmeadows@csuchico.edu

1

The Communication Technology Ecosystem

August E. Grant, Ph.D.*

Communication technologies are the nervous system of contemporary society, transmitting and distributing sensory and control information and interconnecting a myriad of interdependent units. These technologies are critical to commerce, essential to entertainment, and intertwined in our interpersonal relationships. Because these technologies are so vitally important, any change in communication technologies has the potential to impact virtually every area of society.

One of the hallmarks of the industrial revolution was the introduction of new communication technologies as mechanisms of control that played an important role in almost every area of the production and distribution of manufactured goods (Beniger, 1986). These communication technologies have evolved throughout the past two centuries at an increasingly rapid rate. This evolution shows no signs of slowing, so an understanding of this evolution is vital for any individual wishing to attain or retain a position in business, government, or education.

The economic and political challenges faced by the United States and other countries since the beginning of the new millennium clearly illustrate the central role these communication systems play in our society. Just as the prosperity of the 1990s was credited to advances in technology, the economic challenges that followed were linked as well to a major downturn in the technology sector. Today, communication technology is seen by many as a tool for making more efficient use of a wide range of resources including time and energy.

Communication technologies play as big a part in our private lives as they do in commerce and control in society. Geographic distances are no longer barriers to relationships thanks to the bridging power of communication technologies. We can also be entertained and informed in ways that were unimaginable a century ago thanks to these technologies—and they continue to evolve and change before our eyes.

This text provides a snapshot of the state of technologies in our society. The individual chapter authors have compiled facts and figures from hundreds of sources to provide the latest information on more than two dozen communication technologies. Each discussion explains the roots and evolution, recent developments, and current status of the technology as of mid-2014. In discussing each technology, we address them from a systematic perspective, looking at a range of factors beyond hardware.

The goal is to help you analyze technologies and be better able to predict which ones will succeed and which ones will fail. That task is harder to achieve than it sounds. Let's look at Google for an example of how unpredictable technology can be.

The Google Tale

As this book goes to press in mid-2014, Google is the most valuable media company in the world in terms of market capitalization (the total value of all

* J. Rion McKissick Professor of Journalism, School of Journalism and Communications, University of South Carolina (Columbia, South Carolina).

shares of stock held in the company). To understand how Google attained that lofty position, we have to go back to the late 1990s, when commercial applications of the Internet were taking off. There was no question in the minds of engineers and futurists that the Internet was going to revolutionize the delivery of information, entertainment, and commerce. The big question was how it was going to happen.

Those who saw the Internet as a medium for information distribution knew that advertiser support would be critical to its long-term financial success. They knew that they could always find a small group willing to pay for content, but the majority of people preferred free content. To become a mass medium similar to television, newspapers, and magazines, an Internet advertising industry was needed.

At that time, most Internet advertising was banner ads, horizontal display ads that stretched across most of the screen to attract attention, but took up very little space on the screen. The problem was that most people at that time accessed the Internet using slow, dial-up connections, so advertisers were limited in what they could include in these banners to about a dozen words of text and simple graphics. The dream among advertisers was to be able to use rich media, including full-motion video, audio, animation, and every other trick that makes television advertising so successful.

When broadband Internet access started to spread, advertisers were quick to add rich media to their banners, as well as create other types of ads using graphics, video, and sound. These ads were a little more effective, but many Internet users did not like the intrusive nature of rich media messages.

At about the same time, two Stanford students, Sergey Brin and Larry Page, had developed a new type of search engine, Google, that ranked results on the basis of how often content was referred to or linked from other sites, allowing their computer algorithms to create more robust and relevant search results (in most cases) than having a staff of people indexing Web content. What they needed was a way to pay for the costs of the servers and other technology.

According to Vise & Malseed (2006), their budget did not allow Google to create and distribute rich media ads. They could do text ads, but they decided to do them differently from other Internet advertising, using computer algorithms to place these small text ads on the search results that were most likely to give the advertisers results. With a credit card, anyone could use this "AdWords" service, specifying the search terms they thought should display their ads, writing the brief ads (less than 100 characters total—just over a dozen words), and even specifying how much they were willing to pay every time someone clicked on their ad. Even more revolutionary, the Google team decided that no one should have to pay for an ad unless a user clicked on it.

For advertisers, it was as close to a no-lose proposition as they could find. Advertisers did not have to pay unless a person was interested enough to click on the ad. They could set a budget that Google computers could follow, and Google provided a control panel for advertisers that gave a set of measures that was a dream for anyone trying to make a campaign more effective. These measures indicated not only the overall effectiveness of the ad, but also the effectiveness of each message, each keyword, and every part of every campaign.

The result was remarkable. Google's share of the search market was not that much greater than the companies that had held the number one position earlier, but Google was making money—lots of money—from these little text ads. Wall Street investors noticed, and, once Google went public, investors bid up the stock price, spurred by increases in revenues and a very large profit margin. Today, Google is involved in a number of other ventures designed to aggregate and deliver content ranging from text to full-motion video, but its little text ads are still the primary revenue generator.

In retrospect, it was easy to see why Google was such a success. Their little text ads were effective because of context—they always appeared where they would be the most effective. They were not intrusive, so people did not mind the ads on Google pages, and later on other pages that Google served ads to through its "content network." Plus advertisers had a degree of control, feedback, and accountability that no advertising medium had ever offered before (Grant & Wilkinson, 2007).

So what lessons should we learn from the Google story? Advertisers have their own set of lessons, but there are a separate set of lessons for those wishing to understand new media. First, no matter how insightful, no one is ever able to predict whether a technology will succeed or fail. Second, success can be due as much to luck as to careful, deliberate planning and investment. Third, simplicity matters—there are few advertising messages as simple as the little text ads you see when doing a Google search.

The Google tale provides an example of the utility of studying individual companies and industries, so the focus throughout this book is on individual technologies. These individual snapshots, however, comprise a larger mosaic representing the communication networks that bind individuals together and enable them to function as a society. No single technology can be understood without understanding the competing and complementary technologies and the larger social environment within which these technologies exist. As discussed in the following section, all of these factors (and others) have been considered in preparing each chapter through application of the "technology ecosystem." Following this discussion, an overview of the remainder of the book is presented.

The Communication Technology Ecosystem

The most obvious aspect of communication technology is the hardware—the physical equipment related to the technology. The hardware is the most tangible part of a technology system, and new technologies typically spring from developments in hardware. However, understanding communication technology requires more than just studying the hardware. One of the characteristics of today's digital technologies is that most are based upon computer technology, requiring instructions and algorithms more commonly known as "software."

In addition to understanding the hardware and software of the technology, it is just as important to understand the content communicated through the technology system. Some consider the content as another type of software. Regardless of the terminology used, it is critical to understand that digital technologies require a set of instructions (the software) as well as the equipment and content.

The hardware, software, and content must also be studied within a larger context. Rogers' (1986) definition of "communication technology" includes some of these contextual factors, defining it as "the hardware equipment, organizational structures, and social values by which individuals collect, process, and exchange information with other individuals" (p. 2). An even broader range of factors is suggested by Ball-Rokeach (1985) in her media system dependency theory, which suggests that communication media can be understood by analyzing dependency relations within and across levels of analysis, including the individual, organizational, and system levels. Within the system level, Ball-Rokeach identifies three systems for analysis: the media system, the political system, and the economic system.

These two approaches have been synthesized into the "Technology Ecosystem" illustrated in Figure 1.1. The core of the technology ecosystem consists of the hardware, software, and content (as previously defined). Surrounding this core is the organizational infrastructure: the group of organizations involved in the production and distribution of the technology. The next level moving outwards is the system level, including the political, economic, and media systems, as well as other groups of individuals or organizations serving a common set of functions in society. Finally, the individual users of the technology cut across all of the other areas, providing a focus for understanding each one. The basic premise of the technology ecosystem is that all areas of the ecosystem interact and must be examined in order to understand a technology.

(The technology ecosystem is an elaboration of the "umbrella perspective" (Grant, 2010) that was explicated in earlier editions of this text to illustrate the elements that need to be studied in order to understand communication technologies.)

Adding another layer of complexity to each of the areas of the technology ecosystem is also helpful. In order to identify the impact that each individual characteristic of a technology has, the factors within each area of the ecosystem may be identified as "enabling," "limiting," "motivating," and "inhibiting," depending upon the role they play in the technology's diffusion.

Enabling factors are those that make an application possible. For example, the fact that the coaxial cable used to deliver traditional cable television can carry dozens of channels is an enabling factor at the hardware level. Similarly, the decision of policy makers to allocate a portion of the spectrum for cellular telephony is an enabling factor at the system level (political system). One starting point to use in examining any technology is to make a list of the underlying factors from each area of the technology ecosystem that make the technology possible in the first place.

Figure 1.1
The Communication Technology Ecosystem

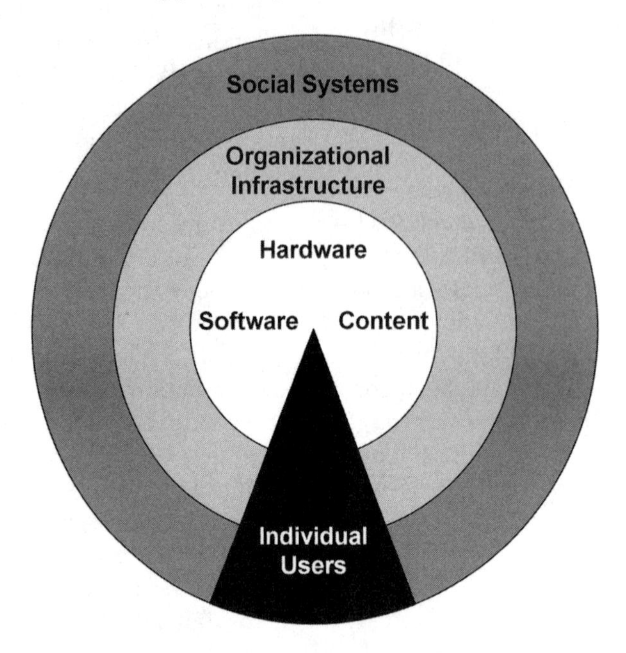

Source: A. E. Grant

Limiting factors are the opposite of enabling factors; they are those factors that create barriers to the adoption or impacts of a technology. A great example is related to the cable television illustration in the previous paragraph. Although coaxial cable increased the number of television programs that could be delivered to a home, most analog coaxial networks cannot transmit more than 100 channels of programming. To the viewer, 100 channels may seem to be more than is needed, but to the programmer of a new cable television channel unable to get space on a filled-up cable system, this hardware factor represents a definite limitation. Similarly, the fact that the policy makers discussed above initially permitted only two companies to offer cellular telephone service in each market was a system-level limitation on that technology. Again, it is useful to apply the technology ecosystem to create a list of factors that limit the adoption, use, or impacts of any specific communication technology.

Motivating factors are a little more complicated. They are those factors that provide a reason for the adoption of a technology. Technologies are not adopted just because they exist. Rather, individuals, organizations, and social systems must have a reason to take advantage of a technology. The desire of local telephone companies for increased profits, combined with the fact that growth in providing local telephone service is limited, is an organizational factor motivating the telcos to enter the markets for new communication technologies. Individual users desiring information more quickly can be motivated to adopt electronic information technologies. If a technology does not have sufficient motivating factors for its use, it cannot be a success.

Inhibiting factors are the opposite of motivating ones, providing a disincentive for adoption or use of a communication technology. An example of an inhibiting factor at the content level might be a new electronic information technology that has the capability to update information more quickly than existing technologies, but provides only "old" content that consumers have already received from other sources. One of the most important inhibiting factors for most new technologies is the cost to individual users. Each potential user must decide whether the cost is worth the service, considering his or her budget and the number of competing technologies. Competition from other technologies is one of the biggest barriers any new (or existing) technology faces. Any factor that works against the success of a technology can be considered an inhibiting factor. As you might guess, there are usually more inhibiting factors for most technologies than motivating ones.

And if the motivating factors are more numerous and stronger than the inhibiting factors, it is an easy bet that a technology will be a success.

All four factors—enabling, limiting, motivating, and inhibiting—can be identified at the system, organizational, content, and individual user levels. However, hardware and software can only be enabling or limiting; by themselves, hardware and software do not provide any motivating factors. The motivating factors must always come from the messages transmitted or one of the other areas of the ecosystem.

The final dimension of the technology ecosystem relates to the environment within which communication technologies are introduced and operate. These factors can be termed "external" factors, while ones relating to the technology itself are "internal" factors. In order to understand a communication technology or be able to predict the manner in which a technology will diffuse, both internal and external factors must be studied and compared.

Each communication technology discussed in this book has been analyzed using the technology ecosystem to ensure that all relevant factors have been included in the discussions. As you will see, in most cases, organizational and system-level factors (especially political factors) are more important in the development and adoption of communication technologies than the hardware itself. For example, political forces have, to date, prevented the establishment of a single world standard for high-definition television (HDTV) production and transmission. As individual standards are selected in countries and regions, the standard selected is as likely to be the product of political and economic factors as of technical attributes of the system.

Organizational factors can have similar powerful effects. For example, as discussed in Chapter 4, the entry of a single company, IBM, into the personal computer business in the early 1980s resulted in fundamental changes in the entire industry, dictating standards and anointing an operating system (MS-DOS) as a market leader. Finally, the individuals who adopt (or choose not to adopt) a technology, along with their motivations and the manner in which they use the technology, have profound impacts on the development and success of a technology following its initial introduction.

Perhaps the best indication of the relative importance of organizational and system-level factors is the number of changes individual authors made to the chapters in this book between the time of the initial chapter submission in March 2014 and production of the final, camera-ready text in May 2014. Very little new information was added regarding hardware, but numerous changes were made due to developments at the organizational and system levels.

To facilitate your understanding of all of the elements related to the technologies explored, each chapter in this book has been written from the perspective of the technology ecosystem. The individual writers have endeavored to update developments in each area to the extent possible in the brief summaries provided. Obviously, not every technology experienced developments in each area of the ecosystem, so each report is limited to areas in which relatively recent developments have taken place.

Why Study New Technologies?

One constant in the study of media is that new technologies seem to get more attention than traditional, established technologies. There are many reasons for the attention. New technologies are more dynamic and evolve more quickly, with greater potential to cause change in other parts of the media system. Perhaps the reason for our attention is the natural attraction that humans have to motion, a characteristic inherited from our most distant ancestors.

There are a number of other reasons for studying new technologies. Perhaps you want to make a lot of money off a new technology—and there is a lot of money to be made (and lost!) on new technologies. If you are planning a career in the media, you may simply be interested in knowing how the media are changing and evolving, and how those changes will affect your career.

Or you might want to learn lessons from the failure of new communication technologies so you can avoid failure in your own career, investments, etc. Simply put, the majority of new technologies introduced do not succeed in the market. Some fail because the technology itself was not attractive to consumers (such as the 1980s' attempt to provide AM stereo radio). Some fail because they were far ahead of the market, such as Qube, the first interactive cable television system, introduced in the 1970s.

Others failed because of bad timing or aggressive marketing from competitors that succeeded despite inferior technology.

The final reason we offer for studying new communication technologies is to identify patterns of adoption, effects, economic opportunity, and competition so that we can be prepared to understand, use, and/or compete with the next generation of new media. Virtually every new technology discussed in this book is going to be one of those "traditional, established technologies" in just a few short years, but there will always be another generation of new media to challenge the status quo.

Overview of Book

The key to getting the most out of this book is therefore to pay as much attention as possible to the reasons that some technologies succeed and others fail. To that end, this book provides you with a number of tools you can apply to virtually any new technology that comes along. These tools are explored in the first five chapters, which we refer to as the *Communication Technology Fundamentals*. You might be tempted to skip over these to get to the latest developments about the individual technologies that are making an impact today, but you will be much better equipped to learn lessons from these technologies if you are armed with these tools.

The first of these is the "technology ecosystem" discussed previously that broadens attention from the technology itself to the users, organizations, and system surrounding that technology. To that end, each of the technologies explored in this book provides details about all of the elements of the ecosystem.

Of course, studying the history of each technology can help you find patterns and apply them to different technologies, times, and places. In addition to including a brief history of each technology, the next chapter, A History of Communication Technologies, provides a broad overview of most of the technologies discussed later in the book, allowing a comparison along a number of dimensions: the year each was first introduced, growth rate, number of current users, etc. This chapter anchors the book to highlight commonalties in the evolution of individual technologies, as well as presents the "big picture" before we delve into the details. By focusing on the number of users over time, this chapter also provides the most useful basis of comparison across technologies.

Another useful tool in identifying patterns across technologies is the application of theories related to new communication technologies. By definition, theories are general statements that identify the underlying mechanisms for adoption and effects of these new technologies. Chapter 3 provides an overview of a wide range of these theories and provides a set of analytic perspectives that you can apply to both the technologies in this book and any new technologies that follow.

The structure of communication industries is then addressed in Chapter 4. The complexity of organizational relationships, along with the need to differentiate between the companies that make the technologies and those that sell the technologies, are explored in this chapter. The most important force at the system level of the ecosystem, regulation, is then introduced in Chapter 5.

These introductory chapters provide a structure and a set of analytic tools that define the study of communication technologies in all forms. Following this introduction, the book then addresses the individual technologies.

The technologies discussed in this book are organized into three sections: electronic mass media, computers and consumer electronics, and networking technologies. These three are not necessarily exclusive; for example, Internet video technologies could be classified as either an electronic mass medium or a computer technology. The ultimate decision regarding where to put each technology was made by determining which set of current technologies most closely resembled the technology from the user's perspective. Thus, Internet video was classified with electronic mass media. This process also locates the discussion of a cable television technology—cable modems—in the Broadband and Home Networks chapter in the Networking Technology section.

Each chapter is followed by a brief bibliography. These reference lists represent a broad overview of literally hundreds of books and articles that provide details about these technologies. It is hoped that the reader will not only use these references, but will examine the list of source material to determine the best places to find newer information since the publication of this *Update*.

To help you understand the importance of the people involved in the creation and distribution of these technologies, each of the technology "update" chapters includes a sidebar introducing a person who has made or is expected to continue to make a difference in the creation or evolution of the technology. It is our hope that the stories of these pioneers will help you to identify other individuals who will make a difference in the future, as well as inspiring you to have your own impact on these technologies.

Most of the technologies discussed in this book are continually evolving. As this book was completed, many technological developments were announced but not released, corporate mergers were under discussion, and regulations had been proposed but not passed. Our goal is for the chapters in this book to establish a basic understanding of the structure, functions, and background for each technology, and for the supplementary Internet home page to provide brief synopses of the latest developments for each technology. (The address for the home page is http://www.tfi.com/ctu.)

The final chapter returns to the "big picture" presented in this book, attempting to place these discussions in a larger context, exploring the process of starting a company to exploit or profit from these technologies. Any text such as this one can never be fully comprehensive, but ideally this text will provide you with a broad overview of the current developments in communication technology.

Bibliography

Ball-Rokeach, S. J. (1985). The origins of media system dependency: A sociological perspective. *Communication Research, 12* (4), 485-510.

Beniger, J. (1986). *The control revolution.* Cambridge, MA: Harvard University Press.

Grant, A. E. (2010). Introduction to communication technologies. In A. E. Grant & J. H. Meadows (Eds.) *Communication Technology Update and Fundamentals (12th ed).* Boston: Focal Press.

Grant, A. E. & Wilkinson, J. S. (2007, February). Lessons for communication technologies from Web advertising. Paper presented to the Mid-Winter Conference of the Association of Educators in Journalism and Mass Communication, Reno.

Rogers, E. M. (1986). *Communication technology: The new media in society.* New York: Free Press.

Vise, D. & Malseed, M. (2006). *The Google story: Inside the hottest business, media, and technology success of our time.* New York: Delta.

2

A History of Communication Technology

Dan Brown, Ph.D.*

The history of communication technologies can be examined from many perspectives. Each chapter in this book provides a brief history and a current update of the technology discussed in that chapter, but providing a *big picture* overview is important in studying the growth of respective technologies. This chapter focuses on early developments and brief updates in various communications media. The discussion is organized into categories that cover eras of print media, electronic media, and digital media.

The most useful perspective permits comparisons among technologies across time: numerical statistics of adoption and use of these technologies. To that end, this chapter follows patterns adopted in previous summaries of trends in U.S. communications media (Brown & Bryant, 1989; Brown, 1996, 1998, 2000, 2002, 2004, 2006, 2008, 2010, 2012). Nonmonetary units are favored as more meaningful tools for assessing diffusion of media innovations, although dollar expenditures appear as supplementary measures. A notable exception is the de facto standard of measuring motion picture acceptance in the market: box office receipts.

Government sources are preferred. Although they are frequently based on private reports, they provide some consistency. However, many government reports in recent years offered inconsistent units of measurement and discontinued annual market updates. Readers should use caution in interpreting data for individual years and instead emphasize the trends over several years. One limitation of this government data is the lag time before statistics are reported, with the most recent data being a year or more old. The companion website for this book (www.tfi.com/ctu) reports more detailed statistics than could be printed in this chapter.

New media technologies experienced increasingly rapid development as the 20th century closed. This rapid increase in development is the logical consequence of the relative degree of permeation of technology in recent years versus the lack of technological sophistication of earlier eras. This chapter excludes several media that the marketplace abandoned, such as quadraphonic sound, CB radios, 8-track audiotapes, and 8mm film cameras. Other media that receive mention have already and may yet suffer this fate. For example, audiocassettes and compact discs seem doomed in the face of rapid adoption of newer forms of digital audio recordings. This chapter traces trends that reveal clues about what has happened and what may happen in the use of respective media forms.

To illustrate the growth rates and specific statistics regarding each technology, a large set of tables and figures have been placed on the companion website for this book at www.tfi.com/ctu. Your understanding of each technology will be aided by referring to the website as you read each section.

* Associate Dean of Arts & Sciences, East Tennessee State University (Johnson City, Tennessee).

The Print Era

Printing began in China thousands of years before Johann Gutenberg developed the movable type printing press in 1455 in Germany. Gutenberg's press triggered a revolution that began an industry that remained stable for another 600 years (Rawlinson, 2011).

Printing in the United States grew from a one-issue newspaper in 1690 to become the largest print industry in the world (U.S. Department of Commerce/International Trade Association, 2000). This enterprise includes newspapers, periodicals, books, directories, greeting cards, and other print media.

Newspapers

Publick Occurrences, Both Foreign and Domestick was the first newspaper produced in North America, appearing in 1690 (Lee, 1917). Table 2.1 and Figure 2.1 from the companion website (www.tfi.com/ctu) for this book show that U.S. newspaper firms and newspaper circulation had extremely slow growth until the 1800s. Early growth suffered from relatively low literacy rates and the lack of discretionary cash among the bulk of the population. The progress of the industrial revolution brought money for workers and improved mechanized printing processes. Lower newspaper prices and the practice of deriving revenue from advertisers encouraged significant growth beginning in the 1830s. Newspapers made the transition from the realm of the educated and wealthy elite to a mass medium serving a wider range of people from this period through the Civil War era (Huntzicker, 1999).

The Mexican and Civil Wars stimulated public demand for news by the middle 1800s, and modern journalism practices, such as assigning reporters to cover specific stories and topics, began to emerge. Circulation wars among big city newspapers in the 1880s featured sensational writing about outrageous stories. Both the number of newspaper firms and newspaper circulation began to soar. Although the number of firms would level off in the 20th century, circulation continued to rise.

The number of morning newspapers more than doubled after 1950, despite a 16% drop in the number of daily newspapers over that period. Overall newspaper circulation remained higher at the start of the new millennium than in 1950, although it inched downward throughout the 1990s. Although circulation actually increased in many developing nations, both U.S. newspaper circulation and the number of U.S. newspaper firms are today lower than the respective figures posted in the early 1990s. Many newspapers that operated for decades are now defunct, and many others offer only online electronic versions.

Periodicals

"The first colonial magazines appeared in Philadelphia in 1741, about 50 years after the first newspapers" (Campbell, 2002, p. 310). Few Americans could read in that era, and periodicals were costly to produce and circulate. Magazines were often subsidized and distributed by special interest groups, such as churches (Huntzicker, 1999). The Saturday Evening Post, the longest running magazine in U.S. history, began in 1821 and became the first magazine to both target women as an audience and to be distributed to a national audience. By 1850, nearly 600 magazines were operating.

By early in the 20th century, national magazines became popular with advertisers who wanted to reach wide audiences. No other medium offered such opportunity. However, by the middle of the century, the many successful national magazines began dying in the face of advertiser preferences for the new medium of television and the increasing costs of periodical distribution. Magazines turned to smaller niche audiences that were more effectively targeted. Table 2.2, Figure 2.2, and Figure 2.3 on the companion website (www.tfi.com/ ctu) show the number of American periodical titles by year, revealing that the number of new periodical titles nearly doubled from 1958 to 1960.

Single copy magazine sales were mired in a long period of decline in 2009 when circulation fell by 17.2%. However, subscription circulation fell by only 5.9%. In 2010, the Audit Bureau of Circulation reported that, among the 522 magazine titles monitored by the Bureau, the number of magazine titles in the United States fell by 8.7% (Agnese, 2011).

In 2010, 20,707 consumer magazines were published in North America, reaching a paid circulation of $8.8 billion. Subscriptions accounted for $6.2 billion (71%) of that circulation. During that year, 193 new North American magazines began publishing, but 176 magazines closed. Many print magazines were also available in digital form, and many had eliminated print circulation in favor of digital publishing. In 2009, 81 North American magazines moved online, but the number of additional magazines that went online in 2010 dropped to 28 (Agnese, 2011).

Books

Stephen Daye printed the first book in colonial America, The Bay Psalm Book, in 1640 (Campbell, 2002). Books remained relatively expensive and rare until after the printing process benefited from the industrial revolution. Linotype machines developed in the 1880s allowed for mechanical typesetting. After World War II, the popularity of paperback books helped the industry expand. The current U.S. book publishing industry includes 87,000 publishers, most of which are small businesses. Many of these literally operate as "mom-and-pop desktop operations" (Peters & Donald, 2007, p. 11).

Table 2.3 and Figures 2.3 and 2.4 from the companion website (www.tfi.com/ctu) show new book titles published by year from the late 1800s through 2008. These data show a remarkable, but potentially deceptive, increase in the number of new book titles published annually, beginning in 1997. The U.S. Bureau of the Census reports furnished data based on material from R. R. Bowker, which changed its reporting methods beginning with the 1998 report. Ink and Grabois (2000) explained the increase as resulting from the change in the method of counting titles "that results in a more accurate portrayal of the current state of American book publishing" (p. 508). The older counting process included only books included by the Library of Congress Cataloging in Publication program. This program included publishing by the largest American publishing companies, but omitted such books as "inexpensive editions, annuals, and much of the output of small presses and self-publishers" (Ink & Grabois, 2000, p. 509). Ink and Grabois observed that the U.S. ISBN (International Standard Book Number) Agency assigned more than 10,000 new ISBN publisher prefixes annually.

The most startling development in book publishing in more than a century is clearly the success of electronic or e-books. Books have long been available for reading via computers, but dedicated e-book readers have transformed the reading experience by bringing many readers into the digital era. By the end of 2009, 3.7 million Americans were reading e-books. In 2010, the readership grew to more than 10.3 million, an increase of 178%, and surveys reported by the Book Industry Study Group (BISG) reported that 20% of respondents had stopped buying printed books in favor of e-books within a year. By July 2010, Amazon reported that sales of e-books surpassed that of print hardcover sales for the first time, with "143 e-books sold for every 100 print hardcover books" (Dillon, 2011, p. 5). From mid-December 2011 through January 2012, the proportion of Americans owning both e-book readers and tablet computers nearly doubled from 10% to 19%, with 29% owning at least one of the devices (Rainie, 2012).

Figure 2.1
Communication Technology Timeline

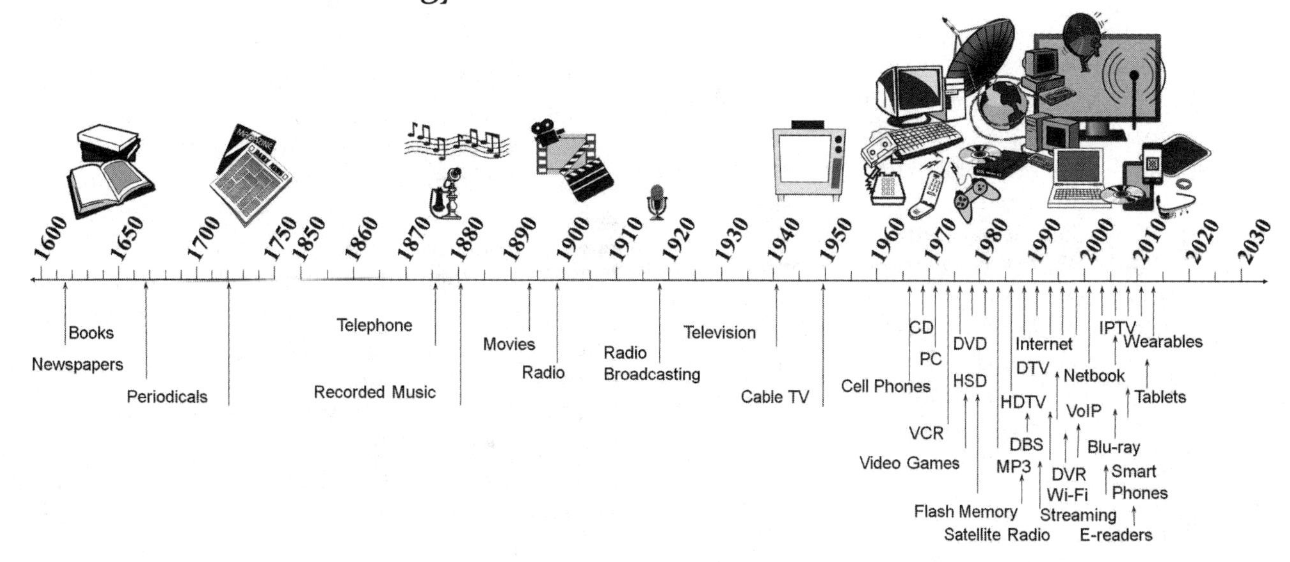

Source: Technology Futures, Inc.

The Electronic Era

The telegraph transitioned from the print era to a new period by introducing a means of sending messages far more rapidly than was previously possible. Soon, Americans and people around the world enjoyed a world enhanced by such electronic media as wired telephones, motion pictures, audio recording, radio, television, cable television, and satellite television.

Telephone

With the telephone, Alexander Graham Bell became the first to transmit speech electronically in 1876. By June 30, 1877, 230 telephones were in use, and the number rose to 1,300 by the end of August, mostly to avoid the need for a skilled interpreter of telegraph messages. The first switching office connected three company offices in Boston beginning on May 17, 1877, reflecting a focus on business rather than residential use during the telephone's early decades. Hotels became early adopters of telephones as they sought to reduce the costs of employing human messengers, and New York's 100 largest hotels had 21,000 telephones by 1909. After 1894, non-business telephone use became ordinary, in part because business use lowered the cost of telephone service. By 1902, 2,315,000 telephones were in service in the United States (Aronson, 1977). Table 2.4 and Figure 2.4 on the companion website (www.tfi.com/ctu) document the growth to near ubiquity of telephones in U.S. households and the expanding presence of wireless telephones.

Wireless Telephones

Guglielmo Marconi sent the first wireless data messages in 1895. The growing popularity of telephony led many to experiment with Marconi's radio technology as another means for interpersonal communication. By the 1920s, Detroit police cars had mobile radiophones for voice communication (ITU, 1999). The Bell system offered radio telephone service in 1946 in St. Louis, the first of 25 cities to receive the service. Bell engineers divided reception areas into cells in 1947, but cellular telephones that switched effectively among cells as callers moved did not arrive until the 1970s. The first call on a portable, handheld cell phone occurred in 1973. However, in 1981, only 24 people in New York City could use their mobile phones at the same time, and only 700 customers could have active contracts. To increase the number of people who could receive service, the Federal Communications Commission (FCC) began offering cellular telephone system licenses by lottery in June 1982 (Murray, 2001). Other countries, such as Japan in 1979 and Saudi Arabia in 1982, operated cellular systems earlier than the United States (ITU, 1999).

The U.S. Congress promoted a more advanced group of mobile communication services in 1993 by creating a classification of commercial mobile services that became known as Commercial Mobile Radio Service. This classification allowed for consistent regulatory oversight of these technologies and encouraged commercial competition among providers (FCC, 2005). By the end of 1996, about 44 million Americans subscribed to wireless telephone services (U.S. Bureau of the Census, 2008).

The new century brought an explosion of wireless telephones, and phones morphed into multipurpose devices (i.e., smartphones) with capabilities previously limited to computers. By the end of 2012, wireless phone penetration in the United States reached 326.4 million subscribers, an approximated penetration rate of 102.2% (Moorman, 2013), easily outdistancing that of wired telephones. *CTIA-The Wireless Association* (as cited in Dezego, 2013) reported that 38.2% of American households had wireless telephones only by the end of 2012. That figure increased from about 10% in 2006. By 2013, worldwide shipments of smartphones exceeded 900 million units, roughly triple the number of units shipped in 2010 (IDC as cited by Amobi, 2013).

Motion Pictures

In the 1890s, George Eastman improved on work by and patents purchased from Hannibal Goodwin in 1889 to produce workable motion picture film. The Lumière brothers projected moving pictures in a Paris café in 1895, hosting 2,500 people nightly at their movies. William Dickson, an assistant to Thomas Edison, developed the kinetograph, an early motion picture camera, and the kinetoscope, a motion picture viewing system. A New York movie house opened in 1894, offering moviegoers several coin-fed kinetoscopes. Edison's Vitascope, which expanded the length of films over those shown via kinetoscopes and allowed larger audiences to simultaneously see the moving images, appeared in public for the first time in 1896. In France in that same year, Georges Méliès started the first motion picture theater. Short movies became part of public entertainment in a variety of American venues by 1900 (Campbell, 2002), and average weekly movie attendance reached 40 million people by 1922.

Average weekly motion picture theater attendance, as shown in Table 2.5 and Figure 2.6 on the companion website (www.tfi.com/ctu), increased annually from the earliest available census reports on the subject in 1922 until 1930. After falling dramatically during the Great Depression, attendance regained growth in 1934 and continued until 1937. Slight declines in the prewar years were followed by a period of strength and stability throughout the World War II years. After the end of the war, average weekly attendance reached its greatest heights: 90 million attendees weekly from 1946 through 1949. After the introduction of television, weekly attendance would never again reach these levels.

Although a brief period of leveling off occurred in the late 1950s and early 1960s, average weekly attendance continued to plummet until a small recovery began in 1972. This recovery signaled a period of relative stability that lasted into the 1990s. Through the last decade of the century, average weekly attendance enjoyed small but steady gains.

Box office revenues, which declined generally for 20 years after the beginning of television, began a recovery in the late 1960s, then began to skyrocket in the 1970s. The explosion continued until after the turn of the new century. However, much of the increase in revenues came from increases in ticket prices and inflation, rather than from increased popularity of films with audiences, and total motion picture revenue from box office receipts declined during recent years, as studios realized revenues from television and videocassettes (U.S. Department of Commerce/International Trade Association, 2000).

As shown in Table 2.5 on the companion website (www.tfi.com/ctu), American movie fans spent an average of 12 hours per person per year from 1993 through 1997 going to theaters. That average stabilized through the first decade of the 21st century (U.S. Bureau of the Census, 2010), despite the growing popularity of watching movies at home with new digital tools. In 2011, movie rental companies were thriving, with *Netflix* boasting 25 million subscribers and *Redbox* having 32,000 rental kiosks in the United States (Amobi, 2011b).

The record-breaking success of *Avatar* as a 3D motion picture triggered a spate of followers who tried to revive the technology that was a brief hit in the 1950s. *Avatar* earned more than $761 million at American box offices and nearly $2.8 billion worldwide.

In the United States, nearly 8,000 of 39,500 theater screens were set up for 3D at the end of 2010, half of them having been installed in that year. The ticket prices for 3D films ran 20-30% higher than that of 2D films, and 3D films comprised 20% of the new films released. Nevertheless, American audiences preferred subsequent 2D films to 3D competitors, although 3D response remained strong outside the United States, where 61% of the world's 22,000 3D screens were installed. Another factor in the lack of success of 3D in America might have been the trend toward viewing movies at home, often with digital playback. In 2010, home video purchases and rentals reached $18.8 billion in North America, compared with only $10.6 billion spent at theaters (Amobi, 2011b).

China offers an example of the globalization of the film industry. Film box office revenues in China increased by 35% in the six years ending in 2010. Total box office earnings in that year reached the equivalent of $10.6 billion, even though China allowed only 20 foreign film releases each year (Amobi, 2011b).

Audio Recording

Thomas Edison expanded on experiments from the 1850s by Leon Scott de Martinville to produce a talking machine or phonograph in 1877 that played back sound recordings from etchings in tin foil. Edison later replaced the foil with wax. In the 1880s, Emile Berliner created the first flat records from metal and shellac designed to play on his gramophone, providing mass production of recordings. The early standard recordings played at 78 revolutions per minute (rpm). After shellac became a scarce commodity because of World War II, records were manufactured from polyvinyl plastic. In 1948, CBS Records produced the long-playing record that turned at 33-1/3 rpm, extending the playing time from three to four minutes to 10 minutes. *RCA* countered in 1949 with 45 rpm records that were incompatible with machines that played other formats. After a five-year war of formats, record players were manufactured that would play recordings at all of the speeds (Campbell, 2002).

The Germans used plastic magnetic tape for sound recording during World War II. After the Americans confiscated some of the tapes, the technology was improved and became a boon for Western audio editing and multiple track recordings that played on bulky reel-to-reel machines. By the 1960s, the reels were encased in cassettes, which would prove to be deadly competition in the 1970s for single song records playing at 45 rpm and long-playing albums playing at 33-1/3 rpm. At first, the tape cassettes were popular in 8-track players. As technology improved, high sound quality was obtainable on tape of smaller width, and 8-tracks gave way to smaller audiocassettes. Thomas Stockholm began recording sound digitally in the 1970s, and the introduction of compact disc (CD) recordings in 1983 decimated the sales performance of earlier analog media types (Campbell, 2002). Tables 2.6 and 2.6A and Figures 2.7 and 2.7A on the companion website (www.tfi.com/ctu) show that total unit sales of recorded music generally increased from the early 1970s through 2008. Although vinyl recordings are hanging on, cassettes vanished after 2000, and CD units sold began a downturn in 2001. Figure 2.8B shows trends in downloaded music.

The 21st century saw an explosion in new digital delivery systems for music. Digital audio players, which had their first limited popularity in 1998 (Beaumont, 2008), hit a new gear of growth with the 2001 introduction of the *Apple iPod,* which increased the storage capacity and became responsible for about 19% of music sales within its first decade. Apple's online iTunes store followed in 2003, soon becoming the world's largest music seller (Amobi, 2009).

Radio

Guglielmo Marconi's wireless messages in 1895 on his father's estate led to his establishing a British company to profit from ship-to-ship and ship-to-shore messaging. He formed a U.S. subsidiary in 1899 that would become the American Marconi Company. Reginald A. Fessenden and Lee De Forest independently transmitted voice by means of wireless radio in 1906, and a radio broadcast from the stage of a performance by Enrico Caruso occurred in 1910. Various U.S. companies and Marconi's British company owned important patents that were necessary to the development of the infant industry, so the U.S. firms formed the Radio Corporation of America (*RCA*) to buy out the patent rights from Marconi.

The debate still rages over the question of who became the first broadcaster among KDKA in Pittsburgh (Pennsylvania), WHA in Madison (Wisconsin), WWJ in Detroit (Michigan), and KQW in San Jose (California). In 1919, Dr. Frank Conrad of Westinghouse broadcast music from his phonograph in his garage in East Pittsburgh. Westinghouse's KDKA in Pittsburgh announced the presidential election returns over the airwaves on November 2, 1920. By January 1, 1922, the Secretary of Commerce had issued 30 broadcast licenses, and the number of licensees swelled to 556 by early 1923. By 1924, RCA owned a station in New York, and Westinghouse expanded to Chicago, Philadelphia, and Boston. In 1922, AT&T withdrew from RCA and started WEAF in New York, the first radio station supported by commercials. In 1923, AT&T linked WEAF with WNAC in Boston by the company's telephone lines for a simultaneous program. This began the first network, which grew to 26 stations by 1925. RCA linked its stations with telegraph lines, which failed to match the voice quality of the transmissions of AT&T. However, AT&T wanted out of the new business and sold WEAF in 1926 to the National Broadcasting Company, a subsidiary of RCA (White, 1971).

The 1930 penetration of radio sets in American households reached 40%, then approximately doubled over the next 10 years, passing 90% by 1947 (Brown, 2006). Table 2.7 and Figure 2.8, on the companion website (www.tfi.com/ctu), show the rapid rate of increase in the number of radio households from 1922 through the early 1980s, when the rate of increase declined. The increases resumed until 1993, when they began to level off.

Although thousands of radio stations were transmitting via the Internet by 2000, Channel1031.com became the first station to cease using FM and move exclusively to the Internet in September 2000 (Raphael, 2000). Many other stations were operating only on the Internet when questions about fees for commercial performers and royalties for music played on the Web arose. In 2002, the Librarian of Congress set royalty rates for Internet transmissions of sound recordings (U.S. Copyright Office, 2003). A federal court upheld the right of the Copyright Office to establish fees on streaming music over the Internet (*Bonneville v. Peters*, 2001).

In March 2001, the first two American digital audio satellites were launched, offering the promise of hundreds of satellite radio channels (Associated

Press, 2001). Consumers were expected to pay about $9.95 per month for access to commercial-free programming that would be targeted to automobile receivers. The system included amplification from about 1,300 ground antennas. By the end of 2003, about 1.6 million satellite radio subscribers tuned to the two top providers, XM and Sirius (Schaeffler, 2004). These two players merged soon before the 2008 stock market crisis, during which the new company, Sirius XM Radio, lost nearly all of its stock value. In 2011, the service was used by 20.5 million subscribers, with its market value beginning to recover (Sirius XM Radio, 2011).

Television

Paul Nipkow invented a scanning disk device in the 1880s that provided the basis from which other inventions would develop into television. In 1927, Philo Farnsworth became the first to electronically transmit a picture over the air. Fittingly, he transmitted the image of a dollar sign. In 1930, he received a patent for the first electronic television, one of many patents for which RCA would be forced, after court challenges, to negotiate. By 1932, Vladimir Zworykin discovered a means of converting light rays into electronic signals that could be transmitted and reconstructed at a receiving device. RCA offered the first public demonstration of television at the 1939 World's Fair.

The FCC designated 13 channels in 1941 for use in transmitting black-and-white television, and the commission issued almost 100 television station broadcasting licenses before placing a freeze on new licenses in 1948. The freeze offered time to settle technical issues, and it ran longer because of U.S. involvement in the Korean War (Campbell, 2002). As shown in Table 2.8 on the companion website (www.tfi.com/ctu), nearly 4,000 households had television sets by 1950, a 9% penetration rate that would escalate to 87% a decade later. Penetration has remained steady at about 98% since 1980. Figure 2.8 illustrates the meteoric rise in the number of households with television by year from 1946 through the turn of the century. In 2010, 288.5 million Americans had televisions, up by 0.8% from 2009, and average monthly time spent viewing reached 158 hours and 47 minutes, an increase of 0.2% from the previous year (Amobi, 2011a).

By the 1980s, Japanese high-definition television (HDTV) increased the potential resolution to more than 1,100 lines of data in a television picture. This increase enabled a much higher-quality image to be transmitted with less electromagnetic spectrum space per signal. In 1996, the FCC approved a digital television transmission standard and authorized broadcast television stations a second channel for a 10-year period to allow the transition to HDTV. As discussed in Chapter 6, that transition made all older analog television sets obsolete because they cannot process HDTV signals (Campbell, 2002).

The FCC (2002) initially set May 2002 as the deadline by which all U.S. commercial television broadcasters were required to be broadcasting digital television signals. Progress toward digital television broadcasting fell short of FCC requirements that all affiliates of the top four networks in the top 10 markets transmit digital signals by May 1, 1999.

Within the 10 largest television markets, all except one network affiliate had begun HDTV broadcasts by August 1, 2001. By that date, 83% of American television stations had received construction permits for HDTV facilities or a license to broadcast HDTV signals (FCC, 2002). HDTV penetration into the home marketplace would remain slow for the first few years of the 21st century, in part because of the high price of the television sets. By fall 2010, 60% of American television households had at least one HDTV set (Nielsen tech survey, 2010), and 3.2 million (9%) televisions were connected to the Internet (Amobi, 2011b).

Although 3D television sets were available in 2010, little sales success occurred. The sets were quite expensive, not much 3D television content was available, and the required 3D viewing glasses were inconvenient to wear (Amobi, 2011b).

During the fall 2011-12 television season, The Nielsen Company reported that the number of households with television in the United States dropped for the first time since the company began such monitoring in the 1970s. The decline to 114.7 million from 115.9 million television households represented a 2.2% decline, leaving the television penetration at 96.7%. Explanations for the reversal of the long-running trend included the economic recession, but the decline could represent a transition to digital access in which viewers were getting TV from devices other than television sets (Wallenstein, 2011).

Cable Television

Cable television began as a means to overcome poor reception for broadcast television signals. John Watson claimed to have developed a master antenna system in 1948, but his records were lost in a fire. Robert J. Tarlton of Lansford (Pennsylvania) and Ed Parsons of Astoria (Oregon) set up working systems in 1949 that used a single antenna to receive programming over the air and distribute it via coaxial cable to multiple users (Baldwin & McVoy, 1983). At first, the FCC chose not to regulate cable, but after the new medium appeared to offer a threat to broadcasters, cable became the focus of heavy government regulation. Under the Reagan administration, attitudes swung toward deregulation, and cable began to flourish. Table 2.9 and Figure 2.9 on the companion website (www.tfi.com/ctu) show the growth of cable systems and subscribers, with penetration remaining below 25% until 1981, but passing the 50% mark before the 1980s ended.

In the first decade of the 21st century, cable customers began receiving access to such options as digital video, video on demand, DVRs, HDTV, and telephone services. The success of digital cable led to the FCC decision to eliminate analog broadcast television as of February 17, 2009. However, in September 2007, the FCC unanimously required cable television operators to continue to provide carriage of local television stations that demand it in both analog and digital formats for three years after the conversion date. This action was designed to provide uninterrupted local station service to all cable television subscribers, protecting the 40 million (35%) U.S. households that remained analog-only (Amobi & Kolb, 2007).

Telephone service became widespread via cable during the early years of the 21st century. For years, some cable television operators offered circuit-switched telephone service, attracting 3.6 million subscribers by the end of 2004. Also by that time, the industry offered telephone services via voice over Internet protocol (VoIP) to 38% of cable households, attracting 600,000 subscribers. That number grew to 1.2 million by July 2005 (Amobi, 2005).

The growth of digital cable in the first decade of the new century also saw the growth of video-on-demand (VOD), offering cable television customers the ability to order programming for immediate viewing. VOD purchases increased by 21% over the previous year in 2010, generating revenue of $1.8 billion (Amobi, 2011b).

Cable penetration declined in the United States after 2000, as illustrated in Figure 2.9 on the companion website (www.tfi.com/ctu). However, estimated use of a combination of cable and satellite television increased steadily over the same period (U.S. Bureau of the Census, 2008).

Worldwide, pay television flourished in the new century, especially in the digital market. From 2009 to 2010, pay TV subscriptions increased from 648 million households to 704 million households. ABI Research estimated that 704 million pay TV subscribers would exist in 2011, about half of whom would be digital television subscribers, and an estimated 225 million households would subscribe to HDTV (HDTV subscribers, 2011).

Direct Broadcast Satellite and Other Cable TV Competitors

Satellite technology began in the 1940s, but HBO became the first service to use it for distributing entertainment content in 1976 when the company sent programming to its cable affiliates (Amobi & Kolb, 2007). Other networks soon followed this lead, and individual broadcast stations (WTBS, WGN, WWOR, and WPIX) used satellites in the 1970s to expand their audiences beyond their local markets by distributing their signals to cable operators around the United States.

Competitors for the cable industry include a variety of technologies. Annual FCC reports distinguish between home satellite dish (HSD) and direct broadcast satellite (DBS) systems. Both are included as MVPDs (multi-channel video program distributors), which include cable television, wireless cable systems called multichannel multipoint distribution services (MMDS), and private cable systems called satellite master antenna television (SMATV). Table 2.10 and Figure 2.10 on the companion website for this book (www.tfi.com/ctu), show trends in home satellite dish, DBS, MMDS, and SMATV (or PCO, Private Cable Operator) subscribers. However, the FCC (2013) noted that little public data was available for the dwindling services of HSD, MMDS, and PCO, citing SNL Kagan conclusions that those services accounted for less than 1% of MVPDs and were expected to continue declining over the coming decade.

Home Video

Although VCRs became available to the public in the late 1970s, competing technical standards slowed the adoption of the new devices. After the longer taping capacity of the VHS format won greater public acceptance over the higher-quality images of the Betamax, the popularity of home recording and playback rapidly accelerated, as shown in Table 2.11 and Figure 2.11 on the companion website (www.tfi.com/ctu).

By 2004, rental spending for videotapes and DVDs reached $24.5 billion, far surpassing the $9.4 billion spent for tickets to motion pictures in that year. During 2005, DVD sales increased by 400% over the $4 billion figure for 2000 to $15.7 billion. However, the annual rate of growth reversed direction and slowed that year to 45% and again the following year to 2%. VHS sales amounted to less than $300 million in 2006 (Amobi & Donald, 2007).

Factors in the decline of VHS and DVD use included growth in cable and satellite video-on-demand services, growth of broadband video availability, digital downloading of content, and the transition to DVD Blu-ray format (Amobi, 2009). The competing new formats for playing high-definition content was similar to the one waged in the early years of VCR development between the Betamax and VHS formats. Similarly, in early DVD player development, companies touting competing standards settled a dispute by agreeing to share royalties with the creator of the winning format. Until early 2008, the competition between proponents of the HD-DVD and Blu-ray formats for playing high-definition DVD content remained unresolved, and some studios were planning to distribute motion pictures in both formats. Blu-ray seemed to emerge the victor in 2008 when large companies (e.g., Time Warner, Wal-Mart, Netflix) declared allegiance to that format. By July 2010, Blu-ray penetration reached 17% of American households (Gruenwedel, 2010), and 170 million Blu-ray discs shipped that year (Amobi, 2011b).

Digital video recorders (DVRs, also called personal video recorders, PVRs) debuted during 2000, and about 500,000 units were sold by the end of 2001 (FCC, 2002). The devices save video content on computer hard drives, allowing fast-forwarding, rewinding, and pausing of live television; retroactive recording of limited minutes of previously displayed live television; automatic recording of all first-run episodes; automatic recording logs; and superior quality to that of analog VCRs. Multiple tuner models allow viewers to watch one program, while recording others simultaneously.

DVR providers generate additional revenues by charging households monthly fees, and satellite DVR households tend to be less likely to drop their satellite subscriptions. Perhaps the most fundamental importance of DVRs is the ability of consumers to make their own programming decisions about when and what they watch. This flexibility threatens the revenue base of network television in several ways, including empowering viewers to skip standard commercials. Amobi (2005) cited potential advertiser responses, such as sponsorships and product placements within programming.

Reflecting the popularity of the DVR, time shifting was practiced in 2010 by 107.1 million American households (up 13.2% from 2009). Time shifted viewing increased by 12.2% in 2010 from 2009 to an average of 10 hours and 46 minutes monthly (Amobi, 2011a).

The Digital Era

The digital era represents a transition in modes of delivery of mediated content. Although the tools of using digital media may have changed, in many cases, the content remains remarkably stable. With other media, such as social media, the digital content fostered new modes of communicating. This section contains histories of the development of computers and the Internet. Segments of earlier discussions could be considered part of the digital era, such as audio recording, HDTV, films on DVD, etc., but discussions of those segments remain under earlier eras.

Computers

The history of computing traces its origins back thousands of years to such practices as using bones as counters (Hofstra University, 2000). Intel introduced the first microprocessor in 1971. The MITS Altair, with an 8080 processor and 256 bytes of RAM (random access memory), sold for $498 in 1975, introducing the desktop computer to individuals. In 1977, Radio Shack offered the TRS80 home computer, and the Apple II set a new standard for personal computing, selling for $1,298. Other companies began introducing personal computers, and, by 1978, 212,000 personal computers were shipped for sale.

Early business adoption of computers served primarily to assist practices such as accounting. When computers became small enough to sit on office desktops in the 1980s, word processing became popular and fueled interest in home computers. With the growth of networking and the Internet in the 1990s, both businesses and consumers began buying computers in large numbers. Computer shipments around the world grew annually by more than 20% between 1991 and 1995 (Kessler, 2007).

By 1997, the majority of American households with annual incomes greater than $50,000 owned a personal computer. At the time, those computers sold for about $2,000, exceeding the reach of lower income groups. By the late 1990s, prices dropped below $1,000 per system (Kessler, 2007), and American households passed the 60% penetration mark in owning personal computers within a couple of years (U.S. Bureau of the Census, 2008).

Table 2.12 and Figure 2.12 on the companion website (www.tfi.com/ctu) trace the rapid and steady rise in American computer shipments and home penetration. By 1998, 42.1% of American households owned personal computers (U.S. Bureau of the Census, 2006). After the start of the 21st century, personal computer prices declined, and penetration increased from 63% in 2000 to 77% in 2008 (Forrester Research as cited in Kessler, 2011). Worldwide personal computer sales increased by 34% from 287 million in 2008 to 385 million in 2011 (IDC as cited Kessler, 2011).

Internet

The Internet began in the 1960s with ARPANET, or the Advanced Research Projects Agency (ARPA) network project, under the auspices of the U.S. Defense Department. The project intended to serve the military and researchers with multiple paths of linking computers together for sharing data in a system that would remain operational even when traditional communications might become unavailable. Early users, mostly university and research lab personnel, took advantage of electronic mail and posting information on computer bulletin boards. Usage increased dramatically in 1982 after the National Science Foundation supported high-speed linkage of multiple locations around the United States. After the collapse of the Soviet Union in the late 1980s, military users abandoned ARPANET, but private users continued to use it and multimedia transmissions of audio and video became possible once this content could be digitized. More than 150,000 regional computer networks and 95 million computer servers hosted data for Internet users (Campbell, 2002).

Penetration and Usage

During the first decade of the 21st century, the Internet became the primary reason that consumers purchased new computers (Kessler, 2007). Cable modems and digital subscriber lines (DSL) telephone line connections increased among home users as the means for connecting to the Internet, as more than half of American households accessed the Internet with high-speed broadband connections.

Tables 2.13 and 2.13 and Figures 2.13 and 2.14 on the companion website (www.tfi.com/ctu) show trends in Internet penetration in the United States. By 2008, 74 million (63%) American households had high-speed broadband access (Kessler, 2011). In June 2013, 91,342,000 fixed broadband subscriptions and 299,447,000 million American subscribers had wireless broadband subscriptions (OECD, 2013).

Synthesis

Older media, including both print and electronic types, have gone through transitions from their original forms into digital media. Many media that originated in digital form have enjoyed explosive growth during the digital age.

Horrigan (2005) noted that the rate of adoption of high-speed Internet approximated that of other electronic technologies. He observed that 10% of the population used high-speed Internet access in just more than five years. Personal computers required four years, CD players needed 4.5 years, cell phones required eight years, VCRs took 10 years, and color televisions used 12 years to reach 10% penetration.

Early visions of the Internet (see Chapter 22) did not include the emphasis on entertainment and information to the general public that has emerged. The combination of this new medium with older media belongs to a phenomenon called *convergence*, referring to the merging of functions of old and new media. By 2002, the FCC (2002) reported that the most important type of convergence related to video content is the joining of Internet services. The report also noted that companies from many business areas were providing a variety of video, data, and other communications services.

Just as media began converging nearly a century ago when radios and record players merged in the same appliance, media in recent years have been converging at a much more rapid pace. As the popularity of print media forms generally declined throughout the 1990s, the popularity of the Internet grew rapidly, particularly with the increase in high-speed broadband connections, for which adoption rates achieved comparability with previous new communications media. Consumer flexibility through the use of digital media became the dominant media consumption theme during the first decade of the new century.

Bibliography

Agnese, J. (2011, October 20). Industry surveys: Publishing & advertising. In E. M. Bossong-Martines (Ed.), *Standard & Poor's industrysSurveys, Vol. 2.*

Amobi, T. N. (2005, December 8). Industry surveys: Broadcasting, cable, and satellite industry survey. In E. M. Bossong-Martines (Ed.), *Standard & Poor's industry surveys, 173* (49), Section 2.

Amobi, T. N. (2009). Industry surveys: Movies and home entertainment. In E. M. Bossong-Martines (Ed.), *Standard & Poor's industry surveys, 177* (38), Section 2.

Amobi, T. N. (2011a). Industry surveys: Broadcasting, cable, and satellite. In E. M. Bossong-Martines (Ed.), *Standard & Poor's industry surveys, Vol. 1.*

Amobi, T. N. (2011b). Industry surveys: Movies & entertainment. In E. M. Bossong-Martines (Ed.), *Standard & Poor's industry surveys, Vol. 2.*

Amobi, T. N. (2013). Industry surveys: Broadcasting, cable, & satellite. In E. M. Bossong-Martines (Ed.), *Standard & Poor's industry surveys, Vol. 1.*

Amobi, T. N. & Donald, W. H. (2007, September 20). Industry surveys: Movies and home entertainment. In E. M. Bossong-Martines (Ed.), *Standard & Poor's industry surveys, 175* (38), Section 2.

Amobi, T. N. & Kolb, E. (2007, December 13). Industry surveys: Broadcasting, cable & satellite. In E. M. Bossong-Martines (Ed.), *Standard & Poor's industry surveys, 175* (50), Section 1.

Aronson, S. (1977). Bell's electrical toy: What's the use? The sociology of early telephone usage. In I. Pool (Ed.). *The social impact of the telephone.* Cambridge, MA: The MIT Press, 15-39.

Associated Press. (2001, March 20). Audio satellite launched into orbit. *New York Times.* Retrieved from http://www.nytimes.com/aponline/national/AP-Satellite-Radio.html?ex=986113045& ei=1&en=7af33c7805ed8853.

Baldwin, T. & McVoy, D. (1983). *Cable communication.* Englewood Cliffs, NJ: Prentice-Hall.

Bonneville International Corp., et al. v. Marybeth Peters, as Register of Copyrights, et al. Civ. No. 01-0408, 153 F. Supp.2d 763 (E.D. Pa., August 1, 2001).

Brown, D. & Bryant, J. (1989). An annotated statistical abstract of communications media in the United States. In J. Salvaggio & J. Bryant (Eds.), *Media use in the information age: Emerging patterns of adoption and consumer use.* Hillsdale, NJ: Lawrence Erlbaum Associates, 259-302.

Brown, D. (1996). A statistical update of selected American communications media. In Grant, A. E. (Ed.), *Communication Technology Update* (5th ed.). Boston, MA: Focal Press, 327-355.

Brown, D. (1998). Trends in selected U. S. communications media. In Grant, A. E. & Meadows, J. H. (Eds.), Communication Technology Update (6th ed.). Boston, MA: Focal Press, 279-305.

Brown, D. (2000). Trends in selected U. S. communications media. In Grant, A. E. & Meadows, J. H. (Eds.), *Communication Technology Update* (7th ed.). Boston, MA: Focal Press, 299-324.

Brown, D. (2002). Communication technology timeline. In A. E. Grant & J. H. Meadows (Eds.), *Communication technology update* (8th ed.) Boston: Focal Press, 7-45.

Brown, D. (2004). Communication technology timeline. In A. E. Grant & J. H. Meadows (Eds.). *Communication technology update* (9th ed.). Boston: Focal Press, 7-46.

Brown, D. (2006). Communication technology timeline. In A. E. Grant & J. H. Meadows (Eds.), *Communication technology update* (10th ed.). Boston: Focal Press. 7-46.

Brown, D. (2008). Historical perspectives on communication technology. In E. Grant & J. H. Meadows (Eds.), *Communication technology update and fundamentals* (11th ed.). Boston: Focal Press. 11-42.

Brown, D. (2010). Historical perspectives on communication technology. In E. Grant & J. H. Meadows (Eds.), *Communication technology update and fundamentals* (12th ed.). Boston: Focal Press. 9-46.

Brown, D. (2012). Historical perspectives on communication technology. In E. Grant & J. H. Meadows (Eds.), *Communication technology update and fundamentals* (13th ed.). Boston: Focal Press. 9-24.

Campbell, R. (2002). *Media & culture.* Boston, MA: Bedford/St. Martins.

Dezego, R. (2013, July). Industry surveys: Telecommunications: Wireline. In E. M. Bossong-Martines (Ed.), Standard & Poor's industry surveys, Vol 2.
Dillon, D. (2011). E-books pose major challenge for publishers, libraries. In D. Bogart (Ed.), *Library and book trade almanac* (pp. 3-16). Medford, NJ: Information Today, Inc.
Federal Communications Commission. (2002, January 14). *In the matter of annual assessment of the status of competition in the market for the delivery of video programming* (eighth annual report). CS Docket No. 01-129. Washington, DC 20554.
Federal Communications Commission. (2005). *In the matter of Implementation of Section 6002(b) of the Omnibus Budget Reconciliation Act of 1993: Annual report and analysis of competitive market conditions with respect to commercial mobile services* (10th report). WT Docket No. 05-71. Washington, DC 20554.
Federal Communications Commission. (2013, July 22). *In the matter of annual assessment of the status of competition in the market for the delivery of video programming* (fifteenth annual report). MB Docket No. 12-203. Washington, DC 20554.
Gruenwedel, E. (2010). Report: Blu's household penetration reaches 17%. *Home Media Magazine, 32*, 40.
HDTV subscribers to total 225 million globally in 2011 (2011, March). *Broadcast Engineering* [Online Exclusive].
Hofstra University. (2000). *Chronology of computing history*. Retrieved from http://www.hofstra.edu/pdf/CompHist_9812tla1.pdf.
Horrigan, J. B. (2005, September 24). *Broadband adoption at home in the United States: Growing but slowing*. Paper presented to the 33rd Annual Meeting of the Telecommunications Policy Research Conference. Retrieved from http://www.pewinternet.org/PPF/r/164/report_display.asp.
Huntzicker, W. (1999). *The popular press, 1833-1865*. Westport, CT: Greenwood Press.
Ink, G. & Grabois, A. (2000). *Book title output and average prices: 1998 final and 1999 preliminary figures*, 45th edition. D. Bogart (Ed.). New Providence, NJ: R. R. Bowker, 508-513.
International Telecommunications Union. (1999). *World telecommunications report 1999*. Geneva, Switzerland: Author.
Kessler, S. H. (2007, August 26). Industry surveys: Computers: Hardware. In E M. Bossong-Martines (Ed.), *Standard & Poor's industry surveys, 175* (17), Section 2.
Kessler, S. H. (2011). Industry surveys: Computers: Consumer services and the Internet. In E. M. Bossong-Martines (Ed.), *Standard & Poor's industry surveys, Vol. 1.*
Lee, J. (1917). *History of American journalism*. Boston: Houghton Mifflin.
Moorman, J. G. (2013, July). Industry surveys: Publishing and advertising. In E. M. Bossong-Martines (Ed.), Standard & Poor's industry surveys, Vol 2.
Murray, J. (2001). *Wireless nation: The frenzied launch of the cellular revolution in America*. Cambridge, MA: Perseus Publishing.
OECD. (2013). Broadband and telecom. Retrieved from http://www.oecd.org/internet/broadband/oecdbroadbandportal.htm
Peters, J. & Donald, W. H. (2007). Industry surveys: Publishing. In E. M. Bossong-Martines (Ed.), *Standard & Poor's industry surveys. 175* (36). Section 1.
Rainie, L. (2012, January 23). Tablet and e-book reader ownership nearly doubles over the holiday gift-giving period. *Pew Research Center's Internet & American Life Project*. Retrieved from http://pewinternet.org/Reports/2012/E-readers-and-tablets.aspx
Raphael, J. (2000, September 4). Radio station leaves earth and enters cyberspace. Trading the FM dial for a digital stream. *New York Times*. Retrieved from http://www.nytimes.com/library/tech/00/ 09/biztech/articles/04radio.html.
Rawlinson, N. (2011, April 28). Books vs ebooks. *Computer Act!ve*. Retrieved from General OneFile database.
Schaeffler, J. (2004, February 2). The real satellite radio boom begins. *Satellite News, 27* (5). Retrieved from Lexis-Nexis.
Sirius XM Radio Poised for Growth, Finally. (2011, May 10). *Newsmax*.
U.S. Bureau of the Census. (2006). *Statistical abstract of the United States: 2006* (125th Ed.). Washington, DC: U.S. Government Printing Office.
U.S. Bureau of the Census. (2008). *Statistical abstract of the United States: 2008* (127th Ed.). Washington, DC: U.S. Government Printing Office. Retrieved from http://www.census.gov/compendia/statab/.
U.S. Bureau of the Census. (2010). *Statistical abstract of the United States: 2008* (129th Ed.). Washington, DC: U.S. Government Printing Office. Retrieved from http://www.census.gov/compendia/statab/.
U.S. Copyright Office. (2003). *106th Annual report of the Register of Copyrights for the fiscal year ending September 30, 2003*. Washington, DC: Library of Congress.
U.S. Department of Commerce/International Trade Association. (2000). *U.S. industry and trade outlook 2000*. New York: McGraw-Hill.
Wallenstein, A. (2011, September 16). Tube squeeze: economy, tech spur drop in TV homes. *Daily Variety*, 3.
White, L. (1971). *The American radio*. New York: Arno Press.

3

Understanding Communication Technologies

Jennifer H. Meadows, Ph.D.*

Today you can do dozens of things that your parents never dreamed of: surfing the Web anytime and anywhere, watching crystal-clear sports on a large high definition television (HDTV) in your home, battling aliens on "distant worlds" alongside game players scattered around the globe, and "Googling" any subject you find interesting. This book was created to help you understand these technologies, but there is a set of tools that will not only help you understand them, but also understand the next generation of technologies.

All of the communication technologies explored in this book have a number of characteristics in common, including how their adoption spreads from a small group of highly interested users to the general public, what the effects of these technologies are upon the people who use them (and on society in general), and how these technologies affect each other.

For more than a century, researchers have studied adoption, effects, and other aspects of new technologies, identifying patterns that are common across dissimilar technologies, and proposing theories of technology adoption and effects. These theories have proven to be valuable to entrepreneurs seeking to develop new technologies, regulators who want to control those technologies, and everyone else who just wants to understand them.

* Professor and Chair, Department of Communication Design, California State University, Chico (Chico, California).

The utility of these theories is that they allow you to apply lessons from one technology to another or from old technologies to new technologies. The easiest way to understand the role played by the technologies explored in this book is to have a set of theories you can apply to virtually any technology you discuss. The purpose of this chapter is to give you those tools by introducing you to the theories.

The technology ecosystem discussed in Chapter 1 is a useful framework for studying communication technologies, but it is not a theory. This perspective is a good starting point to begin to understand communication technologies because it targets your attention at a number of different levels that might not be immediately obvious: hardware, software, content, organizational infrastructure, social systems, and, finally, the user.

Understanding each of these levels is aided by knowing a number of theoretical perspectives that can help us understand the different sections of the ecosystem for these technologies. Theoretical approaches are useful in understanding the origins of the information-based economy in which we now live, why some technologies take off while others fail, the impacts and effects of technologies, and the economics of the communication technology marketplace.

The Information Society and the Control Revolution

Our economy used to be based on tangible products such as coal, lumber, and steel. That has changed. Now, information is the basis of our economy. Information industries include education; research and development; creating informational goods such as computer software, banking, insurance; and even entertainment and news (Beniger, 1986).

Information is different from other commodities like coffee and pork bellies, which are known as "private goods." Instead, information is a "public good" because it is intangible, lacks a physical presence, and can be sold as many times as demand allows without regard to consumption.

For example, if 10 sweaters are sold, then 10 sweaters must be manufactured using raw materials. If 10 subscriptions to an online dating service are sold, there is no need to create new services: 10—or 10,000—subscriptions can be sold without additional raw materials.

This difference actually gets to the heart of a common misunderstanding about ownership of information that falls into a field known as "intellectual property rights." A common example is the purchase of a digital music download. A person may believe that because he or she purchased the music, that he or she can copy and distribute that music to others. Just because the information (the music) was purchased doesn't mean you own the song and performance (intellectual property).

Several theorists have studied the development of the information society, including its origin. Beniger (1986) argues that there was a control revolution: "A complex of rapid changes in the technological and economic arrangements by which information is collected, stored, processed, and communicated and through which formal or programmed decisions might affect social control" (p. 52). In other words, as society progressed, technologies were created to help control information. For example, information was centralized by mass media.

In addition, as more and more information is created and distributed, new technologies must be developed to control that information. For example, with the explosion of information available over the Internet, search engines were developed to help users find relevant information.

Another important point is that information is power, and there is power in giving information away. Power can also be gained by withholding information. At different times in modern history, governments have blocked access to information or controlled information dissemination to maintain power.

Adoption

Why do some technologies succeed while others fail? This question is addressed by a number of theoretical approaches including the diffusion of innovations, social information processing theory, and critical mass theory.

Diffusion of Innovations

The diffusion of innovations, also referred to as diffusion theory, was developed by Everett Rogers (1962; 2003). This theory tries to explain how an innovation is communicated over time through different channels to members of a social system. There are four main aspects of this approach.

First, there is the innovation. In the case of communication technologies, the innovation is some technology that is perceived as new. Rogers also defines characteristics of innovations: relative advantage, compatibility, complexity, trialability, and observability.

So, if someone is deciding to purchase a new mobile phone, characteristics would include the relative advantage over other mobile phones; whether or not the mobile phone is compatible with the existing needs of the user; how complex it is to use; whether or not the potential user can try it out; and whether or not the potential user can see others using the new mobile phone with successful results.

Information about an innovation is communicated through different channels. Mass media is good for awareness knowledge. For example, each new iPhone has Web content, television commercials, and print advertising announcing its existence and its features.

Interpersonal channels are also an important means of communication about innovations. These interactions generally involve subjective evaluations of the innovation. For example, a person might ask some friends how they like their new iPhones.

Rogers (2003) outlines the decision-making process a potential user goes through before adopting an innovation. This is a five-step process.

The first step is knowledge. You find out there is a new mobile phone available and learn about its new features. The next step is persuasion—the formation of a positive attitude about the innovation. Maybe you like the new phone. The third step is when you decide to accept or reject the innovation. Yes, I will get the new mobile phone. Implementation is the fourth step. You use the innovation, in this case, the mobile phone. Finally, confirmation occurs when you decide that you made the correct decision. Yes, the mobile phone is what I thought it would be; my decision is reinforced.

Another stage discussed by Rogers (2003) and others is "reinvention," the process by which a person who adopts a technology begins to use it for purposes other than those intended by the original inventor. For example, mobile phones were initially designed for calling other people regardless of location, but users have found ways to use them for a wide variety of applications ranging from alarm clocks to personal calendars and flashlights.

Have you ever noticed that some people are the first to have the new technology gadget, while others refuse to adopt a proven successful technology? Adopters can be categorized into different groups according to how soon or late they adopt an innovation.

The first to adopt are the innovators. Innovators are special because they are willing to take a risk adopting something new that may fail. Next come the early adopters, the early majority, and then the late majority, followed by the last category, the laggards. In terms of percentages, innovators make up the first 2.5% percent of adopters, early adopters are the next 13.5%, early majority follows with 34%, late majority are the next 34%, and laggards are the last 16%.

Adopters can also be described in terms of ideal types. Innovators are venturesome. These are people who like to take risks and can deal with failure. Early adopters are respectable. They are valued opinion leaders in the community and role models for others. Early majority adopters are deliberate. They adopt just before the average person and are an important link between the innovators, early adopters, and everyone else. The late majority are skeptical. They are hesitant to adopt innovations and often adopt because they pressured. Laggards are the last to adopt and often are isolated with no opinion leadership. They are suspicious and resistant to change. Other factors that affect adoption include education, social status, social mobility, finances, and willingness to use credit (Rogers, 2003).

Adoption of an innovation does not usually occur all at once; it happens over time. This is called the rate of adoption. The rate of adoption generally follows an S-shaped "diffusion curve" where the X-axis is time and the Y-axis is percent of adopters. You can note the different adopter categories along the diffusion curve.

Figure 3.1 shows a diffusion curve. See how the innovators are at the very beginning of the curve, and the laggards are at the end. The steepness of the curve depends on how quickly an innovation is adopted. For example, DVD has a steeper curve than VCR because DVD players were adopted at a faster rate than VCRs.

Also, different types of decision processes lead to faster adoption. Voluntary adoption is slower than collective decisions, which, in turn, are slower than authority decisions. For example, a company may let its workers decide whether to use a new software package, the employees may agree collectively to use that software, or finally, the management may decide that everyone at the company is going to use the software. In most cases, voluntary adoption would take the longest, and a management dictate would result in the swiftest adoption.

Moore (2001) further explored diffusion of innovations and high-tech marketing in *Crossing the Chasm*. He noted there are gaps between the innovators and the early adopters, the early adopters and the early majority, and the early majority and late majority.

For a technology's adoption to move from innovators to the early adopters the technology must show a major new benefit. Innovators are visionaries that take the risk of adopting something new such as UltraHD televisions.

Early adopters then must see the new benefit of UltraHD televisions before adopting. The chasm between early adopters and early majority is the greatest of these gaps. Early adopters are still visionary and want to be change agents. They don't mind dealing with the troubles and glitches that come along with a new technology. Early adopters were likely to use a beta version of a new service like Snapchat.

The early majority, on the other hand, are pragmatists and want to see some improvement in productivity—something tangible. Moving from serving the visionaries to serving the pragmatists is difficult; hence Moore's description of "crossing the chasm."

Finally, there is a smaller gap between the early majority and the late majority. Unlike the early majority, the late majority reacts to the technical demands on the users. The early majority is more comfortable working with technology. So, the early majority would be comfortable using social networking like Instagram but the late majority is put off by the perceived technical demands. The technology must alleviate this concern before late majority adoption.

Figure 3.1
Innovation Adoption Rate

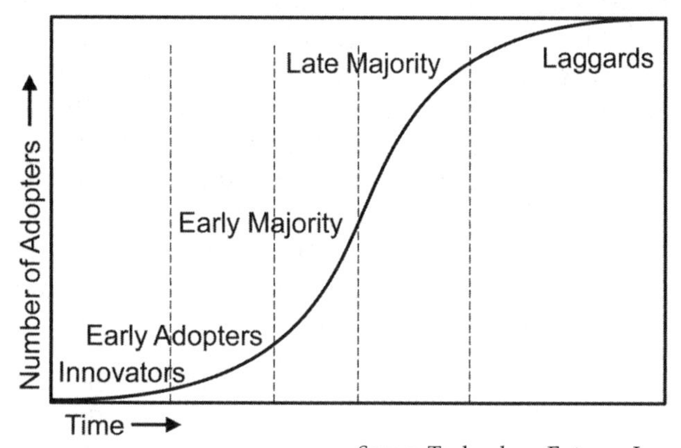

Source: Technology Futures, Inc.

Another perspective on adoption can be found in most marketing textbooks (e.g., Kottler & Keller, 2011): the product lifecycle. As illustrated in Figure 3.2, the product lifecycle extends the diffusion curve to include the maturity and decline of the technology. This perspective provides a more complete picture of a technology because it focuses our attention beyond the initial use of the technology to the time that the technology is in regular use, and ultimately, disappears from market.

Considering the short lifespan of many communication technologies, it may be just as useful to study the entire lifespan of a technology rather than just the process of adoption.

Figure 3.2
Product Lifecycle

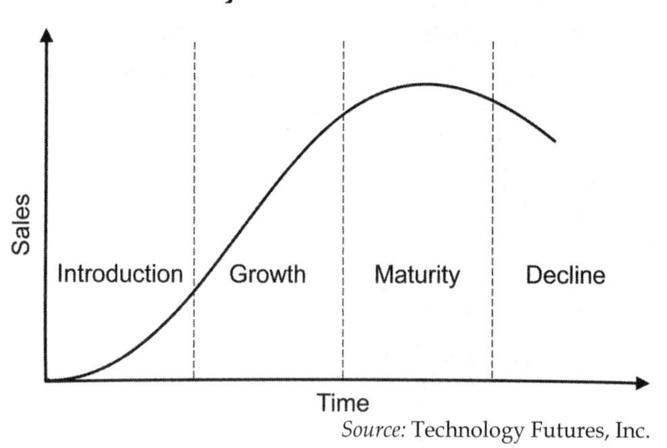

Source: Technology Futures, Inc.

Other Theories of Adoption

Other theorists have attempted to explain the process of adoption as well. Among the most notable perspectives in this regard are the Theory of Planned Behavior (TPB) and the Technology Acceptance Model (TAM). Both of these models emerged from the need to identify factors that can help predict future adoption of a new technology when there is no history of adoption or use of the technology.

The Theory of Planned Behavior (TPB) (Ajzen, 1991; Ajzen & Fishbein, 1980; Fishbein & Ajzen, 1975) presumes that a suitable predictor of future behavior is "behavioral intention," a cognitive rather than behavioral variable that represents an individual's plans for adopting (or not adopting!) an innovation. Behavioral intentions are, in turn, predicted by attitudes toward the innovation and the innovators.

The Technology Acceptance Model (Davis, 1986; Davis, Bagozzi & Warshaw, 1989) elaborates on TPB by adding factors that may predict attitudes toward an innovation and behavioral intentions, including perceived usefulness, perceived ease of use, and external variables.

A substantial body of research has demonstrated the efficacy of these factors in predicting behavioral intentions (e.g., Mathieson, 1991; Jiang, Hsu & Klein, 2000; Chau & Hu, 2002; Park, Lee & Cheong, 2007), but much more research is needed regarding the link between behavioral intentions and actual adoption at a later point in time.

Another theory that expands upon Rogers' diffusion theory is presented in Chapter 4. Grant's pre-diffusion theory identifies organizational functions that must be served before any consumer adoption of the technology can take place.

Critical Mass Theory

Have you ever wondered who had the first e-mail address or the first telephone? Who did he communicate with? Interactive technologies such as telephony and e-mail become more and more useful as more people adopt these technologies. There have to be some innovators and early adopters who are willing to take the risk to try a new interactive technology.

These users are the "critical mass," a small segment of the population that chooses to make big contributions to the public good (Markus, 1987). In general terms, any social process involving actions by individuals that benefit others is known as "collective action." In this case, the technologies become more useful if everyone in the system is using the technology, a goal known as "universal access".

Ultimately, universal access means that you can reach anyone through some communication technology. For example, in the United States, the landline phone system reaches almost everywhere, and everyone benefits from this technology, although a small segment of the population initially chose to adopt the telephone to get the ball rolling. There is a stage in the diffusion process that an interactive medium has to reach in order for adoption to take off. This is the critical mass.

Another conceptualization of critical mass theory is the "tipping point" (Gladwell, 2002). Here is an example. The videophone never took off, in part, because it never reached critical mass. The videophone was not really any better than a regular phone unless the person you were calling also had a videophone. If there were not enough people you knew who had videophones, then you might not adopt it because it was not worth it.

On the other hand, if most of your regular contacts had videophones, then that critical mass of users might drive you to adopt the videophone. Critical mass is an important aspect to consider for the adoption of any interactive technology.

Another good example is facsimile or fax technology. The first method of sending images over wires was invented in the '40s—the 1840s—by Alexander Bain, who proposed using a system of electrical pendulums to send images over wires (Robinson, 1986). Within a few decades, the technology was adapted by the newspaper industry to send photos over wires, but the technology was limited to a small number of news organizations.

The development of technical standards in the 1960s brought the fax machine to corporate America, which generally ignored the technology because few businesses knew of another business that had a fax machine.

Adoption of the fax took place two machines at a time, with those two usually being purchased to communicate with each other, but rarely used to communicate with additional receivers. By the 1980s, enough businesses had fax machines that could communicate with each other that many businesses started buying fax machines one at a time.

As soon as the critical mass point was reached, fax machine adoption increased to the point that it became referred to as the first technology adopted out of fear of embarrassment that someone would ask, "What's your fax number?" (Wathne & Leos, 1993). In less than two years, the fax machine became a business necessity.

Social Information Processing

Another way to look at how and why people choose to use or not use a technology is social information processing. This theory begins by critiquing rational choice models, which presume that people make adoption decisions and other evaluations of technologies based upon objective characteristics of the technology. In order to understand social information processing, you first have to look at a few rational choice models.

One model, social presence theory, categorizes communication media based on a continuum of how the medium "facilitates awareness of the other person and interpersonal relationships during the interaction (Fulk, et al., 1990, p. 118)."

Communication is most efficient when the social presence level of the medium best matches the interpersonal relationship required for the task at hand. For example, a person would break up with another person face to face instead of using a text message.

Another rational choice model is information richness theory. In this theory, media are also arranged on a continuum of richness in four areas: speed of feedback, types of channels employed, personalness of source, and richness of language carried (Fulk, et al., 1990). Face-to-face communications is the highest in social presence and information richness.

In information richness theory, the communication medium chosen is related to message ambiguity. If the message is ambiguous, then a richer medium is chosen. In this case, teaching someone how to dance would be better with a DVD that illustrates the steps rather than just an audio CD that describes the steps.

Social information processing theory goes beyond the rational choice models because it states that perceptions of media are "in part, subjective and socially constructed." Although people may use objective standards in choosing communication media, use is also determined by subjective factors such as the attitudes of coworkers about the media and vicarious learning, or watching others' experiences.

Social influence is strongest in ambiguous situations. For example, the less people know about a medium, then the more likely they are to rely on social information in deciding to use it (Fulk, et al., 1987).

Think about whether you prefer a Macintosh or a Windows-based computer. Although you can probably list objective differences between the two, many of the important factors in your choice are based upon subjective factors such as which one is owned by friends and coworkers, the perceived usefulness of the computer, and advice you receive from people who can help you set up and maintain your computer.

In the end, these social factors probably play a much more important role in your decision than "objective" factors such as processor speed, memory capacity, etc.

Impacts & Effects

Do video games make players violent? Do users seek out social networking sites for social interactions? These are some of the questions that theories of impacts or effects try to answer.

To begin, Rogers (1986) provides a useful typology of impacts. Impacts can be grouped into three dichotomies: desirable and undesirable, direct and indirect, and anticipated and unanticipated.

Desirable impacts are the functional impacts of a technology. For example a desirable impact of e-commerce is the ability to purchase goods and services from your home. An undesirable impact is one that is dysfunctional, such as credit card fraud.

Direct impacts are changes that happen in immediate response to a technology. A direct impact of wireless telephony is the ability to make calls while driving. An indirect impact is a byproduct of the direct impact. To illustrate, laws against driving and using a handheld wireless phone are an impact of the direct impact described above.

Anticipated impacts are the intended impacts of a technology. An anticipated impact of text messaging is to communicate without audio. An unanticipated impact is an unintended impact, such as people sending text messages in a movie theater and annoying other patrons. Often, the desirable, direct, and anticipated impacts are the same and are considered first. Then, the undesirable, indirect, and unanticipated impacts are noted later.

Here is an example using e-mail. A desirable, anticipated, and direct impact of e-mail is to be able to quickly send a message to multiple people at the same time. An undesirable, indirect, and unanticipated impact of e-mail is spam—unwanted e-mail clogging the inboxes of millions of users.

Uses and Gratifications

Uses and gratifications research is a descriptive approach that gives insight into what people do with technology. This approach sees the users as actively seeking to use different media to fulfill different needs (Rubin, 2002). The perspective focuses on "(1) the social and psychological origins of (2) needs, which generate (3) expectations of (4) the mass media or other sources, which lead to (5) differential patterns of media exposure (or engagement in other activities), resulting in (6) needs gratifications and (7) other consequences, perhaps mostly unintended ones" (Katz, et al., 1974, p. 20).

Uses and gratifications research surveys audiences about why they choose to use different types of media. For example, uses and gratifications of television studies have found that people watch television for information, relaxation, to pass time, by habit, excitement, and for social utility (Rubin, 2002).

This approach is also useful for comparing the uses and gratifications between media, as illustrated by studies of the World Wide Web (WWW) and television gratifications that found that, although there are some similarities such as entertainment and to pass time, they are also very different on other variables such as companionship, where the Web was much lower than for television (Ferguson & Perse, 2000). Uses and gratifications studies have examined a multitude of communication technologies including mobile phones (Wei, 2006), radio (Towers, 1987), and satellite television (Etefa, 2005).

Media System Dependency Theory

Often confused with uses and gratifications, media system dependency theory is "an ecological theory that attempts to explore and explain the role of media in society by examining dependency relations within and across levels of analysis" (Grant, et al., 1991, p. 774). The key to this theory is the focus it provides on the dependency relationships that result from the interplay between resources and goals.

The theory suggests that, in order to understand the role of a medium, you have to look at relationships at multiple levels of analysis, including the individual level—the audience, the organizational level, the media system level, and society in general.

These dependency relationships can by symmetrical or asymmetrical. For example, the dependency relationship between audiences and network television is asymmetrical because an individual audience member may depend more on network television to reach his or her goal than the television networks depend on that one audience member to reach their goals.

A typology of individual media dependency relations was developed by Ball-Rokeach & DeFleur (1976) to help understand the range of goals that individuals have when they use the media. There are six dimensions: social understanding, self-understanding, action orientation, interaction orientation, solitary play, and social play.

Social understanding is learning about the world around you, while self-understanding is learning about yourself. Action orientation is learning about specific behaviors, while interaction orientation is about learning about specific behaviors involving other people. Solitary play is entertaining yourself alone, while social play is using media as a focus for social interaction.

Research on individual media system dependency relationships has demonstrated that people have different dependency relationships with different media. For example, Meadows (1997) found that women had stronger social understanding dependencies for television than magazines, but stronger self-understanding dependencies for magazines than television.

In the early days of television shopping (when it was considered "new technology"), Grant, et al. (1991) applied media system dependency theory to the phenomenon. Their analysis explored two dimensions: how TV shopping changed organizational dependency relations within the television industry and how and why individual users watched television shopping programs.

By applying a theory that addressed multiple levels of analysis, a greater understanding of the new technology was obtained than if a theory that focused on only one level had been applied.

Social Learning Theory/Social Cognitive Theory

Social learning theory focuses on how people learn by modeling others (Bandura, 2001). This observational learning occurs when watching another person model the behavior. It also happens with symbolic modeling, modeling that happens by watching the behavior modeled on a television or computer screen. So, a person can learn how to fry an egg by watching another person fry an egg in person or on a video.

Learning happens within a social context. People learn by watching others, but they may or may not perform the behavior. Learning happens, though, whether the behavior is imitated or not.

Reinforcement and punishment play a role in whether or not the modeled behavior is performed. If the behavior is reinforced, then the learner is more likely to perform the behavior. For example, if a student is successful using online resources for a presentation, other students watching the presentation will be more likely to use online resources.

On the other hand, if the action is punished, then the modeling is less likely to result in the behavior. To illustrate, if a character drives drunk and gets arrested on a television program, then that modeled behavior is less likely to be performed by viewers of that program.

Reinforcement and punishment is not that simple though. This is where cognition comes in—learners think about the consequences of performing that behavior. This is why a person may play *Grand Theft Auto* and steal cars in the videogame, but will not then go out and steal a car in real life. Self-regulation is an important factor. Self-efficacy is another important dimension: learners must believe that they can perform the behavior.

Social learning/cognitive theory, then, is a useful framework for examining not only the effects of communication media, but also the adoption of communication technologies (Bandura, 2001).

The content that is consumed through communication technologies contains symbolic models of behavior that are both functional and dysfunctional. If viewers model the behavior in the content, then some form of observational learning is occurring.

A lot of advertising works this way. A movie star uses a new shampoo and then is admired by others. This message models a positive reinforcement of using the shampoo. Cognitively, the viewer then thinks about the consequences of using the shampoo.

Modeling can happen with live models and symbolic models. For example, a person can watch another playing Wii bowling, a videogame where the player has to manipulate the controller to mimic rolling the ball. Their avatar in the game also models the bowling action. The other player considers the consequences of this modeling.

In addition, if the other person had not played with this gaming system, watching the other person play with the Wii and enjoy the experience will make it more likely that he or she will adopt the system. Therefore, social learning/cognitive theory can be used to facilitate the adoption of new technologies and to understand why some technologies are adopted and why some are adopted faster than others (Bandura, 2001).

Economic Theories

Thus far, the theories and perspectives discussed have dealt mainly with individual users and communication technologies. How do users decide to adopt a technology? What impacts will a technology have on a user?

Theory, though, can also be applied to organizational infrastructure and the overall technology market. Here, two approaches will be addressed: the theory of the long tail that presents a new way of looking at digital content and how it is distributed and sold, and the principle of relative constancy that examines what happens to the marketplace when new media products are introduced.

The Theory of the Long Tail

Wired editor Chris Anderson developed the theory of the long tail. While some claim this is not a "theory," it is nonetheless a useful framework for understanding new media markets.

This theory begins with the realization that there are not any huge hit movies, television shows, and albums like there used to be. What counts as a hit TV show today, for example, would be a failed show just 15 years ago.

One of the reasons for this is choice: 40 years ago viewers had a choice of only a few television channels. Today, you could have hundreds of channels of video programming on cable or satellite and limitless amounts of video programming on the Internet.

New communication technologies are giving users access to niche content. There is more music, video, video games, news, etc. than ever before because the distribution is no longer limited to the traditional mass media of over-the-air broadcasting, newspapers, etc. or the shelf space at a local retailer.

The theory states that, "our culture and economy are increasingly shifting away from a focus on a relatively small number of 'hits' at the headend of the demand curve and toward a huge number of niches in the tail" (Anderson, n.d.).

Figure 3.3 shows a traditional demand curve; most of the hits are at the head of the curve, but there is still demand as you go into the tail. There is a demand for niche content and there are opportunities for businesses that deliver content in the long tail.

Both physical media and traditional retail have limitations. For example, there is only so much shelf space in the store. Therefore, the store, in order to maximize profit, is only going to stock the products most likely to sell. Digital content and distribution changes this.

For example, Amazon and Netflix can have huge inventories of hard-to-find titles, as opposed to a bricks-and-mortar video rental store, which has to have duplicate inventories at each location. All digital services, such as the iTunes store, eliminate all physical media. You purchase and download the content digitally, and there is no need for a warehouse to store DVDs and CDs.

Figure 3.3
The Long Tail

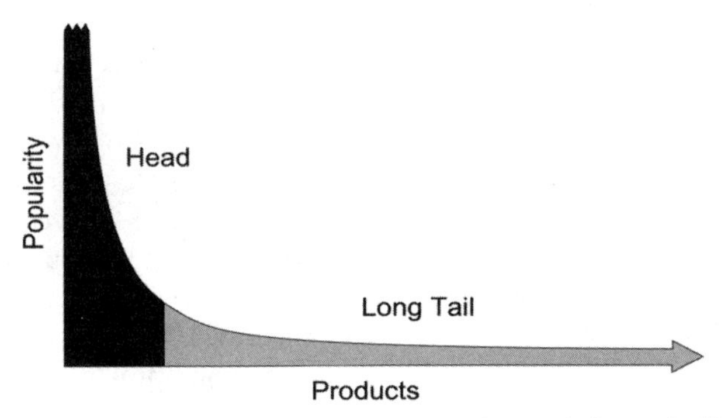

Source: Anderson (n.d.)

Because of these efficiencies, these businesses can better serve niche markets. Taken one at a time, these niche markets may not generate significant revenue but when they are aggregated, these markets are significant.

Anderson (2006) suggests rules for long tail businesses. Make everything available, lower the price, and help people find it. Traditional media are responding to these services. For example, Nintendo is making classic games available for download. Network television is putting up entire series of television programming on the Internet.

The audience is changing, and expectations for content selection and availability are changing. The audience today, Anderson argues, wants what they want, when they want it, and how they want it.

The Principle of Relative Constancy

So now that people have all of this choice of content, delivery mode, etc., what happens to older media? Do people just keep adding new entertainment media, or do they adjust by dropping one form in favor of another?

This question is at the core of the principle of relative constancy, which says that people spend a constant fraction of their disposable income on mass media over time.

People do, however, alter their spending on mass media categories when new services/products are introduced (McCombs & Nolan, 1992). What this means is that, if a new media technology is introduced, in order for adoption to happen, the new technology has to be compelling enough for the adopter to give up something else.

For example, a person who signs up for Netflix may spend less money on movie tickets. A satellite radio user will spend less money purchasing music downloads or CDs. So, when considering a new media technology, the relative advantage it has over existing service must be considered, along with other characteristics of the technology discussed earlier in this chapter. Remember, the money users spend on any new technology has to come from somewhere.

Critical & Cultural Theories

Most of the theories discussed above are derived from sociological perspectives on media and media use, relying upon quantitative analysis and the study of individuals to identify the dimensions of the technologies and the audience that help us understand these technologies. An alternate perspective can be found in critical and cultural studies, which provide different perspectives that help us understand the role of media in society.

Critical and cultural studies make greater use of qualitative analysis, including more analysis of macro-level factors such as media ownership and influences on the structure and content of media.

Marxist scholars, for example, focus on the underlying economic system and the division between those who control and benefit from the means of production and those who actually produce and consume goods and services.

This tradition considers that ownership and control of media and technology can affect the type of content provided. Scholars including Herman & Chomsky (2008) and Bagdikian (2000) have provided a wide range of evidence and analysis informing our understanding of these structural factors.

Feminist theory provides another set of critical perspectives that can help us understand the role and function of communication technology in society. Broadly speaking, feminist theory encompasses a broad range of factors including bodies, race, class, gender, language, pleasure, power, and sexual division of labor (Kolmar & Bartkowski, 2005).

New technologies represent an especially interesting challenge to the existing power relationships in media (Allen, 1999), and application of feminist theory has the potential to explicate the roles and power relationships in these media as well as to prescribe new models of organizational structure that take advantage of the shift in power relationships offered by interactive media (Grant, Meadows, & Storm, 2009).

Cultural studies also address the manner in which content is created and understood. Semiotics, for example, differentiates among "signs," "signifiers," and "signified" (Eco, 1979), explicating the manner in which meaning is attached to content. Scholars such as Hall (2010) have elaborated the processes by which content is encoded and decoded, helping us to understand the complexities of interpreting content in a culturally diverse environment.

Critical and cultural studies put the focus on a wide range of systemic and subjective factors that challenge conventional interpretations and understandings of media and technologies.

Conclusion

This chapter has provided a brief overview of several theoretical approaches to understanding communication technology. As you work through the book, consider theories of adoption, effects, economics, and critical/cultural studies and how they can inform you about each technology and allow you to apply lessons from one technology to others. For more in-depth discussions of these theoretical approaches, check out the sources cited in the bibliography.

Bibliography

Ajzen, I. (1991). The theory of planned behavior. *Organizational Behavior and Human Decision Processes*, 50, 179–211.

Ajzen, I. & Fishbein, M. (1980). *Understanding attitudes and predicting social behavior.* Englewood Cliffs, NJ: Prentice-Hall.

Allen, A. (1999). *The power of feminist theory.* Boulder, CO: Westview Press.

Anderson, C. (n.d.). *About me.* Retrieved from http://www.thelongtail.com/about.html.

Anderson, C. (2006). *The long tail: Why the future of business is selling less of more.* New York, NY: Hyperion.

Ball-Rokeach, S. & DeFleur, M. (1976). A dependency model of mass-media effects. *Communication Research, 3,* 1 3-21.

Bagdikian, B. H. (2000). *The media monopoly* (Vol. 6). Boston: Beacon Press.

Bandura, A. (2001). Social cognitive theory of mass communication. *Media Psychology, 3,* 265-299.

Beniger, J. (1986). The information society: Technological and economic origins. In S. Ball-Rokeach & M. Cantor (Eds.). *Media, audience, and social structure.* Newbury Park, NJ: Sage, pp. 51-70.

Chau, P. Y. K. & Hu, P. J. (2002). Examining a model of information technology acceptance by individual professionals: An exploratory study. *Journal of Management Information Systems,* **18**(4), 191–229.

Davis, F. D. (1986). *A technology acceptance model for empirically testing new end-user information systems: Theory and results.* Unpublished doctoral dissertation, Massachusetts Institute of Technology, Cambridge.

Davis, F. D., Bagozzi, R. P. & Warshaw, P. R. (1989). User acceptance of computer technology: A comparison of two theoretical models. *Management Science,* **35**(8), 982–1003.

Eco, U. (1979). *A theory of semiotics* (Vol. 217). Indiana University Press.

Etefa, A. (2005). *Arabic satellite channels in the U.S.: Uses & gratifications.* Paper presented at the annual meeting of the International Communication Association, New York. Retrieved May 2, 2008 from http://www.allacademic.com/ meta/p14246 index.html.

Ferguson, D. & Perse, E. (2000, Spring). The World Wide Web as a functional alternative to television. *Journal of Broadcasting and Electronic Media. 44* (2), 155-174.

Fishbein, M. & Ajzen, I. (1975). *Belief, attitude, intention and behavior: An introduction to theory and research.* Reading, MA: Addison-Wesley.

Fulk, J., Schmitz, J. & Steinfield, C. W. (1990). A social influence model of technology use. In J. Fulk & C. Steinfield (Eds.), *Organizations and communication technology.* Thousand Oaks, CA: Sage, pp. 117-140.

Fulk, J., Steinfield, C., Schmitz, J. & Power, J. (1987). A social information processing model of media use in organizations. *Communication Research, 14* (5), 529-552.

Gladwell, M. (2002). *The tipping point: How little things can make a big difference.* New York: Back Bay Books.

Grant, A. E., Guthrie, K. & Ball-Rokeach, S. (1991). Television shopping: A media system dependency perspective. *Communication Research, 18* (6), 773-798.

Grant, A. E., Meadows, J. H., & Storm, E. J. (2009). A feminist perspective on convergence. In Grant, A. E. & Wilkinson, J. S. (Eds.) *Understanding media convergence: The state of the field.* New York: Oxford University Press.

Hall, S. (2010). Encoding, decoding1. *Social theory: Power and identity in the global era, 2*, 569.

Herman, E. S., & Chomsky, N. (2008). *Manufacturing consent: The political economy of the mass media*. Random House.

Jiang, J. J., Hsu, M. & Klein, G. (2000). E-commerce user behavior model: An empirical study. *Human Systems Management*, 19, 265–276.Mathieson, K. (1991). Predicting user intentions: Comparing the technology acceptance model with the theory of planned behavior. *Information Systems Research*, **2**, 173–191.

Katz, E., Blumler, J. & Gurevitch, M. (1974). Utilization of mass communication by the individual. In J. Blumler & E. Katz (Eds.). *The uses of mass communication: Current perspectives on gratifications research.* Beverly Hills: Sage.

Kolmar, W., & Bartkowski, F. (Ed.). (2005). *Feminist theory: A reader*. Boston: McGraw-Hill Higher Education.

Kottler, P. and Keller, K. L. (2011) *Marketing Management (14th ed.)* Englewood Cliffs, NJ: Prentice-Hall.

Markus, M. (1987, October). Toward a "critical mass" theory of interactive media. *Communication Research, 14* (5), 497-511.

McCombs, M. & Nolan, J. (1992, Summer). The relative constancy approach to consumer spending for media. *Journal of Media Economics*, 43-52.

Meadows, J. H. (1997, May). *Body image, women, and media: A media system dependency theory perspective.* Paper presented to the 1997 Mass Communication Division of the International Communication Association Annual Meeting, Montreal, Quebec, Canada.

Moore, G. (2001). *Crossing the Chasm*. New York: Harper Business.

Park, N., Lee, K. M. & Cheong, P. H. (2007). University instructors' acceptance of electronic courseware: An application of the technology acceptance model. *Journal of Computer-Mediated Communication*, **13**(1), Retrieved from http://jcmc.indiana.edu/vol13/issue1/park.html.

Robinson, L. (1986). *The facts on fax*. Dallas: Steve Davis Publishing.

Rogers, E. (1962). *Diffusion of Innovations*. New York: Free Press.

Rogers, E. (1986). *Communication technology: The new media in society*. New York: Free Press.

Rogers, E. (2003). *Diffusion of Innovations, 3rd Edition*. New York: Free Press.

Rubin, A. (2002). The uses-and-gratifications perspective of media effects. In J. Bryant & D. Zillmann (Eds.). *Media effects: Advances in theory and research*. Mahwah, NJ: Lawrence Earlbaum Associates, pp. 525-548.

Towers, W. (1987, May 18-21). *Replicating perceived helpfulness of radio news and some uses and gratifications*. Paper presented at the Annual Meeting of the Eastern Communication Association, Syracuse, New York.

Wathne, E. & Leos, C. R. (1993). Facsimile machines. In A. E. Grant & K. T. Wilkinson (Eds.). *Communication technology update: 1993-1994.* Austin: Technology Futures, Inc.

Wei, R. (2006). Staying connected while on the move. *New Media and Society, 8* (1), 53-72.

4

The Structure of the Communication Industries

August E. Grant, Ph.D.*

The first factor that many people consider when studying communication technologies is changes in the equipment and utility of the technology. But, as discussed in Chapter 1, it is equally important to study and understand all areas of the technology ecosystem. In editing the Communication Technology Update for the past 20 years, one factor stands out as having the greatest amount of change in the short term: the organizational infrastructure of the technology.

The continual flux in the organizational structure of communication industries makes this area the most dynamic area of technology to study. "New" technologies that make a major impact come along only a few times a decade. New products that make a major impact come along once or twice a year. Organizational shifts are constantly happening, making it almost impossible to know all of the players at any given time.

Even though the players are changing, the organizational structure of communication industries is relatively stable. The best way to understand the industry, given the rapid pace of acquisitions, mergers, startups, and failures, is to understand its organizational functions. This chapter addresses the organizational structure and explores the functions of those industries, which will help you to understand the individual technologies discussed throughout this book.

* J. Rion McKissick, Professor of Journalism, School of Journalism and Communications, University of South Carolina (Columbia, South Carolina).

In the process of using organizational functions to analyze specific technologies, do not forget to consider that these functions cross national as well as technological boundaries. Most hardware is designed in one country, manufactured in another, and sold around the globe. Although there are cultural and regulatory differences addressed in the individual technology chapters later in the book, the organizational functions discussed in this chapter are common internationally.

What's in a Name? The AT&T Story

A good illustration of the importance of understanding organizational functions comes from analyzing the history of AT&T, one of the biggest names in communication of all time. When you hear the name "AT&T," what do you think of? Your answer probably depends on how old you are and where you live. If you live in Florida, you may know AT&T as your local phone company. In New York, it is the name of one of the leading mobile telephone companies. If you are older than 55, you might think of the company's old nickname "Ma Bell."

The Birth of AT&T

In the study of communication technology over the last century, no name is as prominent as AT&T. The company known today as AT&T is an awkward descendent of the company that once held a monopoly on long-distance telephone service and a near monopoly on local telephone service through the first four decades of the 20th century. The AT&T story is a story of visionaries, mergers, divestiture, and rebirth.

Alexander Graham Bell invented his version of the telephone in 1876, although historians note that he barely beat his competitors to the patent office. His invention soon became an important force in business communication, but diffusion of the telephone was inhibited by the fact that, within 20 years, thousands of entrepreneurs established competing companies to provide telephone service in major metropolitan areas. Initially, these telephone systems were not interconnected, making the choice of telephone company a difficult one, with some businesses needing two or more local phone providers to connect with their clients.

The visionary who solved the problem was Theodore Vail, who realized that the most important function was the interconnection of these telephone companies. Vail led American Telephone & Telegraph to provide the needed interconnection, negotiating with the U.S. government to provide "universal service" under heavy regulation in return for the right to operate as a monopoly. Vail brought as many local telephone companies as he could into AT&T, which evolved under the eye of the federal government as a behemoth with three divisions:

- AT&T Long Lines—the company that had a virtual monopoly on long distance telephony in the United States.
- The Bell System—Local telephone companies providing service to 90% of U.S. subscribers.
- Western Electric—A manufacturing company that made equipment needed by the other two divisions, from telephones to switches. (Bell Labs was a part of Western Electric.)

As a monopoly that was generally regulated on a rate-of-return basis (making a fixed profit percentage), AT&T had little incentive—other than that provided by regulators—to hold down costs. The more the company spent, the more it had to charge to make its profit, which grew in proportion with expenses. As a result, the U.S. telephone industry became the envy of the world, known for "five nines" of reliability; that is, the telephone network was available 99.999% of the time. The company also spent millions every year on basic research, with its Bell Labs responsible for the invention of many of the most important technologies of the 20th century, including the transistor and the laser.

Divestiture

The monopoly suffered a series of challenges in the 1960s and 1970s that began to break AT&T's monopoly control. First, AT&T lost a suit brought by the Hush-a-Phone company, which made a plastic mouthpiece that fit over the AT&T telephone mouthpiece to make it easier to hear a call made in a noisy area (*Hush-a-phone v. AT&T*, 1955; *Hush-a-phone v. U.S.*, 1956). (The idea of a company having to win a lawsuit in order to sell such an innocent item might seem frivolous today, but this suit was the first major crack in AT&T's monopoly armor.) Soon, MCI successfully sued for the right to provide long-distance service between St. Louis and Chicago, allowing businesses to bypass AT&T's long lines (*Microwave Communications, Inc.*, 1969).

Since the 1920s, the Department of Justice (DOJ) had challenged aspects of AT&T's monopoly control, earning a series of consent decrees to limit AT&T's market power and constrain corporate behavior. By the 1970s, it was clear to the antitrust attorneys that AT&T's ownership of Western Electric inhibited innovation, and the DOJ attempted to force AT&T to divest itself of its manufacturing arm. In a surprising move, AT&T proposed a different divestiture, spinning off all of its local telephone companies into seven new "Baby Bells," keeping the now-competitive long distance service and manufacturing arms. The DOJ agreed, and a new AT&T was born (Dizard, 1989).

Cycles of Expansion and Contraction

After divestiture, the leaner, "new" AT&T attempted to compete in many markets with mixed success; AT&T long distance service remained a national leader, but few people bought the overpriced AT&T personal computers. In the meantime, the seven Baby Bells focused on serving their local markets, with most of them named after the region they served. Nynex served New York and states in the extreme northeast, Bell Atlantic served the mid-Atlantic states, BellSouth served the southeastern states, Ameritech served the midwest, Southwestern Bell served the south central states, U S West served a set of western states, and Pacific Telesis served California and the far western states.

Over the next two decades, consolidation occurred among these Baby Bells. Nynex and Bell Atlantic merged to create Verizon. U S West was purchased by Qwest Communication and renamed after its new parent, which was, in turn, acquired by

CenturyLink in 2010. As discussed below, Southwestern Bell was the most aggressive Baby Bell, ultimately reuniting more than half of the Baby Bells.

In the meantime, AT&T entered the 1990s with a repeating cycle of growth and decline. It acquired NCR Computers in 1991 and McCaw Communications (at that time the largest U.S. cellular telephone company) in 1993. Then, in 1995, it divested itself of its manufacturing arm (which became Lucent Technologies) and the computer company (which took the NCR name). It grew again in 1998 by acquiring TCI, the largest U.S. cable television company, renaming it AT&T Broadband, and then acquired another cable company, MediaOne. In 2001, it sold AT&T Broadband to Comcast, and it spun off its wireless interests into an independent company (AT&T Wireless), which was later acquired by Cingular (a wireless phone company co-owned by Baby Bells SBC and BellSouth) (AT&T, 2008).

The only parts of AT&T remaining were the long distance telephone network and the business services, resulting in a company that was a fraction of the size of the AT&T behemoth that had a near monopoly on the telephone industry in the United States just two decades earlier.

Under the leadership of Edward Whitacre, Southwestern Bell became one of the most formidable players in the telecommunications industry. With a visionary style not seen in the telephone industry since the days of Theodore Vail, Whitacre led Southwestern Bell to acquire Baby Bells Pacific Telesis and Ameritech (and a handful of other, smaller telephone companies), renaming itself SBC. Ultimately, SBC merged with BellSouth and purchased what was left of AT&T, then renamed the company AT&T, an interesting case comparable to a child adopting its parent.

Today's AT&T is a dramatically different company with a dramatically different culture than its parent, but the company serves most of the same markets in a much more competitive environment. The lesson is that it is not enough to know the technologies or the company names; you also have to know the history of both in order to understand the role that a company plays in the marketplace.

Functions within the Industries

The AT&T story is an extreme example of the complexity of communication industries. These industries are easier to understand by breaking their functions into categories that are common across most of the segments of these industries. Let's start by picking up the heart of the technology ecosystem introduced in Chapter 1, the hardware, software, and content.

For this discussion, let's use the same definitions used in Chapter 1, with hardware referring to the physical equipment used, software referring to instructions used by the hardware to manipulate content, and content referring to the messages transmitted using these technologies. Some companies produce both hardware and software, ensuring compatibility between the equipment and programming, but few companies produce both equipment and content.

The next distinction has to be made between production and distribution of both equipment and content. As these names imply, companies involved in production engage in the manufacture of equipment or content, and companies involved in distribution are the intermediaries between production and consumers. It is a common practice for some companies to be involved in both production and distribution, but, as discussed below, a large number of companies choose to focus on one or the other.

These two dimensions interact, resulting in separate functions of equipment production, equipment distribution, content production, and content distribution. As discussed below, distribution can be further broken down into national and local distribution. The following section introduces each of these dimensions, which are applied in the subsequent section to help identify the role played by specific companies in communication industries.

One other note: These functions are hierarchical, with production coming before distribution in all cases. Let's say you are interested in creating a new type of telephone, perhaps a "high-definition telephone." You know that there is a market, and you want to be the person who sells it to consumers. But you cannot do so until someone first makes the device. Production always comes before distribution, but you cannot have successful production unless you also have distribution—hence the hierarchy in the model. Figure 4.1 illustrates the general pattern, using the U.S. television industry as an example.

Hardware Path

When you think of hardware, you typically envision the equipment you handle to use a communication technology. But it is also important to note that there is a second type of hardware for most communication industries—the equipment used to make the messages. Although most consumers do not deal with this equipment, it plays a critical role in the system.

Production

Production hardware is usually more expensive and specialized than other types. Examples in the television industry include TV cameras, microphones, and editing equipment. A successful piece of production equipment might sell only a few hundred or a few thousand units, compared with tens of thousands to millions of units for consumer equipment. The profit margin on each piece of production equipment is usually much higher than on consumer equipment, making it a lucrative market for electronics manufacturing companies.

(a) Consumer Hardware

Consumer hardware is the easiest to identify. It includes anything from a digital video recorder (DVR) to a mobile phone or DirecTV satellite dish. A common term used to identify consumer hardware in consumer electronics industries is CPE, which stands for customer premises equipment. An interesting side note is that many companies do not actually make their own products, but instead hire manufacturing facilities to make products they design, shipping them directly to distributors. For example, Microsoft does not manufacture the Xbox 360; Flextronics, Wistron, and Celestica do. As you consider communication technology hardware, consider the lesson from Chapter 1—people are not usually motivated to buy equipment because of the equipment itself, but because of the content it enables, from the pictures recorded on a camera to the conversations (voice and text!) on a wireless phone to the information and entertainment provided by a high-definition television (HDTV) receiver.

(b) Distribution

After a product is manufactured, it has to get to consumers. In the simplest case, the manufacturer sells directly to the consumer, perhaps through a company-owned store or a website. In most cases, however, a product will go through multiple organizations, most often with a wholesaler buying it from the manufacturer and selling it, with a mark-up, to a retail store, which also marks up the price before selling it to a consumer. The key point is that few manufacturers control their own distribution channels, instead relying on other companies to get their products to consumers.

Figure 4.1
Structure of the Traditional Broadcast TV Industry

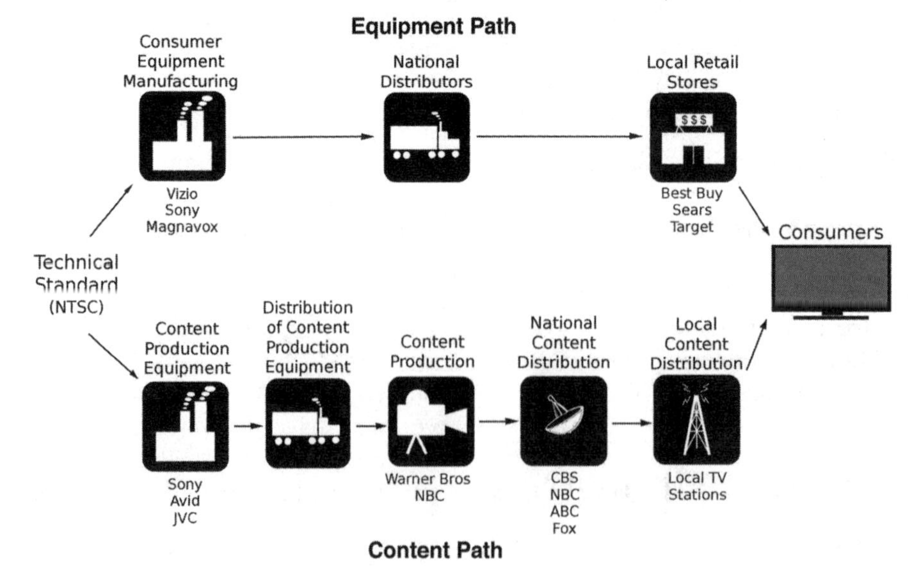

Source: R. Grant & G. Martin, 2012

Content Path: Production and Distribution

The process that media content goes through to get to consumers is a little more complicated than the process for hardware. The first step is the production of the content itself. Whether the product is movies, music, news, images, etc., some type of equipment must be manufactured and distributed to the individuals or companies who are going to create the content. (That hardware production and distribution goes through a similar process to the one discussed above.) The content must then be created, duplicated, and distributed to consumers or other end users.

The distribution process for media content/ software follows the same pattern for hardware. Usually there will be multiple layers of distribution, a national wholesaler that sells the content to a local retailer, which in turn sells it to a consumer.

(c) Disintermediation

Although many products go through multiple layers of distribution to get to consumers, information technologies have also been applied to reduce the complexity of distribution. The process of eliminating layers of distribution is called disintermediation (Kottler & Keller, 2005); examples abound of companies that use the Internet to get around traditional distribution systems to sell directly to consumers.

Netflix is a great example. Traditionally, digital videodiscs (DVDs) of a movie would be sold by the studio to a national distributor, which would then deliver them to hundreds or thousands of individual movie rental stores, which would then rent or sell them to consumers. (Note: The largest video stores buy directly from the studio, handling both national and local distribution.)

Netflix cuts one step out of the distribution process, directly bridging the movie studio and the consumer. (As discussed below, iTunes serves the same function for the music industry, simplifying music distribution.) The result of getting rid of one middleman is greater profit for the companies involved, lower costs to the consumer, or both.

Illustrations: HDTV and HD Radio

The emergence of digital broadcasting provides two excellent illustrations of the complexity of the organizational structure of media industries. HDTV and its distant cousin (HD) radio have had a difficult time penetrating the market because of the need for so many organizational functions to be served before consumers can adopt the technology.

Let's start with the simpler one: HD radio. As illustrated in Figure 4.2, this technology allows existing radio stations to broadcast their current programming (albeit with much higher fidelity), so no changes are needed in the software production area of the model. The only change needed in the software path is that radio stations simply need to add a digital transmitter. The complexity is related to the consumer hardware needed to receive HD radio signals. One set of companies needs to make the radios, another has to distribute the radios to retail stores, other distribution channels, and stores and distributors have to agree to sell them.

The radio industry has therefore taken an active role in pushing diffusion of HD radios throughout the hardware path. In addition to airing thousands of radio commercials promoting HD radio, the industry is promoting distribution of HD radios in new cars (because so much radio listening is done in automobiles).

As discussed in Chapter 9, adoption of HD radio has begun, but has been slow because listeners see little advantage in the new technology. However, if the number of receivers increases, broadcasters will have a reason to begin using the additional channels available with HD. As with FM radio, programming and receiver sales have to *both* be in place before consumer adoption takes place. Also, as with FM, the technology may take decades to take off

The same structure is inherent in the adoption of HDTV, as illustrated in Figure 4.3. Before the first consumer adoption could take place, both programming and receivers (consumer hardware) had to be available. Because a high percentage of primetime television programming was recorded on 35mm film at the time HDTV receivers first went on sale in the United States, that programming could easily be transmitted in high-definition, providing a nucleus of available programming. On the other hand, local news and network programs shot on video required entirely new production and editing equipment before they could be distributed to consumers in high-definition.

As discussed in Chapter 6, the big force behind the diffusion of HDTV and digital TV was a set of regulations issued by the Federal Communications Commission (FCC) that first required stations in the largest markets to begin broadcasting digital signals, then required that all television receivers include the capability to receive digital signals, and finally required that all full-power analog television broadcasting cease on June 12, 2009. In short, the FCC implemented mandates ensuring production and distribution of digital television, easing the path toward both digital TV and HDTV.

Figure 4.2

Structure of the HD Radio Industry

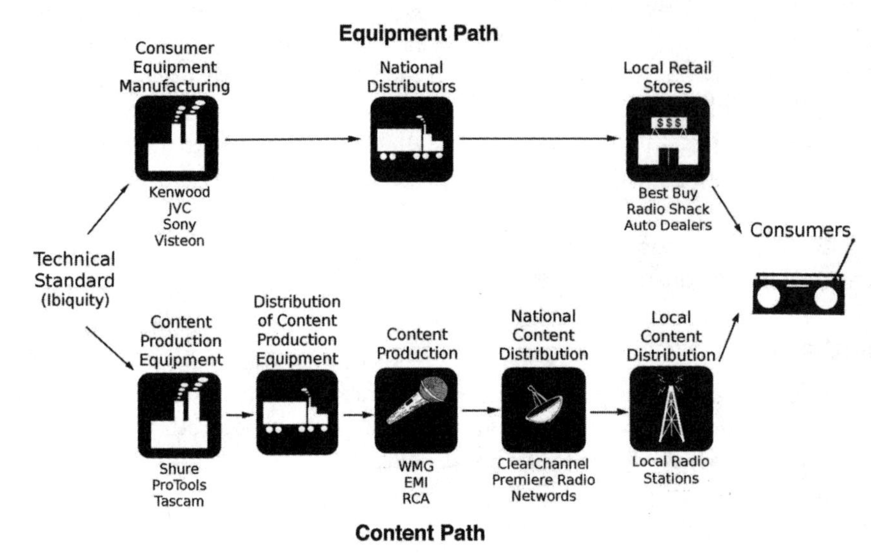

Source: R. Grant & G. Martin, 2012

Figure 4.3

Structure of the HDTV Industry

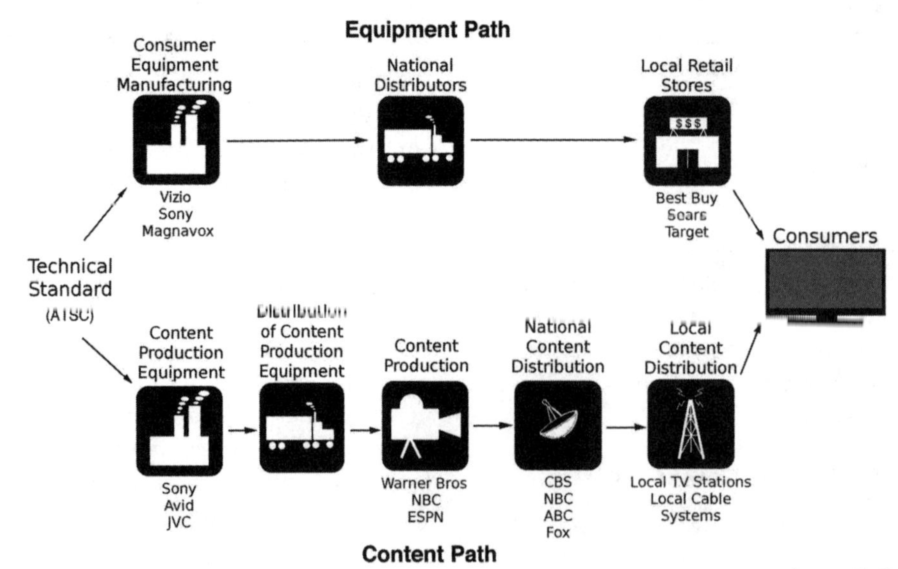

Source: R. Grant & G. Martin, 2012

As you will read in Chapter 6, this cycle will repeat itself over the next few years as the television industry adopts an improved, but incompatible, format known as UltraHDTV (or 4K).

The success of this format will require modifications of virtually all areas of the television system diagrammed in Figure 4.3, with new production equipment manufactured and then used to produce content in the new format, and then distributed to consumers—who are not likely to buy until all of these functions are served.

From Target to iTunes

One of the best examples of the importance of distribution comes from an analysis of the popular music industry. Traditionally, music was recorded on physical media such as CDs and audio tapes and then shipped to retail stores for sale directly to consumers. At one time, the top three U.S. retailers of music were Target, Walmart, and Best Buy.

Once digital music formats that could be distributed over the Internet were introduced in the late 1990s, dozens of start-up companies created online stores to sell music directly to consumers. The problem was that few of these online stores offered the top-selling music. Record companies were leery of the lack of control they had over digital distribution, leaving most of these companies to offer a marginal assortment of music.

The situation changed in 2003 when Apple introduced the iTunes store to provide content for its iPods, which had sold slowly since appearing on the market in 2001. Apple obtained contracts with major record companies that allowed them to provide most of the music that was in high demand.

Initially, record companies resisted the iTunes distribution model that allowed a consumer to buy a single song for $0.99; they preferred that a person have to buy an entire album of music for $13 to $20 to get the one or two songs he or she wanted.

Record company delays spurred consumers to create and use file-sharing services that allowed listeners to get the music for free—and the record companies ended up losing billions of dollars. Soon, the $0.99 iTunes model began to look very attractive to the record companies, and they trusted Apple's digital rights management system to protect their music.

Today, as discussed in Chapter 17, iTunes is the number one music retailer in the United States. The music is similar, but the distribution of music today is dramatically different from what it was when this decade began. The change took years of experimentation, and the successful business model that emerged required cooperation from dozens of separate companies serving different roles in the production and distribution process.

Two more points should be made regarding distribution. First, there is typically more profit potential and less risk in being a distributor than a creator (of either hardware or software) because the investment is less and distributors typically earn a percentage of the value of what they sell. Second, distribution channels can become very complicated when multiple layers of distribution are involved; the easiest way to unravel these layers is simply to "follow the money."

Importance of Distribution

As the above discussion indicates, distributors are just as important to new communication technologies as manufacturers and service providers. When studying these technologies, and the reasons for success or failure, the distribution process (including the economics of distribution) must be examined as thoroughly as the product itself.

Diffusion Threshold

Analysis of the elements in Figure 4.1 reveals an interesting conundrum—there cannot be any consumer adoption of a new technology until all of the production and distribution functions are served, along both the hardware and software paths.

This observation adds a new dimension to Rogers' (2003) diffusion theory, discussed in the previous chapter. The point at which all functions are served has been identified as the "diffusion threshold," the point at which diffusion of the technology can begin (Grant, 1990).

It is easier for a technology to "take off" and begin diffusing if a single company provides a number of different functions, perhaps combining production and distribution, or providing both national and local distribution. The technical term for owning multiple functions in an industry is "vertical integration," and a vertically integrated company has a disproportionate degree of power and control in the marketplace.

Vertical integration is easier said than done, however, because the "core competencies" needed for production and distribution are so different. A company that is great at manufacturing may not have the resources needed to sell the product to end consumers.

Let's consider the next generation of television, UltraHD, again. A company such as Vizio or Samsung might handle the first level of distribution, from the manufacturing plant to the retail store, but they do not own and operate their own stores—that is a very different business. They are certainly not involved in owning the television stations or cable channels that distribute UltraHD programming; that function is served by another set of organizations.

In order for UltraHD to become popular, one organization (or set of organizations) has to make the television receivers, another has to get those televisions into stores, a third has to operate the stores, a fourth has to make UltraHD cameras and technical equipment to produce content, and a fifth has to package and distribute the content to viewers, through the Internet using Internet protocol television (IPTV), cable television systems, or over-the-air.

Most companies that would like to grow are more interested in applying their core competencies by buying up competitors and commanding a greater market share, a process known as "horizontal integration." For example, it makes more sense for a company that makes televisions to grow by making other electronics rather than by buying television stations. Similarly, a company that already owns television stations will probably choose to grow by buying more television stations or cable networks rather than by starting to make and sell television receivers.

The complexity of the structure of most communication industries prevents any one company from serving every needed role. Because so many organizations have to be involved in providing a new technology, many new technologies end up failing.

The lesson is that understanding how a new communication technology makes it to market requires comparatively little understanding of the technology itself compared with the understanding needed of the industry in general, especially the distribution processes.

A "Blue" Lesson

One of the best examples of the need to understand (and perhaps exploit) all of the paths illustrated in Figure 4.1 comes from the earliest days of the personal computer. When the PC was invented in the 1970s, most manufacturers used their own operating systems, so that applications and content could not easily be transferred from one type of computer to other types. Many of these manufacturers realized that they needed to find a standard operating system that would allow the same programs and content to be used on computers from different manufacturers, and they agreed on an operating system called CP/M.

Before CP/M could become a standard, however, IBM, the largest U.S. computer manufacturer—mainframe computers, that is—decided to enter the personal computer market. "Big Blue," as IBM was known (for its blue logo and its dominance in mainframe computers, typewriters, and other business equipment) determined that its core competency was making hardware, and they looked for a company to provide them an operating system that would work on their computers. They chose a then-little-known operating system called MS-DOS, from a small start-up company called Microsoft.

IBM's open architecture allowed other companies to make compatible computers, and dozens of companies entered the market to compete with Big Blue. For a time, IBM dominated the personal computer market, but, over time, competitors steadily made inroads on the market. (Ultimately, IBM sold its personal computer manufacturing business in 2006 to Lenovo, a Chinese company.)

The one thing that most of these competitors had in common was that they used Microsoft's operating systems. Microsoft grew...and grew...and grew. (It is also interesting to note that, although Microsoft has dominated the market for software with its operating systems and productivity software such as *Office*, it has been a consistent failure in most areas of hardware manufacturing. Notable failures include Microsoft's routers and home networking hardware, keyboards and mice, and WebTV hardware. The only major success Microsoft has had in manufacturing hardware is with its Xbox video game system, discussed in Chapter 15.)

The lesson is that there is opportunity in all areas of production and distribution of communication technologies. All aspects of production and distribution must be studied in order to understand communication technologies. Companies have to know their own core competencies, but a company can often improve its ability to introduce a new technology by controlling more than one function in the adoption path.

What are the Industries?

We need to begin our study of communication technologies by defining the industries involved in providing communication-related services in one form or another. Broadly speaking, these can be divided into:

- *Mass media,* including books, newspapers, periodicals, movies, radio, and television.
- *Telecommunications,* including networking and all kinds of telephony (landlines, long distance, wireless, and voice over Internet protocol).
- *Computers,* including hardware (desktops, laptops, tablets, etc.) and software.
- *Consumer electronics,* including audio and video electronics, video games, and cameras.
- *Internet,* including enabling equipment, network providers, content providers, and services.

These industries were introduced in Chapter 2 and discussed in more detail in the individual chapters that follow.

At one point, these industries were distinct, with companies focusing on one or two industries. The opportunity provided by digital media and convergence enables companies to operate in numerous industries, and many companies are looking for synergies across industries.

Table 4.4 lists examples of well-known companies in the communication industries, some of which work across many industries, and some of which are (as of this writing) focused on a single industry. Part of the fun in reading this chart is seeing how much has changed since the book was printed in mid-2014.

There is a risk in discussing specific organizations in a book such as this one; in the time between when the book is written and when it is published, there are certain to be changes in the organizational structure of the industries. For example, as this chapter was being written in early 2014, Netflix had announced plans to be the first to distribute UltraHD television. By the time you read this, other companies will certainly announce their own plans to distribute UltraHD programming to viewers.

Fortunately, mergers and takeovers that revolutionize an industry do not happen that often—only a couple a year! The major players are more likely to acquire other companies than to be acquired, so it is fairly safe (but not completely safe) to identify the major players and then analyze the industries in which they are doing business.

As in the AT&T story earlier in this chapter, the specific businesses a company is in can change dramatically over the course of a few years.

Future Trends

The focus of this book is on changing technologies. It should be clear that some of the most important changes to track are changes in the organizational structure of media industries. The remainder of this chapter projects organizational trends to watch to help you predict the trajectory of existing and future technologies.

Disappearing Newspapers

For decades, newspapers were the dominant mass medium, commanding revenues, consumer attention, and significant political and economic power. Since the dramatic drop in newspaper revenues and subscriptions began about a decade ago, newspaper publishers have been reconsidering their core business. Some prognosticators have even predicted the demise of the printed newspaper completely, forcing newspaper companies to plan for digital distribution of their news and advertisements.

Before starting the countdown clock, it is necessary to define what we mean by a "newspaper publisher." If a newspaper publisher is defined as an organization that communicates and obtains revenue by smearing ink on dead trees, then we can easily predict a steady decline in that business. If, however, a newspaper publisher is defined as an organization that gathers news and advertising messages, distributing them via a wide range of available media, then newspaper publishers should be quite healthy through the century.

Table 4.4

Examples of Major Communication Company Industries, 2014

	TV/Film/Video Production	TV/Film/Video Distribution	Print	Telephone	Wireless	Internet
AT&T		••		•••	•••	••
Comcast	•••	•••		••		•••
Disney	•••	•••	•			••
Gannett	•	•••	•••			
Google		•			•	•••
News Corp.	•••	•••	•••			••
Sony	•••	•				•
Time Warner, Inc.	•••	•	•			•
Time Warner Cable		•••		••		••
Verizon				•••	•••	••
Viacom	•••	•••				••
Yahoo!	•	•				•••

The number of dots is proportional to the importance of this business to each company. *Source:* A. Grant (2014)

The current problem is that there is no comparable revenue model for delivery of news and advertising through new media that approaches the revenues available from smearing ink on dead trees. It is a bad news/good news situation.

The bad news is that traditional newspaper readership and revenues are both declining. Readership is suffering because of competition from the Web and other new media, with younger cohorts increasingly ignoring print in favor of other news sources.

Advertising revenues are suffering for two reasons. The decline in readership and competition from new media are impacting revenues from display advertising. More significant is the loss in revenues from classified advertising, which at one point comprised up to one-third of newspaper revenues.

The good news is that newspapers remain profitable, at least on a cash flow basis, with gross margins of 10% to 20%. This profit margin is one that many industries would envy. But many newspaper companies borrowed extensively to expand their reach, with interest payments often exceeding these gross profits. Stockholders in newspaper publishers have been used to much higher profit margins, and the stock prices of newspaper companies have been punished for the decline in profits.

Some companies reacted by divesting themselves of their newspapers in favor of TV and new media investments. Some newspaper publishers are using the opportunity to buy up other newspapers; consider McClatchy's 2006 purchase of the majority of Knight-Ridder's newspapers (McClatchy, 2008).

Gannett, on the hand, is taking the boldest, and potentially the riskiest, strategy by aggressively transforming both their newspaper and television newsrooms into "Information Centers," where the goal is to be platform agnostic, getting news out in any available medium as quickly as possible. According to former Gannett CEO Craig Dubow, the goal was to deliver the news and content anywhere the consumer is, and then solve the revenue question later (Gahran, 2006).

Gannett's approach was seen by many as a risky one, but it follows the model that has worked for new media in the past—the revenue always follows the audience, and the companies that are first to reach an audience through a new medium are disproportionately likely to profit from their investments.

Advertiser-Supported Media

For advertiser-supported media organizations, the primary concern is the impact of the Internet and other new media on revenues. As discussed above, some of the loss in revenues is due to loss of advertising dollars (including classified advertising), but that loss is not experienced equally by all advertiser-supported media.

The Internet is especially attractive to advertisers because online advertising systems have the most comprehensive reporting of any advertising medium. For example, an advertiser using the Google AdWords system discussed in Chapter 1 gets comprehensive reports on the effectiveness of every message—but "effectiveness" is defined by these advertisers as an immediate response such as a click-through.

As Grant and Wilkinson (2007) discuss, not all advertising is this type of "call-to-action" advertising. There is another type of advertising that is equally important—image advertising, which does not demand immediate results, but rather works over time to build brand identity and increase the likelihood of a future purchase.

Any medium can carry any type of advertising, but image advertising is more common on television (especially national television) and magazines, and call-to-action advertising is more common in newspapers. As a result, newspapers, at least in the short term, are more likely to be impacted by the increase in Internet advertising.

Interestingly, local advertising is more likely to be call-to-action advertising, but local advertisers have been slower than national advertisers to move to the Internet, most likely because of the global reach of the Internet. This paradox could be seen as an opportunity for an entrepreneur wishing to earn a million or two by exploiting a new advertising market (Wilkinson, Grant & Fisher, 2012).

The "Mobile Revolution"

Another important trend that can help you analyze media organizations is the shift toward mobile communication technologies. Companies that are positioned to produce and distribute content and technology that further enable the "mobile revolution" are likely to have increased prospects for growth.

Areas to watch include mobile Internet access (involving new hardware and software, provided by a mixture of existing and new organizations), mobile advertising, new applications of GPS technology, and a host of new applications designed to take advantage of Internet access available anytime, anywhere.

Consumers—Time Spent Using Media

Another piece of good news for media organizations in general is the fact that the amount of time consumers are spending with media is increasing, with much of that increase coming from simultaneous media use (Papper, et al., 2009). Advertiser-supported media thus have more "audience" to sell, and subscription-based media have more prospects for revenue.

Furthermore, new technologies are increasingly targeting specific messages at specific consumers, increasing the efficiency of message delivery for advertisers and potentially reducing the clutter of irrelevant advertising for consumers. Already, advertising services such as Google's Double-Click and Google's AdWords provide ads that are targeted to a specific person or the specific content on a Web page, greatly increasing their effectiveness.

Imagine a future where every commercial on TV that you see is targeted—and is interesting—to you! Technically, it is possible, but the lessons of previous technologies suggest that the road to customized advertising will be a meandering one.

Principle of Relative Constancy

The potential revenue from consumers is limited because they devote a fixed proportion of their disposable income to media, the phenomenon discussed in Chapter 3 as the "Principle of Relative Constancy." The implication is emerging companies and technologies have to wrest market share and revenue from established companies. To do that, they can't be just as good as the incumbents. Rather, they have to be faster, smaller, less expensive, or in some way better so that consumers will have the motivation to shift spending from existing media.

Conclusions

The structure of the media system may be the most dynamic area in the study of new communication technologies, with new industries and organizations constantly emerging and merging. In the following chapters, organizational developments are therefore given significant attention. Be warned, however; between the time these chapters are written and published, there is likely to be some change in the organizational structure of each technology discussed in this book. To keep up with these developments, visit the *Communication Technology Update and Fundamentals* home page at www.tfi.com/ctu.

Bibliography

AT&T. (2008). *Milestones in AT&T history*. Retrieved from http://www.corp.att.com/history/ milestones.html.

Dizard, W. (1989). *The coming information age: An overview of technology, economics, and politics, 2nd ed.* New York: Longman.

Gahran, A. (2006). Gannett "Information Centers"—Good for daily journalism? *Poynter Online E-Media Tidbits*. Retrieved May 4, 2008 from http://www.poynter.org/dg.lts/id.31/aid.113411/column.htm.

Grant, A. E. (1990, April). The "pre-diffusion of HDTV: Organizational factors and the "diffusion threshold. Paper presented to the Annual Convention of the Broadcast Education Association, Atlanta.

Grant, A. E. & Wilkinson, J. S. (2007, February). Lessons for communication technologies from Web advertising. Paper presented to the Mid-Winter Conference of the Association of Educators in Journalism and Mass Communication, Reno.

Hush-A-Phone Corp. v. AT&T, et al. (1955). FCC Docket No. 9189. Decision and order (1955). 20 FCC 391.

Hush-A-Phone Corp. v. United States. (1956). 238 F. 2d 266 (D.C. Cir.). Decision and order on remand (1957). 22 FCC 112.

Kottler, P. & Keller, K. L. (2005). *Marketing management, 12th ed.* Englewood Cliffs, NJ: Prentice-Hall.

McClatchy. (2008). *About the McClatchy Company*. Retrieved from http://www.mcclatchy.com/100/story/ 179.html.

Microwave Communications, Inc. (1969). FCC Docket No. 16509. Decision, 18 FCC 2d 953.

Meyer, P. (2006). *The vanishing newspaper: Saving journalism in the information age*. Columbia, MO: University of Missouri Press.

Papper, R. E., Holmes, M. A. & Popovich, M. N. (2009). Middletown media studies II: Observing consumer interactions with media. In A. E. Grant & J. S. Wilkinson (Eds.) *Understanding media convergence: The state of the field*. New York: Oxford.

Rogers, E. M. (2003). *Diffusion of innovations, 5th ed.* New York: Free Press.

Wilkinson, J.S., Grant, A. E. & Fisher, D. J. (2012). *Principles of convergent journalism (2nd ed.)*. New York: Oxford University Press.

5

Communication Policy & Technology

Lon Berquist, M.A.*

Throughout its history, U.S. communication policy has been shaped by evolving communication technologies. As a new communication technology is introduced into society, it is often preceded by an idealized vision, or Blue Sky perspective, of how the technology will positively impact economic opportunities, democratic participation, and social inclusion. Due, in part, to this perspective, government policy makers traditionally look for policies and regulations that will foster the wide diffusion of the emerging technology. At the same time, however, U.S. policy typically displays a light regulatory touch, promoting a free-market approach that attempts to balance the economic interests of media and communication industries, the First Amendment, and the rights of citizens.

Indeed, much of the recent impetus for media deregulation was directly related to communication technologies as "technological plenty is forcing a widespread reconsideration of the role competition can play in broadcast regulation" (Fowler & Brenner, 1982, p. 209). From a theoretical perspective, some see new communication technologies as technologies of freedom where "freedom is fostered when the means of communication are dispersed, decentralized, and easily available" (Pool, 1983, p. 5). Others fear technologies favor government and private interests and become technologies of control (Gandy, 1989). Still others argue that technologies are merely neutral in how they shape society. No matter the perspective, the purpose of policy and regulation is to allow society to shape the use of communication technologies to best serve the citizenry.

* Telecommunications and Information Policy Institute, University of Texas at Austin (Austin, Texas).

Background

The First Amendment is a particularly important component of U.S. communication policy, balancing freedom of the press with the free speech rights of citizens. The First Amendment was created at a time when the most sophisticated communication technology was the printing press. Over time, the notion of "press" has evolved with the introduction of new communication technologies. The First Amendment has evolved as well, with varying degrees of protection for the traditional press, broadcasting, cable television, and the Internet.

Communication policy is essentially the balancing of national interests and the interests of the communications industry (van Cuilenburg & McQuail, 2003). In the United States, communication policy is often shaped in reaction to the development of a new technology. As a result, policies vary according to the particular communication policy regime: press, common carrier, broadcasting, cable TV, and the Internet. Napoli (2001) characterizes this policy tendency as a "technologically particularistic" approach leading to distinct policy and regulatory structures for each new technology. Thus, the result is differing First Amendment protections for the printed press, broadcasting, cable television, and the Internet (Pool, 1983).

In addition to distinct policy regimes based on technology, scholars have recognized differing types of regulation that impact programming; the industry market and economics; and the transmission and delivery of programming and information. These include content regulation, structural regulation, and technical regulation.

Content regulation refers to the degree to which a particular industry enjoys First Amendment protection. For example, in the United States, the press is acknowledged as having the most First Amendment protection, and there certainly is no regulatory agency to oversee printing. Cable television has limited First Amendment protection, while broadcasting has the most limitations on its First Amendment rights. This regulation is apparent in the type of programming rules and regulations imposed by the Federal Communication Commission (FCC) on broadcast programming that is not imposed on cable television programming.

Structural regulation addresses market power within (horizontal integration) and across (vertical integration) media industries. Federal media policy has long established the need to promote diversity of programming by promoting diversity of ownership. The *Telecommunications Act of 1996* changed media ownership limits for the national and local market power of radio, television, and cable television industries; however, the FCC is given the authority to review and revise these rules. Structural regulation includes limitations or permissions to enter communication markets. For example, the *Telecommunications Act of 1996* opened up the video distribution and telephony markets by allowing telephone companies to provide cable television service and for cable television systems to offer telephone service (Parsons & Frieden, 1998).

Technical regulation needs prompted the initial development of U.S. communication regulation in the 1920s, as the fledgling radio industry suffered from signal interference while numerous stations transmitted without any government referee (Starr, 2004). Under FCC regulation, broadcast licensees are allowed to transmit at a certain power, or wattage, on a precise frequency within a particular market. Cable television systems and satellite transmission also follow some technical regulation to prevent signal interference.

Finally, in addition to technology-based policy regimes and regulation types, communication policy is guided by varying jurisdictional regulatory bodies. Given the global nature of satellites, both international (International Telecommunications Union) and national (FCC) regulatory commissions have a vested interest in satellite transmission.

Regulation of U.S. broadcasting is exclusively the domain of the federal government through the FCC. The telephone industry is regulated primarily at the federal level through the FCC, but also with regulations imposed by state public utility commissions. Cable television, initially regulated through local municipal franchises, is now regulated at the federal level, the local municipal level, and, for some, at the state level (Parsons & Frieden, 1998).

The Evolution of Communication Technologies

Telegraph

Although the evolution of technologies has influenced the policymaking process in the United States, many of the fundamental characteristics of U.S. communication policy were established early in the history of communication technology deployment, starting with the telegraph.

There was much debate on how best to develop the telegraph. For many lawmakers and industry observers, the telegraph was viewed as a natural extension of the Post Office, while others favored government ownership based on the successful European model as the only way to counter the power of a private monopoly (DuBoff, 1984).

In a prelude to the implementation of universal service for the telephone (and the current discussion of a "digital divide"), Congress decreed that, "Where the rates are high and facilities poor, as in this country, the number of persons who use the telegraph freely, is limited. Where the rates are low and the facilities are great, as in Europe, the telegraph is extensively used by all classes for all kinds of business" (Lubrano, 1997, p. 102).

Despite the initial dominance of Western Union, there were over 50 separate telegraph companies operating in the United States in 1851. Interconnecting telegraph lines throughout the nation became a significant policy goal of federal, state, and local governments. No geographic area wanted to be disconnected from the telegraph network and its promise of enhanced communication and commerce.

Eventually, in 1887, the *Interstate Commerce Act* was enacted, and the policy model of a regulated, privately-owned communication system was initiated and formal federal regulation began. Early in the development of communication policy, the tradition of creating communications infrastructure through government aid to private profit-making entities was established (Winston, 1998).

Telephone

Similar to the development of the telegraph, the diffusion of the telephone was slowed by competing, unconnected companies serving their own interests. Although AT&T dominated most urban markets, many independent telephone operators and rural cooperatives provided service in smaller towns and rural areas. Since there was no interconnection among the various networks, some households and businesses were forced to have dual service in order to communicate (Starr, 2004).

As telephone use spread in the early 1900s, states and municipalities began regulating and licensing operators as public utilities, although Congress authorized the Interstate Commerce Commission (ICC) to regulate interstate telephone service in 1910. Primarily an agency devoted to transportation issues, the ICC never became a major historical player in communication policy. However, two important phrases originated with the commission and the related *Transportation Act of 1920*.

The term "common carrier," originally used to describe railroad transportation, was used to classify the telegraph and eventually the telephone (Pool, 1983). Common carriage law required carriers to serve their customers without discrimination.

The other notable phrase utilized in transportation regulation was a requirement to serve the "public interest, convenience, or necessity" (Napoli, 2001). This nebulous term was adopted in subsequent broadcast legislation, and continues to guide the FCC even today.

As telephone use increased, it became apparent that there was a need for greater interconnection among competing operators, or the development of some national unifying agreement. In 1907, AT&T President Theodore Vail promoted a policy with the slogan, "One system, one policy, universal service" (Starr, 2004, p. 207). There are conflicting accounts of Vail's motivations: whether it was a sincere call for a national network available to all, or merely a ploy to protect AT&T's growing power in the telephone industry (Napoli, 2001).

Eventually, the national network envisioned by Vail became a reality, as AT&T was given the monopoly power, under strict regulatory control, to build and maintain local and long distance telephone service throughout the nation. Of course, this regulated monopoly was ended decades ago, but the concept of universal service as a significant component of communication policy remains today.

Broadcasting

While U.S. policymakers pursued an efficient national network for telephone operations, they developed radio broadcasting to primarily serve local markets. Before the federal government imposed regulatory control over radio broadcasting in 1927, the industry suffered from signal interference and an uncertain financial future. The *Federal Radio Act* imposed technical regulation on use of spectrum and power, allowing stations to develop a stable local presence.

Despite First Amendment concerns about government regulation of radio, the scarcity of spectrum was considered an adequate rationale for licensing stations. In response to concerns about freedom of the press, the *Radio Act* prohibited censorship by the Radio Commission, but the stations understood that the power of the commission to license implied inherent censorship (Pool, 1983). In 1934, Congress passed the *Communication Act of 1934*, combining regulation of telecommunications and broadcasting by instituting a new Federal Communications Commission.

The *Communication Act* essentially reiterated the regulatory thrust of the 1927 *Radio Act*, maintaining that broadcasters serve the public interest. This broad concept of "public interest" has stood as the guiding force in developing communication policy principles of competition, diversity, and localism (Napoli, 2001; Alexander & Brown, 2007).

Rules and regulations established to serve the public interest for radio transferred to television when it entered the scene. Structural regulation limited ownership of stations, technical regulation required tight control of broadcast transmission, and indirect content regulation led to limitations on station broadcast of network programming and even fines for broadcast of indecent material (Pool, 1983).

One of the most controversial content regulations was the vague Fairness Doctrine, established in 1949, that required broadcasters to present varying viewpoints on issues of public importance (Napoli, 2001). Despite broadcasters' challenges to FCC content regulation on First Amendment grounds, the courts defended the commission's Fairness Doctrine (*Red Lion Broadcasting v. FCC*, 1969) and its ability to limit network control over programming (*NBC v. United States*, 1943), In 1985, the FCC argued the Fairness Doctrine was no longer necessary given the increased media market competition, due in part to the emergence of new communication technologies (Napoli, 2001).

Historically, as technology advanced, the FCC sought ways to increase competition and diversity in broadcasting with AM radio, UHF television, low-power TV, low-power FM, and more recently, digital television.

Cable Television and Direct Broadcast Satellite

Since cable television began simply as a technology to retransmit distant broadcast signals to rural or remote locations, early systems sought permission or franchises from the local authorities to lay cable to reach homes. As cable grew, broadcasters became alarmed with companies making revenue off their programming, and they lobbied against the new technology.

Early on, copyrights became the major issue, as broadcasters complained that retransmission of their signals violated their copyrights. The courts sided with cable operators, but Congress passed compulsory license legislation that forced cable operators to pay royalty fees to broadcasters (Pool, 1983). Because cable television did not utilize the public airwaves, courts rebuffed the FCC's attempt to regulate cable.

In the 1980s, the number of cable systems exploded and the practice of franchising cable systems increasingly was criticized by the cable industry as cities demanded more concessions in return for granting rights-of-way access and exclusive multi-year franchises. The *Cable Communications Act of 1984* was passed to formalize the municipal franchising process while limiting some of their rate regulation authority. The act also authorized the FCC to evaluate cable competition within markets (Parsons & Frieden, 1998).

After that, cable rates increased dramatically. Congress reacted with the *Cable Television Consumer Protection and Competition Act of 1992*. With the 1992 Cable Act, rate regulation returned with the FCC given authority to regulate basic cable rates.

To protect broadcasters and localism principles, the act included "must carry" and "retransmission consent" rules that allowed broadcasters to negotiate with cable systems for carriage (discussed in more detail in Chapter 7). Although challenged on First Amendment grounds, the courts eventually found that the FCC had a legitimate interest in protecting local broadcasters (*Turner Broadcasting v. FCC*, 1997).

To support the development of direct broadcast satellites (DBS), the 1992 act prohibited cable television programmers from withholding channels from DBS and other prospective competitors. As with cable television, DBS operators have been subject to must-carry and retransmission consent rules.

The *1999 Satellite Home Viewers Improvement Act* (SHIVA) required and, more recently, the *Satellite Home Viewer Extension and Reauthorization Act* (SHVER) reconfirmed that DBS operators must carry all local broadcast signals within a local market if they choose to carry one (FCC, 2005). DBS operators challenged this in court, but as in *Turner Broadcasting v. FCC*, the courts upheld the FCC rule (Frieden, 2005).

Policies to promote the development of cable television and direct broadcast satellites have become important components of the desire to enhance media competition and video program diversity, while at the same time preserving localism principles within media markets.

Convergence and the Internet

The *Telecommunications Act of 1996* was a significant recognition of the impact of technological innovation and convergence occurring within the media and telecommunications industries. Because of that recognition, Congress discontinued many of the cross-ownership and service restrictions that had prevented telephone operators from offering video service and cable systems from providing telephone service (Parsons & Frieden, 1998).

The primary purpose of the 1996 Act was to "promote competition and reduce regulation in order to secure lower prices and higher-quality service for American telecommunications consumers and encourage the rapid deployment of new telecommunications technologies" (*Telecommunications Act of 1996*). Competition was expected by opening up local markets to facilities-based competition and deregulating rates for cable television and telephone service to let the market work its magic. The 1996 Act also opened up competition in the local exchange telephone markets and loosened a range of media ownership restrictions.

In 1996, the Internet was a growing phenomenon, and some in Congress were concerned with the adult content available online. In response, along with passing the act, Congress passed the *Communication Decency Act* (CDA) to make it a felony to transmit obscene or indecent material to minors. The Supreme Court struck down the CDA on First Amendment grounds in *Reno v. ACLU* (Napoli, 2001).

Congress continued to pursue a law protecting children from harmful material on the Internet with the *Child Online Protection Act* (COPA), passed in 1998; however, federal courts have found it, too, unconstitutional due to First Amendment concerns (McCullagh, 2007). It is noteworthy that the courts consider the Internet's First Amendment protection more similar to the press, rather than broadcasting or telecommunications (Warner, 2008).

Similarly, from a regulatory perspective, the Internet does not fall under any traditional regulatory regime such as telecommunications, broadcasting, or cable television. Instead, the Internet is considered an "information service" and therefore not subject to regulation (Oxman, 1999). There are, however, policies in place that indirectly impact the Internet. For example, section 706 of the *Telecommunications Act of 1996* requires the FCC to "encourage the deployment on a reasonable and timely basis of advanced telecommunications capability to all Americans," with advanced telecommunications essentially referring to broadband Internet connectivity (Grant & Berquist, 2000).

Recent Developments

Broadband

In response to a weakening U.S. economy, Congress passed the American Recovery and Reinvestment Act (ARRA) of 2009. Stimulus funds were appropriated for a wide range of infrastructure grants, including broadband, to foster economic development. Congress earmarked $7.2 billion to encourage broadband deployment, particularly in unserved and underserved regions of the country.

The U.S. Department of Agriculture Rural Utility Service (RUS) was provided with $2.5 billion to award Broadband Initiatives Program (BIP) grants, while the National Telecommunications and Information Administration (NTIA) was funded with $4.7 billion to award Broadband Technology Opportunity Program (BTOP) grants. In addition, Congress appropriated funding for the *Broadband Data Improvement Act*, legislation that was approved during the Bush Administration in 2008 but lacked the necessary funding for implementation.

Finally, as part of ARRA, Congress directed the FCC to develop a *National Broadband Plan* that addressed broadband deployment, adoption, affordability, and the use of broadband to advance healthcare, education, civic participation, energy, public safety, job creation, and investment.

While developing the comprehensive broadband plan, the FCC released its periodic broadband progress report, required under section 706 of the *Telecommunications Act of 1996*. Departing from the favorable projections in previous progress reports, the FCC conceded that "broadband deployment to *all* Americans is not reasonable and timely" (FCC, 2010c, p. 3).

In addition, the Commission revised the dated broadband benchmark of 200 Kb/s by redefining broadband as having download speeds of 4.0Mb/s and upload speeds of 1.0 Mb/s. The broadband plan, entitled *Connecting America: National Broadband Plan*, was released in 2010 and prominently declared that "Broadband is the great infrastructure challenge of the early 21st century" (FCC, 2010a, p. 19). The plan offered a direction for government partnership with the private sector to innovate the broadband ecosystem and advance "consumer welfare, civic participation, public safety and homeland security, community development, healthcare delivery, energy independence and efficiency, education, employee training, private sector investment, entrepreneurial activity, job creation and economic growth, and other national purposes" (FCC, 2010a, p. xi).

In preparing the *National Broadband Plan*, the FCC commissioned a number of studies to determine the state of broadband deployment in the United States. In analyzing U.S. broadband adoption in 2010, the FCC determined 65% of U.S. adults used broadband (Horrigan, 2010); however, researchers forecast that regions around the country, particularly rural areas, would continue to suffer from poor broadband service due to lack of service options or slow broadband speeds (Atkinson & Schultz, 2009).

The *National Broadband Plan* highlighted a number of strategies and goals to connect the millions of Americans who do not have broadband at home. Among the strategies were plans to:

- Design policies to ensure robust competition.
- Ensure efficient allocation and use of government-owned and government-influenced assets.
- Reform current universal service mechanisms to support deployment of broadband.
- Reform laws, policies, standards, and incentives to maximize the benefits of broadband for government priorities, such as public education and healthcare.

The targeted goals of the plan through 2020 include 1) at least 100 million homes should have affordable access to download speeds of 100 Mb/s and upload speeds of 50 Mb/s, 2) the United States should lead the world in mobile innovation with the fastest and most extensive wireless network in the world, 3) every American should have affordable access to robust broadband service, and the means and skills to subscribe, 4) every American community should have affordable access to at least 1 Gb/s broadband service to schools, hospitals, and government buildings, 5) every first responder should have access to a nationwide, wireless, interoperable broadband public safety network, and 6) every American should be able to use broadband to track and manage their real-time energy consumption (FCC, 2010a).

The most far-reaching components of the plan included freeing 500 MHz of wireless spectrum, including existing television frequencies, for wireless broadband use, and reforming the traditional telephone Universal Service Fund to support broadband deployment.

The *Broadband Data Improvement Act* requires the FCC to compare U.S. broadband service capability to at least 25 countries abroad. International comparisons of broadband deployment confirm that the U.S. continues to lag in access to broadband capability (FCC, 2012b); and within the U.S. 19 million Americans lack access to wireline broadband service (FCC, 2012 a).

The most recent broadband data from the Organisation for Economic Co-operation and Development (OECD) reports the United States ranked 16th among developed nations for fixed wireline broadband penetration (see Table 5.1). Other OECD studies show the United States ranked 13th for average advertised download speed (44.7 Mb/s), with top-ranked Sweden offering significantly greater broadband speed (136.2 Mb/s), followed by Japan (95 Mb/s), Netherlands (89.7 Mb/s), Portugal (76.3 Mb/s), and Norway (74.5 Mb/s) (OECD, 2013).

Universal Service Reform

For 100 years, universal service policies have served the United States well, resulting in significant telephone subscription rates throughout the country. But telephone service, an indispensable technology for 20th century business and citizen use, has been displaced by an even more essential technology in the 21st century—broadband Internet access.

Just as the universal deployment of telephone infrastructure was critical to foster business and citizen communication in the early 1900s, broadband has become a crucial infrastructure for the nation's economic development and civic engagement. However, nearly a third of Americans have not adopted broadband, and broadband deployment gaps in rural areas remain significant (FCC, 2011b).

In response to uneven broadband deployment and adoption, the FCC has shifted most of the billions of dollars currently subsidizing voice networks in the Universal Service Fund (USF) to supporting broadband deployment. Two major programs were introduced to modernize universal service in the United States: The Connect America Fund, and the Mobility Fund.

Table 5.1
International Fixed Broadband Penetration

Country	Broadband Penetration*
Switzerland	43.8
Netherlands	40.0
Denmark	39.7
Korea	37.1
France	37.0
Norway	36.6
Iceland	35.1
United Kingdom	34.9
Germany	34.5
Belgium	34.0
Canada	32.9
Luxembourg	32.7
Sweden	32.3
Finland	30.5
New Zealand	29.5
United States	29.3

* Fixed Broadband access per 100 inhabitants

Source: OECD (2013)

The Connect America Fund provides up to $4.5 billion annually over six years to fund broadband and high-quality voice in geographic areas throughout the U.S. where private investment in communication is limited or absent. The funding is released to telecommunications carriers in phases, with Phase I releasing $115 million in 2012 to carriers providing broadband infrastructure to 400,000 homes, businesses, and institutions previously without access to broadband. As of March, 2014, the second round of Phase I funding had provided $324 million for broadband deployment to connect over 1.2 million Americans (FCC, 2014b). The Mobility Fund targets wireless availability in unserved regions of the nation by ensuring all areas of the country achieve 3G service, with enhanced opportunities for 4G data and voice service in the future (Gilroy, 2011). In 2012, the Mobility Fund Phase I awarded close to $300 million to fund advanced voice and broadband service for primarily rural areas in over 30 states (Wallsten, 2013).

The Connect America Fund and the Mobility Fund maintain the long tradition of government support for communication development, as intended by Congress in 1934 when they created the FCC to make "available...to all the people of the United States...a rapid, efficient, Nation-wide, and world-wide wire and radio communication service with adequate facilities at reasonable charges" (FCC, 2011a, p. 4). Ultimately the Connect America Fund is expected to connect 18 million unserved Americans to broadband, with the hope of creating 500,000 jobs and generating $50 billion in economic growth (FCC, 2011b).

Spectrum Reallocation

Wireless spectrum, or the "invisible infrastructure," that allows wireless communications, is rapidly facing a deficit as demand for mobile data continues to grow (FCC, 2010b). Mobile wireless connections have increased 160% from 2008 to 2010, while the average data per line has increased 500% (Executive Office of the President, 2012). The increase in wireless data traffic is projected to grow by a factor of 20 by 2015.

Globally, the United States ranks 7th in wireless broadband penetration (See Table 5.2); however, mobile broadband speeds are much greater in most Asian and European countries due to greater spectrum availability and wireless technology (Executive Office of the President, 2012). In response to the U.S. spectrum crunch, the *National Broadband Plan* called for freeing 500 MHz of spectrum for wireless broadband. President Obama also established a National Wireless Initiative for federal agencies to support the FCC in making available the 500MHz of spectrum before 2020 for mobile and fixed wireless broadband (The White House, 2010).

The FCC has begun the steps to repurpose spectrum for wireless broadband service by making spectrum from the 2.3 GHz, Mobile Satellite Service, and TV bands available for mobile broadband service. The Commission will make additional spectrum available for unlicensed wireless broadband by leveraging unused portions of the TV bands, or "white space" that might offer unique solutions for innovative developers of broadband service. Congress supported the effort with passage of the Middle Class Tax Relief and Job Creation Act of 2012, including provisions from the Jumpstarting Opportunity with Broadband Spectrum (JOBS) Act of 2011 (Moore, 2012). The spectrum reallocation will be accomplished through the FCC's authority to conduct incentive auctions where existing license holders, such as broadcasters, will relinquish spectrum in exchange for proceeds that will be shared with the Federal government.

Table 5.2
International Wireless Broadband Penetration

Country	Broadband Penetration*
Australia	114.0
Finland	112.9
Sweden	107.9
Japan	105.3
Korea	102.9
Denmark	102.7
United States	96.0
Estonia	89.1
Norway	86.9
New Zealand	82.5
Luxembourg	82.0
United Kingdom	80.4
Iceland	74.7
Ireland	66.3
Netherlands	66.0

* Wireless Broadband access per 100 inhabitants (Measure includes both standard and dedicated mobile broadband subscriptions)

Source: OECD (2013)

Network Neutrality

In 2005, then-AT&T CEO Edward Whitacre, Jr. created a stir when he suggested Google and Vonage should not expect to use his pipes for free (Yang, 2005). Internet purists insist the Internet should remain open and unfettered, as originally designed, and decried the notion that broadband providers might discriminate by the type and amount of data content streaming through their pipes. Users of Internet service are concerned that, as more services become available via the Web such as video streaming and voice over Internet protocol (VoIP), Internet service providers (ISPs) will become gatekeepers limiting content and access to information (Gilroy, 2008).

In 2007, the FCC received complaints accusing Comcast of delaying Web traffic on its cable modem service for the popular file sharing site BitTorrent (Kang, 2008). Because of the uproar among consumer groups, the FCC held hearings on the issue and ordered Comcast to end its discriminatory network management practices (FCC, 2008).

Although Comcast complied with the order and discontinued interfering with peer-to-peer traffic like BitTorrent, it challenged the FCC's authority in court. In April 2010, the U.S. Court of Appeals for the D.C. Circuit determined that the FCC had failed to show it had the statutory authority to regulate an Internet service provider's network practices and vacated the order (*Comcast v. FCC*, 2010).

Because broadband service is an unregulated information service, the FCC argued it had the power to regulate under its broad "ancillary" authority highlighted in Title I of the Telecommunications Act. The court's rejection of the FCC argument disheartened network neutrality proponents who feared the court's decision would encourage broadband service providers to restrict network data traffic, undermining the traditional openness of the Internet.

In evaluating the court decision, the FCC revisited net neutrality and determined it could establish rules for an open Internet through a combination of regulatory authority. First, since broadband is considered an information service under Title I, the Commission is directed under Section 706 of the *Telecommunications Act of 1996* to take action if broadband capability is not deployed in a reasonable and timely fashion. Second, under Title II of the *Telecommunications Act*, the FCC has a role in protecting consumers who receive broadband over telecommunications services. Third, Title III of the *Telecommunications Act* provides the FCC with the authority to license spectrum used to provide wireless broadband services, and last, Title IV gives the FCC authority to promote competition in video services (Gilroy, 2011).

The FCC Open Internet Rule was adopted in November 2010 and established three basic rules to ensure Internet providers do not restrict innovation on the Internet. The rules required Internet service providers to disclose information about their network management practices and commercial terms to consumers and content providers to ensure *transparency*. Internet service providers were also subject to *no blocking* requirements of lawful content and applications. Finally, to prevent *unreasonable discrimination*, broadband Internet providers could not unreasonably discriminate in transmitting lawful network traffic over a consumer's Internet service.

In January, 2011, both Verizon Communications and Metro PSC filed lawsuits challenging the FCC's Open Internet Rule, and in early 2014, the U.S. Court of Appeals for the D.C. Circuit struck down two vital portions of the rule (*Verizon v. FCC*, 2014).

As in the *Comcast v. FCC* decision, the court determined the FCC lacked legal authority to prevent Internet service providers from blocking traffic, and lacked authority to bar wireline broadband providers from discriminating among Internet traffic. However, the court upheld the transparency rule which requires Internet service providers to disclose how they manage Internet traffic, and affirmed the FCC's general authority to oversee broadband services under Section 706 of the *Telecommunications Act of 1996*.

Rather than appeal the ruling, the FCC plans to propose new rules that consider the Circuit Court's legal rationale (Mazmanian, 2014). In addition, to promote local broadband competition, the FCC will consider ways to counter state laws that prohibit municipalities from providing government managed broadband services.

In April 2014, the FCC proposed new rules designed to protect consumers and allow broadband providers the ability to offer some preferential treatment to content providers.

Under these rules, for example, Netflix could make a deal with Cablevision so their content would stream faster on a dedicated portion of Cablevision's network. Critics are concerned that this preferential treatment to those that can pay may harm smaller, up and coming services distributed over the Internet (Nagesh, 2014)

Competition and Technology Transformation

Technological transformation and policy changes have significantly improved video programming capabilities and competition among multichannel video programming distributors (MVPDs). According to the FCC, 35% of U.S. homes have access to at least four MVPDs which include cable television, direct broadcast satellite, and cable services offered through telephone systems (FCC, 2013).

The FCC's *Fifteenth Report on Video Competition* showed that from 2010 to 2011, cable lost video market share to telephone systems (See Table 5.3). More importantly, the report discovered online video distributors (OVDs) such as Netflix and Hulu have significantly increased video programming options for consumers, with more than 180 million Internet users watching more than 20 hours of online video in June 2012.

Table 5.3

MVPD Market Share

Multichannel Video Programming Distributor	2010	2011
Cable	59.3%	57.4%
DBS	33.1%	33.6%
Telco Cable	6.9%	8.4%

Source: FCC (2013), p. 60

Although MVPDs and OVDs offer an alternative to traditional broadcast television, 9.7% of all U.S. households, approximately 11 million homes, still rely exclusively on over-the-air broadcasting for video news and entertainment. Although the introduction of video competition has gradually decreased broadcast television viewership over time, the FCC competition report shows that the audience share for network affiliates remained steady at 33% for both the 2010-2011 and 2011-2012 television seasons (FCC, 2013).

In 2012, Aereo was launched as an online service in New York City allowing viewers to watch local broadcast stations over the Internet. Soon after, broadcast television networks filed a lawsuit challenging the service for infringing on their copyrights, as Aereo streams live and recorded broadcast signals online without permission.

Aereo argues that their technology, which uses tens of thousands of antennas (one for each unique customer), constitutes a private performance thereby negating any claims of infringement. Because of conflicting district court rulings over the service, the Supreme Court has decided to weigh in on the case (Kravets, 2014).

Legal scholars are curious how the court will eventually approach the case; analogous to the early days of cable television retransmission disputes with broadcasters, or as a permissible remote DVR service (Garner, 2013). (The outcome of this case was pending as this book went to press; for the latest information, see the *Communication Technology Update and Fundamentals* website, www.tfi.com/ctu).

Because much of voice communication has transitioned from circuit-switched phone networks to Internet Protocol (IP) networks, the FCC is interested in exploring the impact this technology transformation is having on policy.

In 2014, the FCC announced a *Technology Transition Order* to research how the transition to IP might affect federal communication regulation; including impacts on public safety, universal service, competition, and consumer protection (FCC, 2014a).

After the FCC Order, AT&T proposed tests in Carbon Hill, Alabama and West Delray Beach, Florida; where the old legacy phone network will be replaced with a new IP based phone system (Hickey, 2014). The FCC will monitor the IP transition and analyze the impact of the new services on customers.

Although the transition to IP based communication services has been occurring for years, it is clear the role of the Internet as a distribution medium for video and voice is accelerating.

Factors to Watch

The FCC's *National Broadband Plan* and Universal Service reform are still early in development, but these efforts confirm a strong national commitment to ubiquitous broadband service for all Americans. Observers will watch to see if these policies result in enhanced broadband deployment and faster Internet speeds for Americans (Kruger, 2013).

As the Federal Communications Commission ponders new rules to sustain an open Internet, they will have to consider recent court decisions before determining the proper role and limits of their authority for regulating broadband service.

Soon after the U.S. Court of Appeals for the D.C. Circuit struck down portions of the FCC's 2010 Open Internet Rule, Netflix announced an agreement with cable giant Comcast that will likely intensify network neutrality debates (Wyatt & Cohen, 2014). The interconnection deal provides Netflix with a faster and more reliable pathway for their streaming videos to reach subscribers on Comcast cable systems.

Comcast has received additional scrutiny as it reported a plan to merge with the cable television operator Time Warner Cable (Kang, 2014). The combined company would garner 38% of the U.S. broadband Internet market, and have over 30 million cable television subscribers, dominating the cable television market with 30% of all U.S. subscribers.

Consumer groups oppose the merger due to concerns about the market power the combined company would achieve (Cooper, 2014). The market share concerns are twofold in that a combined Comcast/Time Warner could exert extraordinary power to control broadband Internet access that impacts network neutrality, as well as controlling a significant share of the MVPD market that might diminish video competition.

The technological changes that have historically prompted regulatory change will again force Congress and the FCC to examine the effectiveness of current policy. The rapid transition to IP based voice and video distribution has the potential to transform the policymaking landscape.

The FCC has already undertaken a series of steps that appreciate the newly emerging communication technology environment. The *Technology Transition Order* may discover a new policy direction as the communication industry increasingly leverages the Internet for video programming distribution and voice communication. Broadcasters are not immune to the expansive nature of the Internet as the emergence of Aereo attests. Finally, the FCC's efforts to expand the use of wireless spectrum recognize that the mobile Internet is quickly becoming a dominant medium for distributing data, voice, and video (Cisco, 2014).

Given this rapidly changing environment, it is not surprising that Congress has expressed renewed interest in revisiting the *Telecommunications Act of 1996.* Written two decades ago, the Act scarcely acknowledged the existence of the Internet with a mere eleven references, and it mentioned broadband only once (Hatten, 2014).

The U.S. House Committee on Energy and Commerce (2014) is issuing a series of whitepapers spelling out its priorities in modernizing the Communications Act as it seeks input from the communications industry, citizens, and other stakeholders.

As communication policy is reshaped to accommodate new technologies, policymakers must continue to explore ways to serve the public interest. However, it is evident that the transformation of technology change is currently outpacing the ability of regulation and policy to adapt.

Bibliography

Alexander, P. J. & Brown, K. (2007). Policy making and policy tradeoffs: Broadcast media regulation in the United States. In P. Seabright & J. von Hagen (Eds.). *The economic regulation of broadcasting markets: Evolving technology and the challenges for policy*. Cambridge: Cambridge University Press.

Atkinson, R.C. & Schultz, I.E. (2009). Broadband in America: Where it is and where it is going (according to broadband service providers). Retrieved from http://broadband.gov/docs/Broadband_in_America.pdf.

The Broadband Data Improvement Act of 2008, Pub. L. 110-385, 122 Stat. 4096 (2008). Retrieved from http://www.ntia.doc.gov/advisory/onlinesafety/BroadbandData_PublicLaw110-385.pdf.

Comcast v. FCC. No. 08-1291 (D.C. Cir., 2010). Retrieved from http://hraunfoss.fcc.gov/edocs_public/attachmatch/DOC-297356A1.pdf.

Cisco, (2014, Feb. 5). Cisco visual network index: Global mobile data traffic forecast update, 2013-2018. Retrieved from http://www.cisco.com/c/en/us/solutions/collateral/service-provider/visual-networking-index-vni/white_paper_c11-520862.pdf.

Cooper, M. (2014). Buyer and bottleneck market power: Make the Comcast-Time Warner merger "unapprovable". Consmer Federation of America. Retrieved from http://www.consumerfed.org/pdfs/CFA-Comcast-TW-Merger-Analysis.pdf.

DuBoff, R. B. (1984). The rise of communications regulation: The telegraph industry, 1844-1880. *Journal of Communication, 34* (3), 52-66..

Executive Office of the President: Council of Economic Advisors (2012, Feb. 12). *The economic benefits of new spectrum for wireless broadband*. Retrieved from http://www.whitehouse.gov/sites/default/files/cea_spectrum_report_2-21-2012.pdf.

Federal Communications Commission. (2005, September 8). *Retransmission consent and exclusivity rules: Report to Congress pursuant to section 208 of the Satellite Home Viewer Extension and Reauthorization Act of 2004*. Retrieved from http://hraunfoss.fcc.gov/edocs_public/attachmatch/DOC-260936A1.pdf.

Federal Communications Commission. (2008, August 20). *Memorandum opinion and order* (FCC 08-183). Broadband Industry Practices. Retrieved from http://hraunfoss.fcc.gov/edocs_public/attachmatch/FCC-08-183A1.pdf.

Federal Communications Commission. (2010a). *Connecting America: National broadband plan*. Retrieved from http://hraunfoss.fcc.gov/edocs_public/attachmatch/DOC-296935A1.pdf.

Federal Communications Commission. (2010b). *Mobile broadband: The benefits of additional spectrum*. Retrieved from http://download.broadband.gov/plan/fcc-staff-technical-paper-mobile-broadband-benefits-of-additional-spectrum.pdf.

Federal Communications Commission. (2010c). *Sixth broadband deployment report* (FCC 10-129). Retrieved from http://hraunfoss.fcc.gov/edocs_public/attachmatch/FCC-10-129A1_Rcd.pdf.

Federal Communications Commission. (2011a). *Bringing broadband to rural America: Update to report on rural broadband strategy*. Retrieved from http://hraunfoss.fcc.gov/edocs_public/attachmatch/DOC-307877A1.pdf.

Federal Communications Commission. (2011b). FCC releases 'Connect America Fund" order to help expand broadband, create jobs, benefit consumers. Press Release. Retrieved from http://hraunfoss.fcc.gov/edocs_public/attachmatch/DOC-311095A1.pdf.

Federal Communications Commission. (2012a). *Eighth broadband deployment report* (FCC 12-90). Retrieved from http://hraunfoss.fcc.gov/edocs_public/attachmatch/FCC-12-90A1.pdf.

Federal Communications Commission. (2012b). *Third international broadband data report* (FCC 12-1334). Retrieved from http://hraunfoss.fcc.gov/edocs_public/attachmatch/DA-12-1334A1.pdf.

Federal Communications Commission. (2013). *Fifteenth report on video competition* (FCC 13-99). Retrieved from http://hraunfoss.fcc.gov/edocs_public/attachmatch/FCC-13-99A1.pdf.

Federal Communication Commission. (2014a). *Technology Transition Order* (fCC14-05). Retrieved from http://transition.fcc.gov/Daily_Releases/Daily_Business/2014/db0131/FCC-14-5A1.pdf.

Federal Communication Commission. (2014b). *Universal service implementation progress report*. Retrieved from http://transition.fcc.gov/Daily_Releases/Daily_Business/2014/db0324/DOC-326217A1.pdf.

Fowler, M. S. & Brenner, D. L. (1982). A marketplace approach to broadcast regulation. *University of Texas Law Review 60* (207), 207-257.

Frieden, R. (2005, April). *Analog and digital must-carry obligations of cable and satellite television operators in the United States*. Retrieved from http://ssrn.com/abstract=704585.

Gandy, O. H. (1989). The surveillance society: Information technology and bureaucratic control. *Journal of Communication, 39* (3), 61-76.
Gardner, E. (2013). TV broadcasters petition for Aereo rehearing at appeals court. *The Hollywood Reporter*. Retrieved from http://www.hollywoodreporter.com/thr-esq/tv-broadcasters-petition-aereo-rehearing-440428.
Gilroy, A. A. (2008, Sept. 16). Net neutrality: Background and issues. *CRS Reports to Congress*. CRS Report RS22444. Retrieved from http://www.fas.org/sgp/crs/misc/RS22444.pdf.
Gilroy, A.A. (2011, June 30). Universal service fund: Background and options for reform. *CRS Reports to Congress*. CRS Report RL33979. Retrieved from http://www.fas.org/sgp/crs/misc/RL33979.pdf.
Grant, A. E. & Berquist, L. (2000). Telecommunications infrastructure and the city: Adapting to the convergence of technology and policy. In J. O. Wheeler, Y, Aoyama & B. Wharf (Eds.). *Cities in the telecommunications age: The fracturing of geographies*. New York: Routledge.
Hatten, J. (2014, Jan. 15). Congress looks to revamp telecom law. *The Hill*. Retrieved from http://thehill.com/blogs/hillicon-valley/technology/195545-congress-looks-to-revamp-telecom-law-for-internet-age.
Hickey, K. (2014, May 18). Cities to test IP-only telephony. *Government Computer News*. Retrieved from http://gcn.com/articles/2014/03/18/ip-telephony-cities.aspx.
Horrigan, J. (2010). *Broadband adoption and use in America* (OBI Working Paper No. 1). Retrieved from http://hraunfoss.fcc.gov/edocs_public/attachmatch/DOC-296442A1.pdf.
Kang, C. (2008, March 28). Net neutrality's quiet crusader. *Washington Post*, D01.
Kang, C. (2014, Feb. 12). Comcast, Time Warner Cable agree to merge I $45 billion deal. *Washington Post*. Retrieved from http://www.washingtonpost.com/business/economy/comcast-time-warner-agree-to-merge-in-45-billion-deal/2014/02/13/7b778d60-9469-11e3-84e1-27626c5ef5fb_story.html.
Kravets, D. (2014, Jan. 10). Supreme Court will soon decide the future of online broadcast TV. *Wired*. Retrieved from http://www.wired.com/2014/01/supreme-court-decide-online-broadcast-tvs-future/.
Kruger, L.G. (2013, March 19). The National Broadband Plan goals: Where do we stand? Congressional Research Service, R43016. Retrieved from http://www.fas.org/sgp/crs/misc/R43016.pdf.
Lubrano, A. (1997). *The telegraph: How technology innovation caused social change*. New York: Garland Publishing.
Mazmanian, A. (2014, Feb. 19). FCC plans new basis for old network neutrality rules. *Federal Computer Week*. Retrieved from http://fcw.com/articles/2014/02/19/fcc-plans-new-basis-for-old-network-neutrality-rules.aspx.
McCullagh, D. (2007, March 22). Net porn ban faces another legal setback. *C/NET News*. Retrieved from http://www.news.com/Net-porn-ban-faces-another-legal-setback/2100-1030_3-6169621.html.
Moore, L.K. (2012, Jan. 5). Spectrum policy in the age of broadband: Issues for Congress. *CRS Reports to Congress*. CRS Report R40674. Retrieved from http://www.fas.org/sgp/crs/misc/R40674.pdf.
Nagesh, G. (2014, April 23). FCC to propose new net-neutrality rules. *The Wall Street Journal*. Retrieved from http://online.wsj.com/news/articles/SB10001424052702304518704579519963416350296
Napoli, P. M. (2001). *Foundations of communications policy: Principles and process in the regulation of electronic media*. Cresskill, NJ: Hampton Press.
National Broadcasting Co. v. United States, 319 U.S. 190 (1949).
Organisation for Economic Co-operation and Development. (2013). *OECD broadband portal*. Retrieved from http://www.oecd.org/sti/broadband/oecdbroadbandportal.htm.
Oxman, J. (1999). *The FCC and the unregulation of the Internet*. OPP Working Paper No. 31. Retrieved from http://www.fcc.gov/Bureaus/OPP/working_papers/oppwp31.pdf.
Parsons, P. R. & Frieden, R. M. (1998). *The cable and satellite television industries*. Needham Heights, MA: Allyn & Bacon.
Pool, I. S. (1983). *Technologies of freedom*. Cambridge, MA: Harvard University Press.
Red Lion Broadcasting Co. v. Federal Communications Commission, 395 U.S. 367 (1969).
Stanton, L. (2012). Action on PIPA, SOPA postponed in wake of protest; Both sides vow to continue work on online copyright bills. *Telecommunications Reports, 78* (3). 3-7.
Starr, P. (2004). *The creation of the media*. New York: Basic Books.
Telecommunications Act of 1996, Pub. L. No. 104-104, 110 Stat. 56 (1996). Retrieved from http://transition.fcc.gov/Reports/tcom1996.pdf.
The White House. (2010, June 28). *Presidential memorandum: Unleashing the wireless broadband revolution*. Retrieved from http://www.whitehouse.gov/the-press-office/presidential-memorandum-unleashing-wireless-broadband-revolution.
Turner Broadcasting System, Inc. v. FCC, 512 U.S. 622 (1997).

U.S. House Committee on Energy and Commerce. (2014, Jan. 8). *Modernizing the Communications Act*. Whitepaper. Retrieved from http://energycommerce.house.gov/sites/republicans.energycommerce.house.gov/files/analysis/CommActUpdate/20140108WhitePaper.pdf.

van Cuilenburg, J. & McQuail, D. (2003). Media policy paradigm shifts: Toward a new communications policy paradigm. *European Journal of Communication, 18* (2), 181-207.

Verizon v. FCC. No. 11-1355 (D.C. Cir., 2014). Retrieved from http://www.cadc.uscourts.gov/internet/opinions.nsf/3AF8B4D938CDEEA685257C6000532062/$file/11-1355-1474943.pdf

Wallsten, S. (2013, April 1). Two cheers for the FCC's mobility fund reverse auction. Technology Policy Institute. Retrieved from https://www.techpolicyinstitute.org/files/wallsten_the%20fccs%20mobility%20reverse%20auction.pdf.

Warner, W. B. (2008). Networking and broadcasting in crisis. In R. E. Rice (Ed.), *Media ownership research and regulation*. Creskill, NJ: Hampton Press.

Winston, B. (1998). *Media technology and society: A history from the telegraph to the Internet*. New York: Routledge.

Wyatt, E. & Cohen, N. (2014, Feb. 23). Comcast and Netflix reach deal on service. *New York Times*. Retrieved from http://www.nytimes.com/2014/02/24/business/media/comcast-and-netflix-reach-a-streaming-agreement.html.

Yang, C. (2005, December 26). At stake: The net as we know it. *Business Week*, 38-39.

II

Electronic Mass Media

6

Digital Television & Video

Peter B. Seel, Ph.D. and
Brian Pauling, Ph.D.*

Why Study Digital Television & Video?

- They are ubiquitous technologies, meaning the glow of digital displays can be seen in remote fishing villages north of the Arctic Circle and in houses mounted on stilts above the Amazon River in the most remote parts of Brazil.
- Digital motion pictures, online video, and over-the-air television programs are media seen globally in multiple languages on displays of all sizes from mobile phones to IMAX "movie" screens.
- Humans like to watch media with moving images, especially if those images have stories that move us emotionally, feature actors we enjoy watching, or televise sporting events such as the Super Bowl or World Cup.

Introduction

As the nations of the world complete their conversion to digital television technology, there are multiple new technological developments on the horizon. Video and television displays are simultaneously getting larger and smaller (in flat screens that exceed five feet in diagonal size and, at the other extreme, with head-mounted digital displays such as the Oculus Rift). There is also improved image resolution in all screen sizes. Program delivery is also shifting from over-the-air broadcast where viewers have content "pushed" to them to an expanding online environment where they "pull" what they want to watch when they want to see it. This global transformation in delivery is affecting every aspect of broadcasting in countries worldwide.

In the United States, this online-delivery process has an evolving name—*Over-The-Top* television—with OTT TV as the obligatory acronym. It should more accurately be called over-the-top video as the term refers to non-broadcast, Internet-streamed video content viewed on a digital display, but the confusion reflects the blurring of the line between broadcast/cable/DBS-delivered "linear" television and that of on-demand streamed content (Seel, 2012). *Over-The-Top* indicates that there is an increasingly popular means of accessing television and video content that is independent of traditional linear over-the-air (OTA) television and multichannel video programming distributors (MVPDs, which include satellite TV services, cable companies, and telco providers). However, much of the content consists of television programs previously broadcast (or cablecast) accessed on websites such as Hulu.com—in addition to streamed movies from providers such as Netflix, Quickflix (in Australia and New Zealand), and Amazon. In addition, the term "Over-The-Top" is unique to the U.S.—in other regions of the world it is called "Connected" video or television. To move past this confusion, we suggest the introduction of an all-encompassing term: **Internet-Delivered Television** or **I-DTV**. The use of this meta-term would subsume all OTT, Connected, and IPTV streamed programming and create a clear

* Seel is a professor in the Department of Journalism and Technical Communication, Colorado State University (Fort Collins, Colorado). Pauling is principal academic staff member at the New Zealand Broadcasting School at the Christchurch Polytechnic Institute of Technology (Christchurch, New Zealand).

demarcation between this mode of reception and broadcast-cable-DBS delivery.

In 2012, for the first time in 12 years, the number of U.S. television households declined with penetration falling from 98.9 percent penetration to 97.1 percent of the 114.7 million TV homes (Nielsen, 2011). However, the number of television households in the U.S. climbed back to 115.5 million households in 2013, an increase of 1.2% (Nielsen, 2013). The Nielsen Company offered several rationales for the 2012 decline in the size of the U.S. television audience; one factor was the then-ongoing global recession that reduced consumer's incomes and their ability to purchase new digital sets. Nielsen researchers also hypothesized that Over-The-Top viewing on multiple displays was a key factor in the drop in the number of television households, especially among younger viewers who have forsaken cable and DBS subscriptions in favor of "free" I-DTV (Nielsen, 2013). While the number of television households increased slightly in 2013, Americans between the ages of 12 and 34 are spending less time watching traditional television in favor of viewing programs streamed on computers and mobile phones (Stelter, 2012). This is a shift with multi-billion-dollar consequences for traditional broadcasters as advertising campaigns target younger viewers watching multimedia online.

The displays to watch all of this content are also changing. Many flat-screen displays larger than 42-inches sold globally in 2014 and beyond will be "smart" televisions that can easily display online media sources as well as websites such as Facebook and Twitter. An example is Samsung's "Smart Hub" display that offers Kinect-type gesture control, voice control, and facial recognition in accessing a panoply of Internet-delivered and MVPD-provided content. Remote controls for these displays can easily toggle between broadcast/cable/DBS content and Internet-streamed videos (see Figure 6.1). Older digital televisions can be easily connected to the Internet with specialized boxes provided by Roku and Apple or via Blu-ray players and most high-end game consoles.

Figure 6.1

Samsung's "Smart Hub" interface screen for its digital television displays.

Source: Samsung

Digital television displays and high-definition video recording are increasingly common features in mobile phones used by five billion of the world's population of 7.14 billion. Digital video cameras have also become more powerful and less expensive. An example is the $300 GoPro HERO3+ (see Figure 6.3) which records video at 1080p and is waterproof down to 141 feet. Digital storage for these cameras has also decreased in price while simultaneously increasing in capacity—a 32 GB SDHC memory card is available in 2014 for about $20, or less than 63 cents per Gigabyte. However, the prevalence of HD-quality video recording in mobile devices is affecting the market for low-end, stand-alone video recorders.

Figures 6.2 and 6.3

Two Contemporary High-Definition Cameras.

The $65,000 Sony F65 CineAlta camera is used to shoot motion pictures and high-end television programs with 4K image resolution (2160 X 4096 pixels).

The $300 GoPro HERO3+ camera captures 1080p HD-quality videos and is also waterproof to 141 feet underwater.

Sources: Sony and GoPro

Sales of digital still cameras fell 43 percent in the first six months of 2013, and the numbers for digital video cameras are also falling (Wakabayshi, 2013). A significant exception is GoPro, which is one of most phenomenal camera success stories in the past decade. GoPro digital video cameras have been used to document dramatic point-of-view footage of snowboarding, surfing, and have been attached to drones for high-definition aerial videography. The ten-year old company made over $500 million in sales in 2012 and the 2013 sales numbers doubled that (Cade, 2013). While the latest 4G mobile phones such as the Apple iPhone5 and Samsung's Galaxy 4 have greatly improved video resolution at 8 and 13 megapixels, respectively—they lack the waterproof housing that is a key selling point for videography in water and snow. However, most mobile phone users do not plan to use them for making surfing or snowboarding videos, and their improved image resolution and convenience (typically always with you) will make them the *defacto* camera of choice for creating spur-of-the-moment digital videos.

The democratization of "television" production generated by the explosion in the number of devices that can record digital video has created a world where 100 hours of video are uploaded *every minute* on the YouTube.com site and over *six billion* hours are viewed there each month. Eighty percent of YouTube viewers live outside the U.S., with content available in 61 languages (YouTube Statistics, 2014). The online distribution of digital video and television programming is an increasingly disruptive force to established broadcasters and program producers. They have responded by making their content available online for free or via subscription as increasing numbers of viewers seek to "pull" digital television content on request rather than watch at times when it is "pushed" as broadcast programming.

Television news programs routinely feature video shot by bystanders, such as the swarm of tornadoes that hit the U.S. Midwest in February of 2012. The increasing ubiquity of digital video recording capability also bodes well for the global free expression and exchange of ideas via the Internet. The expanding "universe" of digital television and video is being driven by improvements in high-definition video cameras for professional production and the simultaneous inclusion of higher-quality video capture capability in mobile phones.

Another key trend is the on-going global conversion from analog to digital television (DTV) technology. The United States completed its national conversion to digital broadcasting in June 2009; Japan and most European nations completed their transitions in 2012; India in 2014; and China plans to do so by 2015. At the outset of high-definition television (HDTV) development in the 1980s, there was hope that one global television standard might emerge, easing the need to perform format conversions for international program distribution. There are now multiple competing DTV standards based on regional affiliations and national political orientation. In many respects, global television has reverted to a "Babel" of competing digital formats reminiscent of the advent of analog color television. However, DTV programming in the widescreen 16:9 aspect ratio is now a commonplace sight in all nations that have made the conversion. The good news for consumers is that digital television displays have become commodity products with prices dropping rapidly each year. In 2014, a consumer in the United States can purchase a 40-inch LCD digital television for $260—falling below the $10 per diagonal inch benchmark that was crossed in 2010.

Background

The global conversion from analog to digital television technology is the most significant change in television broadcast standards since color images were added in the 1960s. Digital television combines higher-resolution image quality with improved multi-channel audio, and new "smart" models include the ability to seamlessly integrate Internet-delivered "television" programming into these displays. In the United States, the Federal Communications Commission (FCC, 1998) defines DTV as "any technology that uses digital techniques to provide advanced television services such as high definition TV (HDTV), multiple standard definition TV (SDTV) and other advanced features and services" (p. 7420). One key attribute of digital technology is "scalability"- the ability to produce audio-visual quality as good (or as bad) as the viewer desires (or will tolerate). The two most common digital display options are:

- HDTV (high-definition television, scalable)
- SDTV (standard-definition television, scalable)

High-definition television (HDTV) represents the highest image and sound quality that can be transmitted through the air. It is defined by the FCC in the United States as a system that provides image quality approaching that of 35 mm motion picture film, that has an image resolution of approximately twice that (1080i or 720p) of analog television, and has a picture aspect ratio of 16:9 (FCC, 1990) (see Table 6.1). SDTV, or standard-definition television, is another type of digital television technology that can be transmitted *along with,* or *instead of,* HDTV. Digital SDTV transmissions offer lower resolution (480p or 480i) than HDTV, and they are available in both narrowscreen and widescreen formats. Using digital video compression technology, it is feasible for U.S. broadcasters to transmit up to five SDTV signals instead of one HDTV signal within the allocated 6 MHz digital channel. The development of multichannel SDTV broadcasting, called "multicasting," is an approach that some U.S. broadcasters at national and local levels have adopted, especially the Public Broadcasting Service (PBS). Many PBS stations broadcast two child-oriented SDTV channels in the daytime along with educational channels geared toward adults. Most public and commercial networks reserve true HDTV programming for evening prime-time hours.

Digital television programming can be accessed via linear over-the-air (OTA) fixed and mobile transmissions, through cable/telco/satellite multichannel video program distributors MVPDs), and through Internet-delivered I-DTV sites. I-DTV is a "pull" technology in that viewers seek out a certain program and watch it in a video stream or as a downloaded file. OTA linear broadcasting is a "push" technology that transmits a digital program to millions of viewers at once. I-DTV, like other forms of digital television, is a scalable technology that can be viewed as lower quality, highly compressed content or programs can be accessed in HDTV quality on sites such as Vimeo.com. I-DTV and its subset IPTV are the subject of Chapter 8 in this text and, as noted above, are a rapidly growing method of accessing digital television programming as more viewers seek to watch their favorite shows on demand.

In the 1970s and 1980s, Japanese researchers at NHK (Japan Broadcasting Corporation) developed two related analog HDTV systems: an analog "Hi-Vision" *production* standard with 1125 scanning lines and 60 fields (30 frames) per second; and an analog "MUSE" *transmission* system with an original bandwidth of 9 MHz designed for satellite distribution throughout Japan. The decade between 1986 and 1996 was a significant era in the diffusion of HDTV technology in Japan, Europe, and the United States. There were a number of key events during this period that shaped advanced television technology and related industrial policies:

- In 1986, the Japanese Hi-Vision system was rejected as a world HDTV production standard by the CCIR, a subgroup of the International Telecommunication Union (ITU). By 1988, a

European research and development consortium, EUREKA EU-95, had created a competing system known as HD-MAC that featured 1250 wide-screen scanning lines and 50 fields (25 frames) displayed per second (Dupagne & Seel, 1998).

- In 1987, the FCC in the United States created the Advisory Committee on Advanced Television Service (ACATS). This committee was charged with investigating the policies, standards, and regulations that would facilitate the introduction of advanced television (ATV) services in the United States (FCC, 1987). See the sidebar about ACATS chairman Richard E. Wiley and his key role in guiding the creation of a U.S. digital television standard.
- U.S. testing of analog ATV systems by ACATS was about to begin in 1990 when the General Instrument Corporation announced that it had perfected a method of digitally transmitting a high-definition signal. Ultimately, the three competitors (AT&T/Zenith, General Instrument/MIT, and Philips/Thomson/Sarnoff) merged into a consortium known as the Grand Alliance and developed a single digital broadcast system for ACATS evaluation (Brinkley, 1997).
- The FCC adopted a number of key decisions during the ATV testing process that defined a national transition process from analog NTSC to an advanced digital television broadcast system.
- In 1990, the Commission outlined a *simulcast* strategy for the transition to an ATV standard (FCC, 1990). This strategy required that U.S. broadcasters transmit *both* the new ATV signal and the existing NTSC signal concurrently for a period of time, at the end of which all NTSC transmitters would be turned off.
- The Grand Alliance system was successfully tested in the summer of 1995, and a U.S. digital television standard based on that technology was recommended to the FCC by the Advisory Committee (Advisory Committee on Advanced Television Service, 1995).
- In May 1996, the FCC proposed the adoption of the *ATSC Digital Television (DTV) Standard* that specified 18 digital transmission variations as outlined in Table 6.1 (FCC, 1996).

In April 1997, the FCC defined how the United States would make the transition to DTV broadcasting and set December 31, 2006 as the target date for the phase-out of NTSC broadcasting (FCC, 1997). In 2005, after it became clear that this deadline was unrealistic due to the slow consumer adoption of DTV sets, it was reset to February 17, 2009 for the cessation of analog full-power television broadcasting (*Deficit Reduction Act*, 2005).

However, as the February 17, 2009 analog shutdown deadline approached, it was apparent that millions of over-the-air households with analog televisions had not purchased the converter boxes needed to continue watching broadcast programs. This was despite a widely publicized national coupon program that made the boxes almost free to consumers. Cable and satellite customers were not affected as provisions were made for the digital conversion at the cable headend or with a satellite set-top box. Neither the newly inaugurated Obama administration nor members of Congress wanted to invite the wrath of millions of disenfranchised analog television viewers, so the shut-off deadline was moved by an act of Congress 116 days to June 12, 2009 (*DTV Delay Act*, 2009).

Table 6.1

U.S. Advanced Television Systems Committee (ATSC) DTV Formats

Format	Active Lines	Horizontal Pixels	Aspect Ratio	Picture Rate*
HDTV	1080 lines	1920 pixels/line	16:9	60i, 30p, 24p
HDTV	720 lines	1280 pixels/line	16:9	60p, 30p, 24p
SDTV	480 lines	704 pixels/line	16:9 or 4:3	60i, 60p, 30p, 24p
SDTV	480 lines	640 pixels/line	4:3	60i, 60p, 30p, 24p

*In the picture rate column, "i" indicates interlace scan in television *fields*/second with two fields required per frame and "p" is progressive scan in *frames*/second.

Source: *ATSC*

Recent Developments

There are two primary areas affecting the diffusion of digital television and video in the United States—an ongoing battle over the digital television spectrum between broadcasters, regulators, and mobile telecommunication providers—and the diffusion of new digital technologies such as 4K-resolution DTV, mobile DTV, thin-screen displays such as OLED, and the advent of smart televisions. The spectrum battle is significant in the context of what is known as the *Negroponte Switch*, which describes the conversion of "broadcasting" from a predominantly over-the-air service to one that is now wired for many American households—and the simultaneous transition of telephony from a traditional wired service to a wireless one for an increasing number of users (Negroponte, 1995).

The growth of wireless broadband services has placed increasing demands on the available radio frequency spectrum—and broadcast television is a significant user of that spectrum. The advent of digital television in the United States made this conflict possible in that the assignment of new DTV channels demonstrated that spectrum assigned to television could be "repacked" at will without the adjacent-channel interference problems presented by analog transmission. The required separation between analog channels is much less of an issue with digital transmissions and DTV channels can be "packed" more tightly within the radio-frequency bands assigned to television. It appears that U.S broadcasters are a victim of their own success in making more efficient digital use of the terrestrial television spectrum, and that this success invited additional attempts at repacking it (Seel, 2011).

The DTV Spectrum Battle

At the 2011 National Association of Broadcasters conference in Las Vegas, then-FCC Chairman Julius Genachowski gave a speech to an audience of U.S. broadcast executives who were decidedly unenthusiastic about his message (Seel, 2011). As part of the Obama administration's National Broadband Plan, the FCC proposed that U.S. television broadcast networks and their affiliate stations voluntarily surrender additional DTV broadcast spectrum for auction to wireless telecommunication providers. The interesting twist in this offer is that the federal government proposed to share a portion of the auction revenue with stations that would voluntarily surrender their spectrum. Chairman Genachowski told the National Association of Broadcasters audience that the auction of 120 MHz of the 294 MHz television spectrum used by U.S broadcasters had the potential to raise up to $30 billion, a guesstimate since the need for wireless spectrum varies widely between U.S. cities (Seel, 2011).

Some context is necessary to understand why the federal government was seeking the return of digital spectrum that it had just allocated in the 2009 U.S. DTV switchover. As a key element of the transition, television spectrum between former channels 52-70 was auctioned off for $20 billion to telecommunication service providers, and the existing 1,500 broadcast stations were "repacked" into lower terrestrial broadcast frequencies (Seel & Dupagne, 2010). The substantial auction revenue was not overlooked by federal officials and as the U.S. suffered massive budget deficits in the global recession of 2008-2010, they proposed auctioning more of the television spectrum as a way to increase revenue without raising taxes in a recession. The outcome of this legislative process is addressed in the Current Status section below.

Mobile Digital Television

The creation of a national mobile standard for digital television broadcasting in the United States was proposed by the industry consortium Open Mobile Video Coalition (OMVC) in 2007, and the ATSC (the same organization that codified the DTV standard in 1995) assumed a similar role for a mobile standard (Advanced Television Systems Committee, 2009). After two years of testing in the lab and in the field, the A/153 ATSC Mobile DTV Standard (mDTV) was approved on October 15, 2009 by a vote of the ATSC membership (O'Neal, 2009). The mDTV standard uses the same 8VSB modulation scheme as the DTV terrestrial broadcast standard, but incorporates decoding technology specifically added to improve mobile reception. Local television broadcasters in the U.S. can include these mobile signals along with their regular DTV transmissions.

By the time of the 2013 NAB show in Las Vegas in April, 130 stations in 46 markets were transmitting a mobile DTV signal—almost a doubling in diffusion over a two-year period (Nakashima, 2013). The problem is that many mobile phone users lack the digital device or chip set needed to view mDTV images—a significant factor inhibiting the consumer adoption of mobile television technology.

Table 6.2

Average U.S. Retail Prices of LCD, LCD-LED, Plasma, and 3D displays, 2009-2013

Display Sizes (diagonal)	Average retail price in 2009	Average retail price in 2011	Average retail price in 2013
32-inch LCD TV	$424 (720p)	$300 (720p)	$230 (720p)
40-42 inch LCD-LED TV	n/a	$610 (1080p)	$450 (1080p)
46-47 inch LCD-LED TV	n/a	$950 (1080p)	$590 (1080p)
50-55 inch LCD-LED TV	n/a	$1208 (1080p)	$730 (1080p)
42-inch plasma TV	$600 (720p) $800 (1080p)	$450 (720p) $650 (1080p)	$ 500 (720p) $ 1,300 (1080p)
50-55 inch plasma TV	$900 (720p) $1,300 (1080p)	$700 (1080p)	$ 770 (1080p)
50-55 inch 3D plasma TV	n/a	$1,150 (1080p)	$1,350 (1080p)
60-inch 3D plasma TV	n/a	$2250 (1080p)	$1,700 (1080p)
55-inch OLED TV (Samsung and LG)	n/a	n/a	$ 8,000 (1080p)

Sources: U.S. retail surveys by P.B. Seel for all data. LCD-LED and 3D models were not widely available until 2010 and new model large-screen OLED displays won't be available until late in 2014. n/a = set not available.

Current Status

United States

Receiver Sales. Consumers are in a dominant position when buying a new television set. Prices are down, screen sizes are larger, and many new sets are "smart" models with a host of features more typically found in a computer than in a traditional television. A high-quality 55-inch television that sold for $1,200 in 2011 could be purchased in 2014 for less than $750. Similar dramatic price reductions have occurred in conventional LCD and plasma models as shown in Table 6.2. Consumers are buying sets with improved quality at less than half the cost compared to a few years earlier. What is good news for consumers has not been greeted warmly by television manufacturers. Profits are down for both display manufacturers and retailers as HDTV sets have become a commodity item. Increases in manufacturing capacity combined with downward price pressures introduced by new manufacturers such as Vizio have reduced profit margins for many well-known brands.

HDTV Adoption and Viewing. U.S. DTV penetration in 2011 was estimated at 79 million television households (of 116 million total) or 68 percent, up from 38 percent in 2009 (Briel, 2012; Leitchman, 2009)—and one third of these households have *more than four television sets* (Gorman, 2012). By October of 2012, 75 percent of U.S. households had at least one HDTV set, although their actual viewing of digital television programs still lagged (McAdams, 2012). As consumers retire their older analog televisions, their only choice at this point is to purchase a DTV model—so the percentage of TV homes with digital televisions will continue to climb toward 100 percent. The conversion is being propelled by the rapidly falling prices of DTV displays as shown in Table 6.2.

DTV Spectrum. On February 17, 2012, the U.S. Congress passed the *Middle Class Tax Relief and Job Creation Act of 2012* (*Middle Class*, 2012) and it was quickly signed into law by President Barack Obama (Eggerton, 2012). The *Act* authorized the Federal Communication Commission to reclaim DTV spectrum assigned to television broadcasters and auction it off to the highest bidder (most likely wireless broadband providers) over the coming decade. As noted above, broadcasters who voluntarily surrender some (or all) of their assigned spectrum will share in the auction revenue (Eggerton, 2012). It is an unusual auction, in that a "reverse" auction must first take place in which television broadcasters offer their DTV spectrum, followed by a "forward" auction in which wireless service providers will bid on it.

However, most network broadcasters in the U.S. are adamant that they won't surrender their spectrum without a legal battle, so this litigation could take several years to resolve. Former FCC Chairman Julius Genachowski sought to have the auction take place in June 2014, but incoming Chairman Thomas Wheeler announced on taking office in November 2013 that the spectrum auction would be delayed until the middle of 2015. He stated that, "These imperatives are balanced with the recognition that we have but one chance to get this right" (Reardon, 2013).

Digital Television Technology

Display Types. Consumers have many technological options for digital television displays, and standard sets sold today include higher-resolution screens (1080p vs. 720p) with UltraHD (2160p) displays going on sale in 2014—and many with more interactive (smart) features. The dominant display technologies are:

- *Liquid Crystal Display (LCD) and LCD-LED models*—LCDs work by rapidly switching color crystals off and on. A key technology is LED (light emitting diode) backlit LCDs that use this alternate light source instead of Cold Cathode Fluorescent (CCFL) backlighting. The use of LEDs for screen backlighting provides higher image contrast ratios with richer blacks and brighter highlights than CCFL models. Sony has decided to produce its new UltraHD sets (with 3840 X 2160 pixels) in 2014 using LCD-LED technology—due to problems manufacturing large OLED displays (Bachman, 2013).
- *Organic Light Emitting Diode (OLED)* – The Sony Corporation introduced remarkably bright and sharp OLED televisions in 2008 that had a display depth of 3 mm–about the thickness of three credit cards. The suggested retail price in Japan at that time for the ultra-thin 11-inch model was $2,200. In 2013, the company dissolved its partnership with Panasonic that had been created to find ways to reduce the cost of manufacturing large OLED displays and make them more durable (the O for "organic" term in OLED means that its color displays may fade over time) (Bachman, 2013). Korean manufacturers LG and Samsung have decided to stake their futures on OLED technology as a way to improve their profit margins. One of the most talked-about technologies at the 2014 Consumer Electronics Show was LG's 55-inch OLED display that dazzled observers with its curved screen (see Figure 6.4). Ideally this technology would emulate the very large circular motion picture screens used for the three-projector Cinerama system popular in the 1960s. However, there are major drawbacks to using a curved display for a relatively small television set in that the "sweet spot" to take advantage of the immersive effect would be limited to just one viewer in a typical home setting. Curved digital displays would need to be wall-sized for a large group to appreciate the immersive effect (Morrison, 2013). Consumers will likely see OLED technology used in their mobile phone screens long before buying an expensive home display. Samsung and other manufacturers are presently deploying AMOLED (active-matrix organic light-emitting diode) versions of these high-definition displays for their touch-screen phones.

Figure 6.4

LG Electronics exhibited curved 3D OLED displays at the 2014 Consumer Electronics Show in Las Vegas.

Source: LG Electronics.

- *Plasma Displays* – Plasma gas is used in these sets as a medium in which tiny color elements are switched off and on in milliseconds. Compared with early LCD displays, plasma sets offered wider viewing angles, better color fidelity and brightness, and larger screen sizes, but these advantages have diminished over the past decade. To the dismay of plasma TV fans, the lone Japanese manufacturer Panasonic announced that is ceasing production of these sets in 2014, so plasma's days may be numbered (Katzmeier, 2013).

- *Digital Light Processing (DLP) projectors*—Developed by Texas Instruments, DLP technology utilizes hundreds of thousands of tiny micromirrors mounted on a 1-inch chip that can project a very bright and sharp color image. This technology is used in a three-chip system to project digital versions of "films" in movie theaters. For under $3,000, a consumer can create a digital home theater with a DLP projector, a movie screen, and a multichannel surround-sound system.
- *UltraHD Television/Video Production and Display*—Just as most U.S. consumers have completed their home adoption of an HDTV display, television manufacturers are introducing displays, cameras, and editing equipment for UltraHD television—the production and distribution of images with 8,294,400 pixels (3840 wide X 2160 high) also known as 4K. These images actually have four times the number of pixels of a conventional 1920w X 1080h HDTV display. One problem with UltraHD technology is that the data processing and storage requirements are four times greater than with 2K HDTV. Just as this appeared to be a significant problem for UltraHD program production and distribution, a new compression scheme has emerged to deal with this issue: High-Efficiency Video Coding (HEVC). It uses advanced compression algorithms in a new global H.265 DTV standard adopted in January of 2013 by the International Telecommunications Union. It superseded their H.264 AVC standard and uses half the bit rate of MPEG-4 coding making it ideal for HEVC production and transmission (Butts, 2013).

Programming. All network, cable, and satellite programming in the United States is now produced in widescreen HDTV. There are some legacy 4:3 productions still being telecast (and there will be for years to come given each network's extensive analog archives), but it is typically being shown on widescreen displays with pillarbox graphics at the left and right side of the screen. As cable operators and satellite television providers phase out their up-converted analog simulcasts, this will free up a great deal of bandwidth now occupied by narrow-screen simulcasts of HDTV programs.

The Aereo Controversy

Aereo, a U.S. company founded in 2012, is a unique Over-The-Top MVPD service that provides a remote, over-the-air antenna to each subscriber. These antennas are similar to those any consumer can purchase to receive OTA broadcasts. The Aereo antenna facilitates the reception of live and streamed programs from local broadcasters, but the company does not pay retransmission fees to them, as U.S. cable and satellite companies are required to do. Aereo technology allows their subscribers to cut the cable and satellite "cord" and receive multiple OTA channels at a lower monthly cost. Local and national broadcasters strenuously objected to the Aereo model and sued the company for copyright infringement. Two federal courts upheld Aereo's technology model saying that, because each subscriber has his or her own antenna, the company is not violating broadcaster copyrights. The battle lines are clearly drawn between U.S. broadcasters (who derive more than $3 billion in annual income from retransmission rights) and Aereo and its allies who would like to encourage greater cord cutting. The U.S. Supreme Court heard the case in April 2014 (as this book is going to press) and will likely issue a decision that will profoundly affect the future of over-the-air broadcasting in the United States. Broadcast executives from CBS and Fox have stated that they would move their network's content from free over-the-air telecasts to a paid subscription model if Aereo prevails before the Supreme Court (Johnson, 2014). For the court decision, see the *Communication Technology Update and Fundamentals* companion website, www.tfi.com/ctu.

The Global Digital Transition

The period from 2012 to 2016 is a crucial one for the global conversion of analog to digital television transmission. France made the transition in 2011 and Japan completed their digital switchover in 2012. By the end of 2014, South Korea, New Zealand, Australia, and the United Kingdom will all switch to digital-only broadcasting (see Table 6.3).

Table 6.3
Global DTV Standards and Switch-Off Dates of Analog Terrestrial Television

Asia/Oceania	Year	Standard	Europe	Year	Standard	Americas	Year	Standard
Taiwan	2010	DVB-T	Netherlands	2006	DVB-T	United States	2009	ATSC
Japan	2012	ISDB-T	Sweden	2007	DVB-T	Canada	2011	ATSC
South Korea	2012	ATSC	Germany	2008	DVB-T	Mexico	2022p	ATSC
New Zealand	2013	DVB-T	Switzerland	2009	DVB-T	Panama	N.D.	DVB-T
Australia	2013	DVB-T	Spain	2010	DVB-T	Columbia	N.D.	DVB-T
Philippines	2015p	DVB-T	France	2011	DVB-T	Brazil	2016p	SBTVD
China	2015p	DMB-TH	United Kingdom	2012	DVB-T	Peru	N.D.	SBTVD

p = projected date. N.D. = No analog termination date set. The SBTVD standards used in Brazil and Peru are South American variants of the Japanese ISDB-T terrestrial DTV standard. DMB-TH is a unique DTV standard developed in China.

Sources: DigiTAG and TV Technology

Japan

The Japan Broadcasting Corporation (NHK) exhibited its 8K next-generation (beyond 4K) HDTV system called Super Hi-Vision (SHV) at the 2013 NAB Show in Las Vegas (see Figure 6.5). They set up a theater in their exhibit area showing breathtaking SHV footage of the 2012 summer Olympic Games shot in London with their prototype 8K camera (Seel, 2013). SHV's technical characteristics include a video format resolution of 7,680 (wide) X 4,320 (high) pixels, an aspect ratio of 16:9, a progressive scanning frequency of 60 frames, and a 22.2 multichannel audio system with a top layer of nine channels, a middle layer of 10 channels, a bottom layer of three channels, and two low-frequency effects channels (Shishikui, Fujita, & Kubota, 2009; Sugawara, 2008). SHV has 33 million pixels per frame—16 times that of HDTV. Beyond the pixel count, this advanced technology "produces three-dimensional spatial sound that augments the sense of reality and presence" (Shishikui et al., 2009, p. 1). These very high resolution displays might be used for large screen projection in corporate or educational settings or would allow home viewers to sit closer to 55-inch and larger displays for a greater sense of immersion or telepresence.

Figure 6.5
8K Television

At the 2013 NAB Show, NHK exhibited a prototype 8K Super Hi-Vision camera made by Hitachi and displayed its output on a 33-million pixel monitor shown in the background at right.

Source: P.B. Seel

Europe

Using the Digital Video Broadcasting-Terrestrial (DVB-T) standard, European broadcasters have focused for many years on delivering lower-definition SDTV programs to provide more digital viewing options to their audiences. But by the mid-2000s, they realized the long-term inevitability of providing widescreen HDTV services, driven in part by the growing popularity of flat-panel displays and Blu-ray disc players in Europe (DigiTAG, 2007).

In 2008, the Digital Video Broadcasting (DVB) Project adopted the technical parameters for the second-generation terrestrial transmission system called DVB-T2. Besides the United Kingdom, Finland launched DVB-T2 services in 2010, and other European countries are conducting DVB-T2 trials. The United Kingdom has avoided the trap of setting a DTV conversion date and then having to postpone it (as the U.S. did twice) by defining two key benchmarks: availability of DTV coverage and affordability of receivers by consumer (Starks, 2011). This process took more than 14 years and yielded a sliding end-date timetable from 2008 to 2012, with the latter date defining the actual switchover and one that coincided with other European nations by an E.U. directive.

New Zealand as an Over-the-Top Case Study

New Zealand is interesting from many perspectives. Its geographical isolation—the most remote western democracy, its small population—just over 4 million, its history—with a unicameral parliament, short election cycle, and no written constitution of substance. It has moved in one lifetime from what Americans would call a socialist state to the most deregulated and open economy in the world.

These changes have affected radio and television broadcasting in the country. Initially broadcasting was state controlled (based on a British 'BBC' model), now it is almost totally private and foreign-owned. The geography and difficult terrain meant that radio was the predominant medium well into the 1960s. New Zealand was one of the last developed countries to adopt a television service. It has a larger share of the advertising spend of any jurisdiction (Pauling, 2013). All television service has been by terrestrial transmission, except for two small attempts at wiring heavy population areas in the cities of Wellington and Christchurch. The small population means that expensive services such as transport, utilities, and broadcasting do not have competitive markets. Thus, there are dominant players. This dominance is helped by the lack of legislation or regulation to control monopoly or duopoly behavior. For example, there is no equivalent of the FCC, or its British regulatory counterpart, Offcom.

However, New Zealanders (Kiwis) are quick to adapt. For example, at the digital switchover on December 13, 2013, 96 percent of households were capable of receiving a digital signal. Forty-four percent of those homes were tuned to free-to-air (terrestrial, satellite or, for a small number, cable) television, another 44 percent were digitally tuned in to a pay-to-view provider (Sky TV), and 12 percent were tuned to both (Beatson, 2013).

The dominance of Sky TV permits it to make contracts with program suppliers that prevent others from easily entering the market. The two other major players, the free-to-air television networks TVNZ and TV3, neatly split the left-overs between them. These three services provide most of New Zealand's television viewing. Thus, programs are scheduled to suit the networks—not Kiwi audiences. Major television series produced in the U.K. or Europe can take up to two years to screen in New Zealand. However, viewers may be forcing a change.

New technology options for young digital-savvy Kiwis are driving a change in viewing culture, which challenges this model. Already 'time-shifting' (using DVRs to record and watch shows when they want to and fast-forwarding through the advertisements) is common and there is an increasing appetite for on-demand "Connected" (or Over-the-Top, as it is called in the U.S.) content that is perfectly suited for the Internet (Norris & Pauling, 2008). The major networks response to this is to increase the capacity of their Internet-only, catch-up services and to use the Internet for 'new' television activity. For example, a local comedy program *Auckland Daze* aimed at a youth audience is distributed as an Internet-only series. It proved so popular that the new series is also airing on traditional broadcast channels. TV3, which has the screening rights for the U.S. -produced drama *House of Cards*, is not only broadcasting the series weekly over 27 weeks, but is streaming all the first 12 episodes collectively on its website. Thus, the television series viewing experience formerly available only on DVD, is now available online in New Zealand.

The 12-month delay in broadcasting this series means that many of the younger audience have already seen the program. Young Kiwis are increasingly shunning the traditional broadcast services and are just as likely to download a television show or spend an evening browsing video clips on the web. (Beatson, 2013). One group of young people when asked which television shows they were viewing, mentioned only programs not yet aired on New

Zealand television. This cohort is downloading its "television" content using proxy servers and (sometimes illegal) torrents. This may explain why the broadcast networks are increasingly using the Internet as an alternate distribution medium.

For example, Coliseum Sports Media is providing an Internet-based sports program providing premium content streamed to any screen platform (phone, computer, tablet, or smart television set). It is a subscription service aimed at European Premier League football fans, of which there are many in New Zealand. Secondly, Telecom New Zealand, the dominant telecommunications entity, has announced the launch of "ShowMeTV" providing movies and television programs to subscribers over the Internet—as a direct competitor to Sky satellite-delivered content.

This has led one commentator to say that Internet-delivered television was becoming a "busy space" in New Zealand, and with the new TVNZ and TV3 services, along with "Quickflix" streamed content (a local version of Netflix), "it will come down to who has the best-quality content" (Galpin as cited in Fletcher, 2014). If the major New Zealand distributors of programs were to release content at the same time as U.S. or European providers, would young people bother to do a work-around? Should broadcasters release content online at the same time as on-air?

To add further impetus to these changes, the New Zealand government is funding a rollout of high-speed fiber-to-the-home (FTTH) service that will reach more than 830,000 homes and businesses across the country by 2019. Delivering the highest data speeds that can support services such as Internet-delivered video and television and high-definition video conferencing is possible (Chorus, 2014)

Television viewing culture is changing, particularly among young people where the concept of time shifting is changing attitudes towards televised content. There a growing demand for the "pull" delivery model, where an episode is just a download or an Internet stream away. Improvements in broadband connectivity in New Zealand—and the government is pouring $1.5 billion into improving high-speed access—will only enhance on-demand access.

Factors to Watch

The global diffusion of DTV technology will evolve over the second decade of the 21st century. In the United States, the future inclusion of mDTV receivers and HD-quality video recording capability in most mobile phones may enhance this diffusion on a personal and familial level. On a global scale, the more-developed nations of the world will have phased out analog television transmissions as they complete their transition to digital broadcasting. A significant number of nations will have completed the digital transition by 2016. In the process, digital television and video will influence what people watch, especially as smart televisions, tablet computers, and 4G phones will offer easy access to Internet-delivered content that is more interactive than traditional television, as the New Zealand case study shows.

These trends will be key areas to watch between 2014 and 2016:

- *Evolving UltraHD and 8K DTV technologies*—The continued development of higher-resolution video production and display technologies will enhance the viewing experience of television audiences around the world. While there are substantial manufacturing, data processing/storage, and transmission issues with UltraHD (and certainly 8K) DTV technologies, these will be resolved in the coming decade. The roll-out of fiber-to-the-home (FTTH) in many nations will enhance the online delivery of these high-resolution programs. When combined with ever-larger, high-resolution displays in homes, audiences can have a more immersive viewing experience (with a greater sense of telepresence) by sitting closer to their large screens.
- *Cord-Shavers and Cord-Nevers*—As noted above, the number of younger television viewers seeking out alternative viewing options on the Internet is steadily increasing. Expect to see more main stream television advertisers placing ads on popular Internet multimedia sites such as YouTube to reach this key demographic. In 2013, Antennas Direct sold 600,000 units of flat DTV antennas in the U.S. that allow viewers to bypass cable and satellite services by pulling in free over-the-air signals from local television broadcasters (Nakashima, 2013). With a smart television, an Internet connection, and a small antenna for

receiving over-the-air digital signals, viewers can see significant amounts of streamed and broadcast content without paying monthly cable or satellite fees. MVPDs such as cable and satellite companies are concerned about this trend as it enables these viewers to "cut the cable" or "toss the dish," so these companies are seeking unique programming such as premium sports events (the Super Bowl and the NCAA Basketball Tournament) to hold on to their customer base.

- *Two-Screen Viewing*—This is a growing trend where viewers watch television programs while accessing related content (via Tweets, texts, and related social media sites) on a mobile device, or via picture-in-picture on their DTV. Samsung's Smart Hub technology would facilitate this type of multitasking while watching television. Producers of all types of television programing are planning to reach out to dual-screen viewers with streams of texts and supplementary content while shows are on the air.
- *Mobile DTV*—Given the relatively modest investment of $100,000 needed by local stations in major cities to transmit mDTV, it is expected that this technology will become a service that most urban U.S. broadcasters can be expected to offer. The investment will be offset by increased revenue from advertising targeted to mobile DTV viewers. The dramatic uptake from 70 stations with mDTV service in April 2011 to 130 stations in 46 markets transmitting it by April of 2013 is indicative of its rapid adoption by larger market stations. However, wider adoption by users is dependent on manufacturers including an mDTV reception chip in newer model 5G smart phones.
- *ATSC 2.0 and 3.0*—The Advanced Television Systems Committee (ATSC) in the U.S. is investigating new technologies for on-demand programming, and the simulcast delivery of Internet content as part of the development of ATSC 2.0 standards that *are compatible* with the present DTV broadcast system. For example, a large-screen smart television could display conventional broadcast programming surrounded by additional windows of related Internet-delivered content around it (See Figure 6.6). The group is also investigating the long-term creation of "next-generation" ATSC 3.0 DTV standards that would be *incompatible* with the present digital broadcast system in the U.S. and would require yet another DTV transition at some point in the distant future.

Figure 6.6
The Future of Television

The future of television is a blended display of broadcast content—in this case, a severe weather alert—combined with an Internet-delivered run-down of the local newscast for random access and other streamed media content around the broadcast image. All viewed on a tablet.

Photo Source: P.B. Seel.

The era of the blended television-computer or "tele-computer" has arrived. Most large-screen televisions are now "smart" models capable of displaying multimedia content from multiple sources: broadcast, cablecast, DBS, and all Internet-delivered motion media. As the typical home screen has evolved toward larger (and smarter) models, it is also getting thinner with the arrival of new OLED models. The image quality of new high-definition displays is startling and will improve further with the delivery of UltraHD and 8K super-HDTV models in the near future. Television, which has been called a "window on the world," will literally resemble a large window in the home for viewing live sports, IMAX-quality motion pictures in 2D and 3D, and any video content that can be streamed over the Internet. On the other end of the viewing scale, mobile customers with 5G phones will have access to this same diverse content on their high-definition AMOLED screens, thanks to Wi-Fi and mDTV technology. Human beings are inherently visual creatures and this is an exciting era for those who like to watch motion media programs of all types in high-definition at home and away.

Digital Television & Video Visionary: Richard E. Wiley

At 4 p.m. on Sunday, May 20, 1993, Washington communications attorney **Richard E. Wiley** received a disturbing phone call. As chairman of the Advisory Committee for Advanced Television (ACATS), Wiley was managing the efforts of a national coalition of U.S. television technical advisors, television manufacturers, television network engineers, and government representatives in the creation of a new national standard for the Federal Communications Commission (FCC) for the production and transmission of digital television signals. It was a project with national multi-billion-dollar consequences, not only for the broadcasting and manufacturing partners, but for the 100 million U.S. television homes which would need to adopt the new digital transmission technology proposed by the FCC (Dupagne & Seel, 1998).

During ACATS testing of new broadcast television technologies in 1992, General Instruments (G.I.) shocked competing companies with a new solution for digital production and transmission, while the other entrant's submissions had been analog. After this development, the competitors (including Philips, Thomson, NBC's Sarnoff Labs, Zenith, AT&T, and the Massachusetts Institute of Technology) were urged to form a "Grand Alliance" consortium which would submit a single digital television standard to the ACATS for testing and eventual national adoption.

The call to Wiley on that Sunday afternoon indicated that technical negotiations had broken down among the Grand Alliance partners, who were convening in a meeting room at the Grand Hotel downtown. A press conference had been scheduled the following Monday, May 21st at FCC headquarters to announce the new national DTV standard, and the Sunday phone call made this unlikely. The fundamental disagreement was over whether the new national DTV standard was going to include legacy interlace scanning to mollify traditional broadcasters and manufacturers—or include progressive scanning, which appealed to the computer industry and three of the Grand Alliance partners: M.I.T., Zenith, and AT&T. These partners argued that progressive scanning was the ideal future-oriented technology for DTV, as it eliminated the artifacts introduced in TV signals with interlaced scanning, such as that used by the analog NTSC standard in place at that time. The interlace advocates argued that there was an enormous investment in interlaced television technology in the U.S. and the ACATS could not just wave a technical wand and make it disappear overnight.

Wiley drove over to the hotel where the group was meeting and used his formidable negotiating skills to find a solution to this fundamental technical impasse. He started by getting them to consent to the basic technical issues (e.g. audio transmission) that they agreed upon. Then he forged a compromise on the scanning issue: the new DTV standard could include provisions for both interlace *and* progressive scanning in varied resolutions at 720 and 1080 lines. Ultimately the U.S. DTV standard, as adopted by the FCC in 1996, included 18 options for digital broadcasting in both interlaced and progressive scanning (see table 6.1).

Wiley's job was not quite finished in forging this crucial compromise on DTV scanning on that Sunday night. On the day of the Monday press conference at the FCC, he supervised a seven-hour conference call among the Grand Alliance partners before all agreed to the key points addressed in the formal press release. As Chairman of the ACATS advisory committee over a nine-year period, he later supervised the testing of the new DTV technology and celebrated with his committee the U.S. digital switchover at midnight on June 12, 2009.

For his central role in forging the crucial agreement that led to the present DTV standard adopted by the U.S., Canada, Mexico, and Korea, Richard Wiley was awarded an Emmy from the Academy of Television Arts and Sciences and received numerous other awards from broadcasting organizations in the United States and abroad. He had previously served as Chairman of the Federal Communications Commission from 1970 to 1971—and is widely viewed as one of the most influential communications attorneys in the U.S. as the founder of the Wiley Rein law firm. He put his considerable negotiation skills to good use on that Sunday night in May of 1993, when he forged a key agreement that led to the digital television service that Americans and other global viewers now access every day.

Bibliography

Advanced Television Systems Committee. (2009, October 15). *A/153: ATSC mobile DTV standard, parts 1–8.* Washington, DC: Author.

Advisory Committee on Advanced Television Service. (1995). *Advisory Committee final report and recommendation.* Washington, DC: Author.

Bachman, J. (2013, December 30). Why the TV of the future might die before it was really launched. *Bloomberg Businessweek.* Retrieved from, http://www.businessweek.com/articles/ 2013-12-30/why-the-tv-of-the-future-might-die-before-it-was-really-launched.

Beatson, D. (2013, December 1). *The Digital Switchover – and the Marginalised Majority*, Kiwiboomers. Retrieved from, http://kiwiboomers.co.nz/the-digital-switchover-and-the-marginalised-majority.
Briel. R. (2012, January 13). Standard definition switch off looms. *Broadband TV News*. Retrieved from, http://www.broadbandtvnews.com/2012/01/13/standard-definition-switch-off-looms/.
Brinkley, J. (1997). *Defining vision: The battle for the future of television.* New York: Harcourt Brace & Company.
Butts, T. (2013, January 25). Next-gen HEVC video standard approved. *TV Technology*. Retrieved from, http://www.tvtechnology.com/article/nexgen-hevc-video-standard-approved/217438 .
Cade, D. L. (2013). A fascinating look at how GoPro became the bestselling camera in the world. *Petapixel.com*. Retrieved from, http://petapixel.com/2013/11/12/fascinating-look-gopro-became-popular-camera-world/
Chorus (2014). Ultra-*Fast Broadband, building a world-class network for New Zealand*. Retrieved from, http://www.chorus.co.nz/ultrafast-broadband.
Deficit Reduction Act of 2005. Pub. L. No. 109-171, § 3001-§ 3013, 120 Stat. 4, 21 (2006).
DigiTAG. (2007). *HD on DTT: Key issues for broadcasters, regulators and viewers.* Geneva: Author. Retrieved from, http://www.digitag.org/HDTV_v01.pdf.
DTV Delay Act of 2009. Pub. L. No. 111-4, 123 Stat. 112 (2009).
Dupagne, M., & Seel, P. B. (1998). *High-definition television: A global perspective.* Ames: Iowa State University Press.
Eggerton, J. (2012, February 27). Spectrum auctions: What now? *Broadcasting & Cable. 142*, 9. pp. 8, 9.
Federal Communications Commission. (1987). Formation of Advisory Committee on Advanced Television Service and Announcement of First Meeting, 52 Fed. Reg. 38523.
Federal Communications Commission. (1990). Advanced Television Systems and Their Impact Upon the Existing Television Broadcast Service (*First Report and Order*), 5 FCC Rcd. 5627.
Federal Communications Commission. (1996). Advanced Television Systems and Their Impact Upon the Existing Television Broadcast Service (*Fifth Further Notice of Proposed Rule Making*), 11 FCC Rcd. 6235.
Federal Communications Commission. (1997). Advanced Television Systems and Their Impact Upon the Existing Television Broadcast Service (*Fifth Report and Order*), 12 FCC Rcd. 12809.
Federal Communications Commission. (1998). Advanced Television Systems and Their Impact Upon the Existing Television Broadcast Service (Memorandum Opinion and Order on Reconsideration of the Sixth Report and Order), 13 FCC Rcd. 7418.
Federal Communications Commission. (2008). DTV Consumer Education Initiative (*Report and Order*, MB Docket No. 07-148). Retrieved from, http://hraunfoss.fcc.gov/edocs_public/attachmatch/FCC-08-56A1.pdf.
Federal Communications Commission. (1997). Advanced Television Systems and Their Impact Upon the Existing Television Broadcast Service (*Fifth Report and Order*), 12 FCC Rcd. 12809.
Fletcher, H. (2014, February 22). More choices for viewers as Telecom takes on Sky. *New Zealand Herald*. Retrieved from, http://www.nzherald.co.nz/ business/news/article.cfm?c_id=3&objectid=11207551.
Gorman, B. (2012, January 6). 1 in 3 U.S. TV households own 4 or more TVs. *TV By the Numbers*. Retrieved from, http://tvbythenumbers.zap2it.com/2012/01/06/almost-1-in-3-us-tv-households-own-4-or-more-tvs/115607/.
Johnson, T. (2014, January 10). Supreme Court to hear Aereo case. *Variety*. Retrieved from,http://variety.com/2014/biz/news/supreme-court-to-hear-aereo-case-1201037308/#.
Katzmeier, D. (2013, November 6). RIP Panasonic plasma TVs: Reactions from industry experts. *CNET*. Retrieved from, http://reviews.cnet.com/8301-33199_7-57610395-221/rip-panasonic-plasma-tvs-reactions-from-industry-experts/.
Leichtman Research Group. (2009, November 30). *Nearly half of U.S. households have an HDTV set* (Press release). Retrieved from, http://www.leichtmanresearch.com/press/ 113009release.html.
McAdams, D. (2012, October 17). HDTV adoption surpases 75 percent. *TV Technology*. Retrieved from, http://www.tvtechnology.com/news/0086/hdtv-adoption-surpasses--percent/215958.
Merli, J. (2011, April 6). Mobile DTV proponents all business at NAB. *TV Technology. 29*, 8. pp. 1, 26.
Middle Class Tax Relief and Job Creation Act of 2012. Pub. L. No. 112-96.
Morrison, (2013). Curved OLED HDTVs are a bad idea (for now). *CNet Reviews*. Retrievedfrom, http://reviews.cnet.com/8301-33199_7-57589082-221/curved-oled-hdtv-screens-are-a-bad-idea-for-now/.
Nakashima, R. (2013, April 7). Broadcasters worry about "zero-TV' households. *Associated Press*. Retrieved from, http://tv.yahoo.com/news/broadcasters-worry-zero-tv-homes-154357101--finance.html.
Negroponte, N. (1995). *Being Digital*. New York: Alfred A. Knopf.
Nielsen. (2011, May 3). Nielsen estimates number of U.S. television homes to be 114.7 million.
Nielsen Wire. Retrieved from, http://blog.nielsen.com/nielsenwire/media_entertainment/nielsen-estimates-number-of-u-s-television-homes-to-be-114-7-million/.

Nielsen. (2013, May 7). Nielsen Estimates 115.6 Million TV Homes in the U.S., Up 1.2%. Retrieved from, http://www.nielsen.com/us/en/newswire/2013/nielsen-estimates-115-6-million-tv-homes-in-the-u-s---up-1-2-.html.

Norris, P. & Pauling, B. (2008), *The Digital Future and Public Broadcasting,* New Zealand on Air, Wellington.

O'Neal, J. E. (2009, November 4). Mobile DTV standard approved. *TV Technology, 27,* 23.

Pauling, B. (2013) *New Zealand, A Radio Paradise?*, in R. J. Hand & M. Traynor, (eds.), *Radio in Small Nations.* Cardiff: University of Wales Press.

Reardon, M. (2013, December 6). FCC delays broadcast spectrum auction until 2015. CNET. Retrieved from, http://news.cnet.com/8301-13578_3-57614832-38/fcc-delays-broadcast-spectrum-auction-until-2015/.

Seel. P. B. (2011). Report from NAB 2011: Future DTV spectrum battles and new 3D, mobile, and IDTV technology. *International Journal of Digital Television. 2,* 3. 371-377.

Seel. P. B. (2012). The 2012 NAB Show: Digital television goes over the top. International Journal of Digital Television, 3, 3. 357-360.

Seel, P. B., & Dupagne, M. (2010). Advanced television and video. In A. E. Grant & J. H. Meadows (Eds.), *Communication Technology Update* (12th ed., pp. 82-100). Boston: Focal Press.

Shishikui, Y., Fujita, Y., & Kubota, K. (2009, January). Super Hi-Vision—The star of the show! *EBU Technical Review,* 1-13. Retrieved from, http://www.ebu.ch/en/technical/trev/trev_2009-Q0_SHV-NHK.pdf.

Starks, M. (2011). Editorial. *International Journal of Digital Television. 2,* 2. pp. 141-143.

Stelter, B. (2012, February 9), Youth are watching, but less often on TV. *New York Times.* pp. B1, B9.

Sugawara, M. (2008, Q2). Super Hi-Vision—Research on a future ultra-HDTV system. *EBU Technical Review,* 1-7. Retrieved from, http://www.ebu.ch/fr/technical/trev/trev_2008-Q2_nhk-ultra-hd.pdf.

Wakabayashi, D. (2013, July 30). The point-and-shoot camera faces its existential moment. *The Wall Street Journal.* Retrieved from, http://online.wsj.com/news/articles/ SB10001424127887324251504578580263719432252

YouTube Statistics. (2014). YouTube Inc. Retrieved from, http://www.youtube.com/yt/press/statistics.html.

7

Multichannel Television Services

Paul Driscoll, Ph.D. and
Michel Dupagne, Ph.D.*

Why Study Multichannel Television Services?

- About 86% of television households subscribe to a multichannel video service.
- Three providers (Comcast, DirecTV, and DISH) serve about 56% of all multichannel video subscribers.
- While the number of basic cable subscribers continued to decline in 2013, the number of video subscribers served by direct broadcast satellite and telephone companies rose moderately.

Introduction

Until the early 1990s, most consumers who sought to receive multichannel television service enjoyed few options other than subscribing to their local cable television operator. Satellite television reception through large dishes was available nationwide in the early 1980s, but this technology lost its luster soon after popular networks decided to scramble their signals.

Multichannel multipoint distribution service (MMDS), using microwave technology and dubbed wireless cable, existed in limited areas of the country in the 1980s, but has never been able to succeed in becoming a major market player. So for all intents and purposes, the cable industry operated as a de facto monopoly with little or no competition for multichannel video service during the 1980s.

The market structure of subscription-based multichannel television began to change when DirecTV launched its direct broadcast satellite (DBS) service in 1994 and an additional option arrived when DISH delivered satellite signals to its first customers in 1996. Another watershed moment occurred when Verizon introduced its fiber-to-the-home FiOS service in 2005 and again when AT&T started deploying U-verse in 2006.

While few cable overbuilders (i.e., two wired cable systems overlapping and competing with one another for the same video subscribers) have ever existed, the multichannel video programming distributor (MVPD) marketplace became increasingly competitive during the second decade of the 2000s. The Federal Communications Commission (FCC) reported in 2013 that 35% of U.S. homes were able to subscribe to at least four MVPDs in 2011, up from 33% in 2010 (Federal Communications Commission [FCC], 2013a).

In 2014, the MVPD industry faced a very different landscape from earlier decades, with a near market saturation that has plateaued around 86%, a subscriber base that is approaching parity between cable MVPDs and DBS, or telco (short for telephone company) MVPDs, a growing cord-cutting activity, and a possible existential threat to

* Driscoll is Associate Professor and Vice Dean for Academic Affairs, Department of Journalism and Media Management, University of Miami (Coral Gables, Florida).
Dupagne is Professor in the same department.

the MVPD business model from over-the-top (OTT) providers including Netflix and Hulu.

This chapter will first provide background information about multichannel television services to situate the historical, regulatory, and technological context of the industry. Special emphasis will be given to the cable industry. The other sections of the chapter will describe and discuss recent developments, major issues, and trends affecting the MVPD industry.

Definitions

Before we delve into the content proper, it is important to delineate the boundaries of this chapter and define key terms that are relevant to multichannel television services. Because definitions often vary from source to source, thereby creating further confusion, we will rely as much as possible on the legal framework of U.S. Federal regulations and laws to define these terms.

The *Cable Television Consumer Protection and Competition Act of 1992* defines a multichannel video programming distributor (MVPD) as "a person such as, but not limited to, a cable operator, a multichannel multipoint distribution service, a direct broadcast satellite service, or a television receive-only satellite program distributor, who makes available for purchase, by subscribers or customers, multiple channels of video programming" (p. 244). For the most part, MVPD service refers to multichannel video service offered by cable operators (e.g., Comcast, Time Warner Cable), DBS providers (DirecTV, DISH), and telephone companies (e.g., AT&T's U-verse, Verizon's FiOS). This chapter will focus on these distributors. As noted above, MMDS and wireless cable providers still exist, but account for a negligible share of the total multichannel video market—estimated at 0.6% at the end of 2012 (Olgeirson, 2013a).

Pay television designates a category of television services that offer programs uninterrupted by commercials for an additional fee on top of the basic MVPD service (FCC, 2013b). These services primarily consist of premium channels, pay-per-view (PPV), and video on demand (VOD). Subscribers pay a monthly fee to receive such premium channels as HBO and Starz. In the case of PPV, they pay a per-unit charge for ordering a movie or another program that is scheduled at a *specified* time. VOD, on the other hand, allows viewers to order any program from a video library at *any given time* in return for an individual charge or a monthly subscription fee. The latter type of VOD service is called subscription VOD or SVOD. We should note that the terms "pay television" and "MVPD" are often used interchangeably to denote a television service for which consumers pay a fee, as opposed to "free over-the-air television," which is available to viewers at no cost.

But perhaps the most challenging definitional issue that confronts regulators is whether over-the-top (OTT) providers could qualify as MVPDs if they offer multiple channels of video programming for purchase, like traditional MVPDs do. The FCC (2012) solicited comments on this very question in a March 2012 *Public Notice*, but has yet to take action as of April 2014. However, in April 2010, the Commission's Media Bureau denied Sky Angel's program access complaint against Discovery Communications on the basis that the company was not an MPVD and did not provide "its subscribers with a transmission path" (FCC, 2010, p. 3883). Sky Angel, which delivered religious and family-friendly programming over IP (Internet Protocol), ceased its OTT/IPTV operations in 2014. Readers will notice that the definition of an MPVD neither contains the words "transmission path" nor implies that such transmission path be owned by a qualified MVPD. In addition, the statutory phrase "such as, but not limited to" could be construed as intentionally illustrative.

These arguments notwithstanding, the FCC will not view an OTT provider like YouTube, which supplies more than 100 paid channels to Internet viewers, as an MVPD. Instead, the Commission would classify this video provider as an online video distributor (OVD), which "offers video content by means of the Internet or other Internet Protocol (IP)-based transmission path provided by a person or entity other than the OVD" (FCC, 2013a, p. 10499). Unlike MVPDs, OVDs may not file program access complaints with the FCC and are not subject to retransmission consent rules. Therefore, OTT providers will not be addressed in much detail in this chapter and instead will be covered in Chapter 8.

Background

The Early Years

While a thorough review of the history, regulation, and technology of the multichannel video industry is beyond this chapter (see Baldwin & McVoy [1988], Parsons [2008], Parsons & Frieden [1998] for more information), a brief overview is necessary to understand the broad context of this technology.

As TV broadcasting grew into a new industry in the late 1940s and early 1950s, many households were unable to access programming because they lived too far from local stations' transmitter sites or because geographic obstacles blocked reception of terrestrial signals. Without access to programming, consumers weren't going to purchase TV receivers, a problem that bedeviled both the local stations seeking viewers and appliance stores eager to profit from set sales.

The solution was to erect a central antenna capable of capturing the signals of local market stations, amplify, and distribute them through copper wire to prospective viewers for a fee. Thus, cable extended local stations' reach and provided an incentive to purchase a set. The first non-commercial Community Antenna TV (CATV) service was established in 1949 in Astoria, Oregon, but multiple communities claim to have pioneered this retransmission technology, including a commercial system launched in Lansford, Pennsylvania, in 1950 (Besen & Crandall, 1981).

Local TV broadcasters initially valued cable's ability to extend their household reach, but tensions rose when cable operators began using terrestrial microwave links to import programming from stations located in "distant markets," increasing the competition for audiences that local stations rely on to sell advertising. Once regarded as a welcome extension to their over-the-air TV signals, broadcasters increasingly viewed cable as a threat and sought regulatory protection from the government.

Evolution of Federal Communications Commission Regulations

At first, the FCC showed little interest in regulating cable TV. But given cable's growth in the mid-1960s, the FCC, sensitive to TV broadcasters' statutory responsibilities to serve their local communities and promote local self-expression, adopted rules designed to protect over-the-air TV broadcasting (FCC, 1965). Viewing cable as only a supplementary service to over-the-air broadcasting, regulators mandated that cable systems carry local TV station signals and placed limits on the duplication of local programming by distant station imports (FCC, 1966). The Commission saw such rules as critical to protecting TV broadcasters, especially struggling UHF-TV stations, although actual evidence of such harm was largely nonexistent. Additional cable regulations followed in subsequent years, including a program origination requirement and restrictions on pay channels' carriage of movies, sporting events, and series programming (FCC, 1969, 1970).

The FCC's robust protection of over-the-air broadcasting began a minor thaw in 1972, allowing an increased number of distant station imports depending on market size, with fewer imports allowed in smaller TV markets (FCC, 1972a). The 1970 pay cable rules were also partially relaxed. Throughout the rest of the decade, the FCC continued its deregulation of cable, sometimes in response to court decisions.

Although the U.S. Supreme Court held in 1968 that the FCC had the power to regulate cable television as reasonably ancillary to its responsibility to regulate broadcast television, the scope of the FCC's jurisdiction under the *Communications Act of 1934* remained murky (*U.S. v. Southwestern Cable Co.*, 1968). The FCC regulations came under attack in a number of legal challenges. For instance, in *Home Box Office v. FCC* (1977), the Court of Appeals for the D.C. Circuit questioned the scope of the FCC's jurisdiction over cable and rejected the Commission's 1975 pay cable rules, designed to protect broadcasters against possible siphoning of movies and sporting events by cable operators.

In the mid-1970s, cable began to move beyond its community antenna roots with the use of domestic satellites to provide additional programming in the form of superstations, cable networks, and pay channels. These innovations were made possible by the FCC's 1972 *Open Skies Policy*, which allowed qualified companies to launch and operate domestic satellites (FCC, 1972b). Not only did the advent of satellite television increase the amount of programming available to households well beyond the coverage area of local TV broadcast channels, but it also provided cable operators with an opportunity to fill their mostly 12-channel cable systems.

Early satellite-delivered programming in 1976 included Ted Turner's "Superstation" WTGC (later WTBS) and the Christian Broadcasting Network, both initially operating from local UHF-TV channels. In 1975, Home Box Office (HBO), which had begun as a pay programming service delivered via terrestrial microwave links, kicked off its satellite distribution offering the "Thrilla in Manila" heavyweight title fight featuring Mohammad Ali against Joe Frasier.

Demand for additional cable programming, offering many more choices than local television stations and a limited number of distant imports, stimulated cable's growth, especially in larger cities. According to the National Cable & Telecommunications Association (NCTA, 2014), there were 28 national cable networks by 1980 and 79 by 1990. This growth started an alternating cycle where the increase in the number of channels led local cable systems to increase their capacity, which in turn stimulated creation of new channels, leading to even higher channel capacity, etc.

Formal congressional involvement in cable TV regulation began with the *Cable Communications Policy Act of 1984* (Cable Act thereafter). Among its provisions, the Act set national standards for franchising and franchise renewals, clarified the roles of Federal, state, and local governments, and freed cable rates except for systems that operated without any "effective competition," defined as TV markets having fewer than three over-the-air local TV stations. Cable rates soared, triggering an outcry by subscribers.

The quickly expanding cable industry was increasingly characterized by a growing concentration of ownership. Congress reacted to this situation in 1992 by subjecting more cable systems to rate regulation of their basic and expanded tiers, and instituting retransmission consent options for stations not opting for must-carry status without remuneration.

The *Cable Television Consumer Protection and Competition Act* (Cable Act thereafter) of 1992 also required that MVPDs make their programming available at comparable terms to satellite and other services. The passage of the *Telecommunications Act of 1996* also reflected a clear preference for competition over regulation. It rolled back much of the rate regulation put in place in 1992 and opened the door for telephone companies to enter the video distribution business.

Distribution of TV Programming by Satellite

By the late 1970s and early 1980s, satellites were being used to distribute TV programming to cable systems, expanding the line-up of programs available to subscribers. But in 1981, about 11 million people lived in rural areas where it was not economical to provide cable; five million residents had no TV at all, and the rest had access to only a few over-the-air stations (FCC, 1982).

One solution for rural dwellers was to install a Television Receive Only (TVRO) satellite dish. Sometimes called "BUGS" (Big Ugly Dishes) because of the 8-to-12-foot diameter dishes needed to capture signals from low-powered satellites, TVROs were initially the purview of hobbyists and engineers. In 1976, H. Taylor Howard, a professor of electrical engineering at Stanford University, built the first homemade TVRO and was able to view HBO programming for free (his letter of notice to HBO having gone unanswered) (Owen, 1985). The backyard earth station movement was born.

Interest in accessing free satellite TV grew in the first half of the 1980s, and numerous companies began to market satellite dish kits. From 1985 to 1995, two to three million dishes were purchased (Dulac & Goodwin, 2006). However, consumer interest in home satellite TV stalled in the late-1980s, when TV programmers began to scramble their satellite feeds.

In 1980, over the strenuous objections of over-the-air broadcasters, the FCC began to plan for direct broadcast satellite (DBS) TV service (FCC, 1980). In September 1982, the Commission authorized the first commercial operation. At the 1983 Regional Administrative Radio Conference, the International Telecommunication Union (ITU), the organization that coordinates spectrum use internationally, awarded the United States eight orbital positions of 32 channels each (Duverney, 1985). The first commercial attempt at DBS service began in Indianapolis in 1983, offering five channels of programming, but ultimately all of the early efforts failed, given the high cost of operations, technical challenges, and the limited number of desirable channels available to subscribers. (Cable operators also used their leverage to dissuade programmers from licensing their products to DBS services.) Still, the early effort demonstrated the technical feasibility of using medium-power Ku-band satellites (12.2 to 12.7 GHz downlink) and foreshadowed a viable DBS service model.

In 1994, the first successful DBS providers were DirecTV, a subsidiary of Hughes Corporation, and U.S. Satellite Broadcasting (USSB) offered by satellite TV pioneer Stanley Hubbard. Technically competitors, the two services offered largely complementary programming and together launched the first high-power digital DBS satellite capable of delivering over 200 channels of programming to a much smaller receiving dish (Crowley, 2013).

DirecTV took over USSB in 1998. EchoStar (now the DISH Network) was established in the United States in 1996 and became DirecTV's primary competitor. In 2005, the FCC allowed EchoStar to take over some of the satellite channel assignments of Rainbow DBS, a failed 2003 attempt by Cablevision Systems Corporation to offer its Voom package of 21 high-definition (HD) TV channels via satellite. In 2003, the U.S. Department of Justice blocked the attempted merger between DirecTV and EchoStar, but there is some speculation that such a merger might again be revisited.

Video Compression

The large increase in the number of available programming channels over satellite was possible due to advances in digital video compression and the development of the MPEG-1 (Moving Picture Experts Group) standard in 1993 and an improved MPEG-2 standard introduced in 1995. Developed by the ITU and the MPEG group, video compression dramatically reduces the number of bits needed to store or transmit a video signal by removing redundant or irrelevant information from the data stream. Satellite signals compressed using the MPEG-2 format allowed up to eight channels of programming to be squeezed into one analog channel. Video compression yields increased capacity and allows satellite carriage of high-definition TV channels.

Impressive advances in compression efficiency have continued, including the adoption by DBS operators of MPEG-4 Advanced Video Coding (AVC) that reduces by half the bit rate needed compared with MPEG-2 (Crowley, 2013). High Efficiency Video Coding (HEVC), approved in 2013 by the ITU and MPEG, would double the program carrying capacity compared with MPEG-4 compression. Once deployed, HEVC would enable DBS providers to carry additional high-definition channels and even UltraHD channels (Crowley, 2013). Other efficiencies captured through a variety of technical advances, including improvements in digital modulation, error correction, and satellite frequency reuse, suggest continuing increases in satellite capacity.

Direct Broadcast Satellite Expansion

DBS systems proved popular with the public but faced a number of obstacles that allowed cable to remain the dominant provider of multichannel television service. One competitive disadvantage of DBS operators over their cable counterparts was that subscribers were unable to view their local-market TV channels without either disconnecting the TV set from the satellite receiver or installing a so-called A/B switch that made it possible to flip between over-the-air and satellite signals.

In 1988, Congress passed the *Satellite Home Viewer Act* that allowed the importation of distant network programming (usually network stations in New York or Los Angeles) to subscribers unable to receive the networks' signals from their over-the-air local network affiliates. Congress followed up in 1999 with the *Satellite Home Viewer Improvement Act* (SHVIA) that afforded satellite companies the opportunity (although not a requirement) to carry the signals of local market TV stations to all subscribers living in that market. Providing local station signals into specific local markets is known as "local into local service" in industry jargon. SHIVA also mandated that DBS operators carrying one local station carry all other local TV stations requesting carriage in a local market, a requirement known as "carry one, carry all."

Today, almost all U.S. TV households subscribing to DBS service are able to receive their local stations via satellite. DirecTV provides local broadcast channels in HD to 96% of U.S. households while DISH Network supplies local broadcast channels in HD to 97% of U.S. households (FCC, 2013a). Most local station signals are delivered from the satellite using spot beam technology that transmits signals into specific local TV markets. Satellites have dozens of feedhorns used to send their signals to earth, from those that cover the entire continental United States (CONUS) to individual spot beams serving local markets.

Cable System Architecture

Early cable TV operators built their systems based on a tree and branch design using coaxial cable, a type of copper wire suitable for transmission of radio frequencies (RF). As shown in Figure 7.1, a variety of video signals are received at the cable system's "headend," including satellite transmissions of cable networks, local TV station signals (delivered either by wire or from over-the-air signals received by headend antennas), and feeds from any public, educational, and governmental (PEG) access channels the operator may provide. Equipment at the headend processes these incoming signals and translates them to frequencies used by the cable system.

Cable companies use a modulation process called quadrature amplitude modulation (QAM) to encode and deliver TV channels. The signals are sent to subscribers through a trunk cable that, in turn, branches out to thinner feeder cables, which terminate at "taps" located in subscribers' neighborhoods. Drop cables run from the taps to individual subscriber homes. Because electromagnetic waves attenuate (lose power) as they travel, the cable operator deploys amplifiers along the path of the coaxial cable to boost signal strength. Most modern TV receivers sold in North America contain QAM tuners, although set-top boxes provided by the cable system are usually used because of the need to descramble encrypted signals.

Figure 7.1

Tree-and-Branch Design

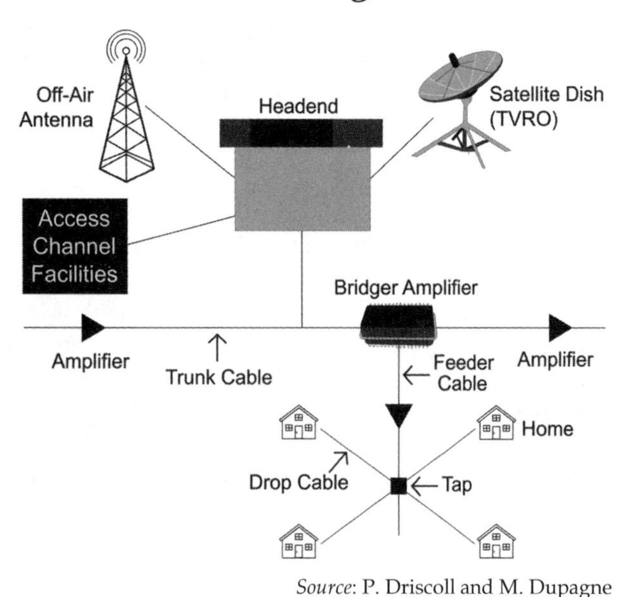

Source: P. Driscoll and M. Dupagne

Over the years, the channel capacity of cable systems has increased dramatically, from 12-22 channel systems through the mid-1970s to today's systems capable of carrying more than 300 TV channels and bi-directional communication. Modern cable systems have 750 MHz of radio frequency capacity (Crowley, 2013). This capacity allows not only plentiful television channel choices, but also enables cable companies to offer a rich selection of pay-per-view (PPV) and video-on-demand (VOD) offerings. It also makes possible high-speed Internet and voice over Internet telephony (VoIP) services to businesses and consumers. Most cable operators have moved to all-digital transmission, although a few still provide some analog channels, usually on the basic tier.

As shown in Figure 7.2, modern cable systems are hybrid fiber-coaxial (HFC) networks. Transmitting signals using light over glass or plastic, fiber optic cables can carry information long distances with much less attenuation than copper coaxial, greatly reducing the need for expensive amplifiers and offering almost unlimited bandwidth with high reliability. A ring of fiber optic cables is built out from the headend, and fiber optic strands are dropped off at each distribution hub. From the hubs, the fiber optic strands are connected to optical nodes, a design known as fiber to the node or fiber to the neighborhood (FTTN). From the nodes, the optical signals are transduced back to electronic signals and delivered to subscriber homes using coaxial cable. The typical fiber node can serve 500 to 2000 homes. Depending on the number of homes served and distance from the optical node, a coaxial trunk, feeder, and drop arrangement may be deployed.

Figure 7.2

Hybrid Fiber-Coaxial Design

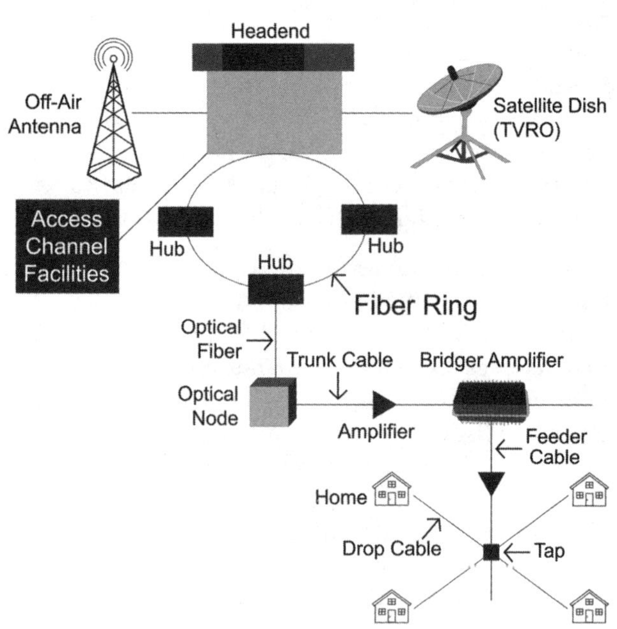

Source: P. Driscoll and M. Dupagne

Rollouts of HFC cable networks and advances in compression have added enormous information capacity compared to systems constructed with only copper cable, boosting the number of TV channels available for live viewing, and allowing expansive VOD offerings and digital video recorder (DVR) service. Additionally, given the HFC architecture's bi-directional capabilities, cable companies are able to offer consumers both telephone and broadband data transmission services, generating additional revenues.

Some MVPDs, such as Verizon with its FiOS service, extend fiber optic cable directly to the consumer's residence, a design known as fiber to the home (FTTH) or fiber to the premises (FTTP). FiOS uses a passive optical network (PON), a point-to-multipoint technology that optically splits a single fiber optic beam sent to the node to multiple fibers that run to individual homes, ending at an optical network terminal (ONT) attached to the exterior or interior of the residence (Corning, 2005). The ONT is a transceiver that converts light waves from the fiber optic strand to electrical signals carried through the home using appropriate cabling for video, broadband (high-speed Internet), and telephony services.

Although the actual design of FTTH systems may vary, fiber has the ability to carry data at much higher speeds compared to copper networks. For example, Google Fiber boasts up to a blazing one gigabit (one billion bits) of data per second. Fiber to the home may be the ultimate in digital abundance, but it is expensive to deploy and may actually provide higher data rates than most consumers currently need.

In addition, technological advances applied to existing HFC systems have significantly increased available bandwidth, making a transition to all-fiber networks a lower priority for existing cable systems in the United States. Going forward, additional efficiencies may be realized by advances in QAM-modulation and possibly transcoding QAM-modulated signals into Internet Protocol (IP) signals, allowing switched video capability that would deliver to the consumer only the channels being watched or recorded.

Direct Broadcast Satellite System Architecture

As shown in Figure 7.3, DBS systems utilize satellites in geostationary orbit, located at 22,236 miles (35,785 km) above the equator, as communication relays. This distance is often rounded off to 22,300 miles. At that point in space, the satellite's orbit matches the earth's rotation (although they are not travelling at the same speed.) As a result, the satellite's coverage area or "footprint" remains constant over a geographic area. In general, footprints are divided into those that cover the entire continental United States, certain portions of the United States (because of the satellite's orbital position), or more highly focused spot beams that send signals into particular TV markets.

DBS operators send or "uplink" data from the ground to a satellite at frequencies in the gigahertz band, or billions of cycles per second. Each satellite is equipped with receive/transmit units called transponders that capture the uplink signals and convert them to different frequencies to avoid interference. The signal is then sent back to earth or "downlinked."

Figure 7.3

Basic Direct Broadcast Satellite Technology

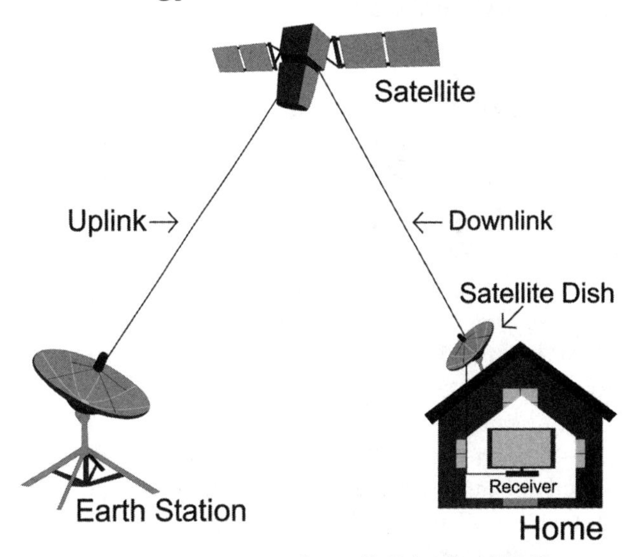

Source: P. Driscoll and M. Dupagne

A small satellite dish attached to the consumer's home collects the signals from the satellite. A low-noise block converter (LBN) is used to amplify the signals by as much as a million times and translate the high frequency satellite signal to the lower frequencies used by TV receivers. The very high radio frequencies used in satellite communications generate extremely short wavelengths, sometimes leading to temporary signal interference in rainy weather when waves sent from the satellite are absorbed and scattered by raindrops.

Unlike cable service that features bi-directional data flows, DBS is a one-way service that relies on broadband connections to offer its VOD services. But with its nationwide footprint, DBS service is available in rural areas where the infrastructure cost of building a cable system would be prohibitive.

Recent Developments

Statewide Video Franchising

Prior to 2005, the required process of cable franchising (i.e., authorizing an entity to construct and operate a cable television system in an area) was generally handled by local governments. The *Cable Act of 1984* did not mandate that the franchising authority be a local body, but subjecting cable operators to local jurisdiction made sense since they used public rights-of-way to deploy their infrastructure.

On the other hand, cable operators have long argued that local franchising requirements are onerous and complex, which often involve negotiations and agreements with multiple local governments (e.g., county and city). "For example, one industry representative indicated that a single Florida county had as many as 26 different local franchising entities" (Office of Program Policy Analysis & Government Accountability, 2009, p. 2).

In 2005, Texas became the first state to enact a law shifting the responsibilities of cable/video franchise requirements from the local to the state government (see Parker, 2011). More than 20 other states followed suit. In those states, it is typically the department of state or the public utilities commission that issues cable franchises (sometimes called certificates) to wireline MVPDs (National Conference of State Legislatures [NCSL], 2013). Other state agencies may share other cable television service duties (e.g., quality complaints) that previously fell within the purview of local governments. As of February 2013, at least 23 states awarded cable/video franchising agreements (NCSL, 2013).

Few studies have investigated the impact of statewide cable franchising reforms on competition. Concerns remain about the amount of franchise fees, funding and regulatory oversight of public, educational, and governmental (PEG) access channels, and the possibility and effectiveness of anti-redlining provisions to prevent service discrimination in low-income areas. The FCC (2013a) reported that some states did away or curtailed the requirements for PEG channels whose funding was often a key provision in local franchise agreements.

A la Carte Pricing

A la carte MVPD service would allow subscribers to order individual channels as a customized programming package instead of relying on standard bundled packages or tiers. This issue has attracted considerable policy interest, especially in the mid-2000s, because consumers have often complained to their elected officials or policymakers about the high cost of cable service in relation to other services. Indeed, the FCC (2013c) has documented for nearly two decades that cable rate increases outpace the general inflation rate. It reported that the average price of expanded basic cable service (cable programming service tier) grew by a compound average annual rate of 6.1% from 1995 to 2012. In contrast, the consumer price index for all items rose by only 2.4% during those 17 years.

The Commission also found that the number of channels available in the expanded package soared from 44 in 1995 to 150 in 2012, or by an annual rate of 5.8% (FCC, 2013c). But the per-channel price hardly changed during that time (0.2% annually), which bolstered the case of the cable industry for legitimate price hikes. But in absolute terms, cable and other MVPD bills continue to increase far above inflation, due largely to climbing license fees (paid by distributors to non-broadcast programmers for carriage) and retransmission fees (paid by operators to broadcasters for carriage).

According to the NPD Group, a market research company, the monthly MVPD subscription fare averaged $86 in 2011 and was forecast to reach $123 in 2015 (Diallo, 2013). SNL Kagan (2013), another leading research firm, estimated that ESPN alone would return more than $7 billion in license fees in 2014 to its parent company Disney. The firm predicted that ESPN's license fee per subscriber (sub) per month would grow from $5.54 in 2013 to $7.40 in 2017.

On paper, a bundling strategy makes more economic sense than a pure components strategy (a la carte) when the marginal cost of channel delivery is low, which is often the case with popular channels, and when consumers price channel packages similarly (Hoskins, McFadyen, & Finn, 2004). Empirical studies about the benefits of a la carte pricing have been mixed and have failed to resolve this debate (see Ramachandran & Launder, 2013).

A study from Needham & Co. indicated that unbundling could result in the elimination of $80 to $113 billion in consumer value, $45 billion in television advertising, and 1.4 million jobs (James, 2013). While a majority of consumers (73% in one survey) would favor customizing their channel packages to suit their individual needs (Barthold, 2013), it is unclear whether they are fully cognizant of the economic implications that a la carte programming would entail. Critics have often pointed out that a la carte programming would not necessarily reduce cable bills because programmers would have to raise their license fees substantially to compensate for declining viewership and advertising revenue. In addition, channel diversity would likely be compromised (Nocera, 2007).

Bundling is hardly an insignificant pricing matter for MVPDs. Not only does bundling play a key role in monetizing MVPDs' video content, but it can also influence the entire revenue stream of cable operators who offer triple-play packages.

Triple play, which refers to the combination of video, voice, and data products, is beneficial to both customers who can save money, and some MVPDs, particularly cable operators, who can secure a competitive advantage over other MVPDs and counter the effect of cord cutting or slow growth in the video segment. SNL Kagan (2013) reported that the penetration of triple play provided by six multiple system operators (MSO) jumped from 21% in the second quarter of 2008 to 42% in the second quarter of 2013.

Adapting to Changes in the Linear Viewing Model

MVPDs continue to adapt their technology and content distribution offerings to accommodate subscribers' shifting patterns of TV consumption. Today, the average adult watches nearly three and a half hours per week of time-shifted TV and seven out of 10 identify themselves as binge viewers (McCord, 2014; Nielsen, 2014).

The trend toward delayed viewing is one factor disrupting the TV ecosystem, especially for advertising revenues and program ratings. The digital video recorder (DVR), first introduced in 1999, is now present in 55% of television households that subscribe to an MVPD (Leichtman Research Group, 2013). Although available as stand-alone units, DVRs are usually integrated into the increasingly powerful set-top boxes provided by cable and DBS operators.

The versatility of DVRs has continued to expand and the available features are sometimes used by MVPDs to distinguish themselves from competitors. For example, the DISH Network's high-end DVR service, the "Hopper," boasts simultaneous recording of up to eight program channels; recording and viewing functionality for all TVs in a household; eight days of automatic recording of prime-time network programming from up to four local stations; automatic ad skipping on playback of some network TV shows; and remote access on mobile devices for out-of-home viewing of live and recorded TV (using Slingbox™ technology) (EchoStar Communications Corp., 2014).

Cablevision Systems offers a remote-storage DVR (RS-DVR) service that allows subscribers to record up to 15 programs simultaneously and store them in the "cloud" (remote servers at the system's headend) eliminating the expense of providing DVR units directly to the household (Cablevision Systems Corp., 2014). Even without a DVR unit, about 70% of MVPD subscribers have access to a huge inventory of VOD and SVOD offerings, including previously aired episodes of programs from channels subscribed to by the consumer (Holloway, 2014).

Another sign of possible upheaval for MVPDs is the growing success of over-the-top (OTT) video services available through a host of streaming media player dongles, boxes, and smart TVs, including Chromecast, Roku, Boxee, Apple TV, and Sony PlayStation. The rising popularity of broadband-delivered video subscription services, such as Netflix, Hulu Plus, and Amazon Prime, have led to concerns that customers, especially younger customers, will be "cutting the cord" on MVPD subscriptions (or never begin subscribing) in favor of more affordable services (Lafayette, 2014). However, for those satisfied with their existing MVPD service, Netflix has announced that it will make its service available for carriage on cable TV systems (Bray, 2014). (For more on OTT, see Chapter 8.)

Retransmission Consent

There are few other issues in the MVPD business that have created more tensions between distributors and broadcasters and angered more subscribers than retransmission consent negotiations going awry.

Table 7.1
Annual Broadcast Retransmission Fee Projections by Medium, 2012-2014

Medium	2012	2013	2014
Cable ($ millions)	1,373.2	1,851.8	2,342.4
Average cable fee/sub/month ($)	2.00	2.78	3.60
DBS ($ millions)	788.2	1,070.6	1,433.3
Average DBS fee/sub/month ($)	1.93	2.61	3.50
Telco ($ millions)	225.8	382.7	512.1
Average telco fee/sub/month ($)	2.04	3.00	3.55
Total retransmission fees ($ millions)	2,387.2	3,305.1	4,287.8
Average fee/sub/month ($)	1.98	2.74	3.56

Note. All figures are estimates as of November 2013. The average fee per month per subscriber for each video medium is obtained by dividing the amount of annual retransmission fee for the medium by the average number of subscribers for that medium. The average number of subscribers, which is calculated by taking the average of the subscriber count from the previous year and the subscriber count of the current year, is meant to estimate the number of subscribers at any given time during the year and reflect better the retransmission fee charged during the year instead of at the year-end. The average estimates refer to the fees paid to television stations on behalf of each subscriber, not to the payment per station.

Source: SNL Kagan. Reprinted with permission.

As noted above, the *Cable Act of 1992* allowed local broadcasters to seek financial compensation for program carriage from MVPDs, a process known as retransmission consent. While most of these agreements are concluded with little fanfare (Lenoir, 2014), some negotiations between parties can degenerate into protracted, mercurial, "who-will-blink-first" disputes, which could lead to high-profile blackouts when the MVPD decides to remove the signal of the broadcaster from its line-up.

The economic stakes are high. Broadcasters claim that they deserve fair rates for their popular programs and increasingly consider retransmission fees as a second revenue stream. Television stations are expected to collect more than $4 billion in broadcast retransmission fees in 2014 (see Table 7.1). On the other hand, the MVPDs, who are fully aware that excessive customer bills would intensify cord cutting, have attempted (generally with little success) to rein in broadcasters' requests for higher retransmission fees.

A case in point was the nasty, public retransmission consent fight that pitted CBS against Time Warner Cable in 2013. About 3.2 million cable subscribers were deprived of CBS programming from August 2 to September 2 (Farrell, 2013). Who won eventually? Industry observers generally agree that CBS CEO Leslie Moonves emerged relatively unscathed from the contractual quarrel, having secured an estimated fee increase to $2 per subscriber, per month, that will total $1 billion by 2017 (Carter, 2013). Moonves has viewed retransmission consent as a growth strategy for the broadcast network.

Current Status

TV Everywhere

According to the FCC (2013a), TV Everywhere (TVE) describes "an MVPD initiative, which allows subscribers of certain services to access video programming on a variety of fixed and mobile Internet-connected devices" (p. 10499). In theory, TVE was designed to provide a subscriber with the opportunity to view live (linear) programs or on-demand (non-linear) video content on such mobile devices as smartphones and tablets inside and outside the home. The subscriber would just need to authenticate (i.e., log in) through an application from the MVPD to confirm his or her identity as an MVPD subscriber.

But after five years, the implementation of TVE has yet to reach its full potential, due to perceived usability and contractual deficiencies. For instance, John Skipper, President of ESPN and Co-Chairman of Disney Media Networks, complained about the "clunkiness" of the authentication process: "We need a single system and we need a system that is automatic" (Dreier, 2014). But former Turner Broadcasting Vice Chairman Andy Heller contended that "the real stumbling block has been deals" and that some distributors and programmers have used the TVE content negotiations as a tool for revisiting the conditions of other contracts (Flint, 2012). By the end of 2013, MVPDs reported various degrees of TVE deployment. But even Time Warner Cable, considered to be a proactive player on the TVE front, provided only a dozen linear channels (and no on-demand content) for out-of-the-home viewing (Olgeirson, 2013b).

Decline of MVPD Subscription

SNL Kagan reported that collectively the cable, DBS, and telco operators experienced their first full-year video subscription drop in 2013, with a loss of 251,000 subscribers (Olgeirson, Lenoir, & Young, 2014). Specifically, cable operators lost about 2 million subscribers in 2013 while DBS and telephone companies added 170,000 and 1.6 million video subscribers, respectively (see Table 7.2). The Leichtman Research Group also confirmed the loss of video subscribers (estimated at 105,000) in 2013 (Baumgartner, 2014a).

Altogether, these estimates point to a growing cord-cutting phenomenon. For 2014, SNL Kagan predicted that the MVPD industry would continue to shed video customers with little or no prospect of growth (Table 7.2). The percentage of TV households who subscribe to an MVPD is expected to decline to about 84% in 2014, down from 87% in 2010.

The ranking of the top 10 MVPDs did not change dramatically in 2013, compared to previous years. The top three providers served about 56% of all MVPD subscribers, and all 10 operators listed in Table 7.3 accounted for a 91% share of the total MVPD market. If we calculate the market shares of these MVPDs, the concentration ratio for the top four would be 67% and would easily exceed the 50% threshold for high concentration (Hoskins et al., 2004). We should note that Table 7.3 does not reflect the possible changes caused by the proposed merger between Comcast and Time Warner Cable announced in early 2014, but not complete when this book was published.

Factors to Watch

Mobile Video Wireless Services

There has been explosive growth in accessing video on mobile devices. A survey by Frank N. Magid Associates found that 71% of tablet users and 45% of smartphone users watch long-form TV, movies, and sports on their devices. Tablet use increased an estimated 56% in 2013 to 9.7 million units while wireless carriers provided service to more than 200 million smartphones in the United States. Total mobile video advertising revenues from all sources soared to an estimated $800 million in 2013, up from $322 million in 2012 (SNL Kagan, 2014).

Table 7.2

U.S. Multichannel Video Industry Benchmarks (in Millions), 2010-2014

Category	2010	2011	2012	2013	2014
Basic cable subscribers	59.8	58.0	56.4	54.4	53.5
Digital service cable subscribers	44.7	46.0	46.8	46.9	47.8
High-speed data cable subscribers	44.4	47.3	50.3	52.7	54.2
Voice service cable subscribers	23.5	24.7	25.8	27.6	26.5
DBS subscribers	33.4	33.8	34.1	34.3	34.1
Telco video subscribers	6.9	8.5	9.9	11.5	12.7
All multichannel video subscribers	100.8	101.0	101.0	100.9	100.8
TV households	115.9	116.4	117.4	118.0	119.3
Occupied households	117.1	117.6	118.6	119.2	120.5

Note. All figures are estimates.

Source: SNL Kagan. Reprinted with permission.

Table 7.3

Top 10 MVPDs in the United States Based on Number of Subscribers (in Thousands), as of December 2013

Company	Subscribers
Comcast	21,690
DIRECTV	20,253
DISH	14,057
Time Warner Cable	11,393
AT&T (U-verse)	5,460
Verizon (FiOS)	5,262
Charter Communications	4,342
Cox Communications	4,302
Cablevision Systems	2,813
Bright House Networks	1,903
Total U.S. MVPD market	100,906

Source: SNL Kagan. Reprinted with permission.

MVPDs, through their TV Everywhere efforts, remain large players in the mobile video marketplace, providing customers with app-based access to live linear TV feeds of selected cable networks, video on demand, subscription video on demand, and extensive programming from premium services. As noted earlier, once mobile users have authenticated their MVPD subscription status, they should be able to access eligible TV content on their mobile devices.

Netflix, Hulu Plus, Amazon Prime, YouTube, and other streaming services remain potent competitors in the mobile market, providing consumers with attractive pricing and growing original programming efforts. Additionally, HBO has announced plans to offer some of its older programming on Amazon Prime beginning in May 2014. It will be the first time that HBO has made its programming available to consumers without a premium subscription through an MVPD.

More potential disruption is on the horizon. The DISH Network plans to launch a subscription broadband TV service in summer 2014, and Sony is developing a broadband service that will include live TV, on-demand content, and DVR service (Baumgartner, 2014b). It remains to be seen how the rapid changes in both technology and business models will impact the mobile marketplace and MVPD core services.

MVPD Consolidation

Since the news of the proposed merger between Comcast and Time Warner Cable in February 2014, there has been rampant speculation that this event could usher in a new wave of MVPD consolidation. Rumors have circulated about another attempted merger between DirecTV and DISH, and a possible acquisition of DirecTV by AT&T. Of course, there is no guarantee that U.S. authorities (the Department of Justice and the FCC) will approve any of these mergers, and any merger is likely be subject to some conditions (e.g., perhaps in the form of net neutrality provisions) in the best case.

In late April 2014, Comcast announced that it would divest itself of 3.9 million subscribers through a complex agreement with Charter Communications (Rubin, 2014). This decision can be viewed as a strategic play to assuage regulators' concerns about the combination of the top two cable operators. With the planned divestment, Comcast would be able to reduce its market share to less than 30%.

Cord Cutting

For several years, it was unclear whether cord cutting (i.e., cancelling MVPD service in favor of alternatives) was a real problem for the multichannel video industry. Some healthy skepticism was not uncommon among distributors, programmers, and analysts. But by 2013, few doubted that cord cutting was adversely affecting the revenue of the MVPD industry, even though the decline in MVPD subscriptions remained small. Leading analyst Craig Moffett predicted that "Pay TV is unmistakably declining and the rate of penetration decline is accelerating" (Wallenstein, 2013). He viewed high subscription prices and low-cost OTT alternatives as two key factors driving subscriber decline. Not surprisingly, cord cutters tend to be younger and subscribe to OTT services like Netflix and Hulu (Kafka, 2013).

The next logical question becomes: What can MPVDs do about cord cutting? It is worth noting that Comcast reported that it added 24,000 basic video subscribers in the first quarter of 2014, due in part to targeted promotions and the accelerated rollout of the user-friendly X1 boxes (Farrell, 2014). But it is far too early to conclude that subscriber losses have stabilized and that the industry has turned the corner.

The following strategies to combat cord cutting are often mentioned in the trade press: promoting triple-play bundling to reduce churn, increasing broadband prices to compensate for revenue losses in the video segment, and offering low-cost bundled packages to woo back cord cutters (Manjoo, 2014). In late 2013, Comcast introduced a $39.99 (promotional price) "Internet Plus" package that includes local channels, 25 Mbps broadband service, HBO, XFINITY Streampix, and XFINITY On Demand.

Multichannel TV Visionary: John C. Malone

At age 73 (he was born on March 7, 1941), Liberty Media Chairman **John C. Malone** is often regarded as the elder statesman in the cable industry, but he remains one of the most forceful and cryptic personalities in the business. During a career that has spanned more than 40 years, Malone has been characterized and castigated as a ruthless, conniving, rapacious Darth Vader (allegedly attributed to Al Gore) figure. As the CEO of Tele-Communications Inc. (TCI), he transformed the company into a lean, mean profit machine with a strong focus on expansion and did not hesitate to use strong-arm tactics to intimidate cities during franchise renewals (Robichaux, 2002).

His depiction as the evil Darth Vader is, of course, a caricature, but Malone has readily admitted that he plays hardball and refuses to be bullied. Malone is also renowned as "a man of science" (Auletta, 1994) who has a doctoral degree in operations research from Johns Hopkins University. He has applied his analytical mind to cable finance to become one of the most consummate dealmakers in the industry. His financial genius was clearly on display when he tried to convince (to no avail) AT&T's Michael Armstrong "to issue a separate tracking stock for AT&T's cable business" (Robichaux, 2002, p. 246) to insulate cable assets that are valued in cash flow terms from the long-distance phone assets that are valued in earning terms. Yet, for all his financial ingenuity, it is hard to understand why such a captain of the industry would care so little about customer service when cable television represents a fundamental service to end-users. Malone blasted the Cable Act of 1992 as "the most un-American thing I've ever experienced" (Robichaux, 2002, p. 130), as if he shared no responsibility for the passage of the legislation.

One can fault Malone on his management style and tactics toward perceived foes, but his business acumen and vision are without equal in the industry. Months before the proposed merger between Comcast and Time Warner Cable, he spoke about the necessity of consolidation "to help the industry prevent further video customer erosion by unifying around a common user interface and a full array of programming available online, on all devices" (Sherman & Lee, 2014). SNL Kagan ranked Liberty Media as the most profitable media company in 2013.

Bibliography

Auletta, K. (1994, February 7). John Malone: Flying solo. *The New Yorker*. Retrieved from http://www.newyorker.com

Baldwin, T. F., & McVoy, D. S. (1988). *Cable communication* (2nd ed.). Englewood Cliffs, NJ: Prentice Hall.

Barthold, J. (2013, October 1). Consumers say a la carte important, but not worth cutting the cord. *FierceCable*. Retrieved from http://www.fiercecable.com.

Baumgartner, J. (2014a, March 14). Top U.S. pay-TV providers lost 105,000 subs in 2013. *Multichannel News*. Retrieved from http://www.multichannel.com.

Baumgartner, J. (2014b, April 23). Report: Dish eyeing summer launch of OTT TV service. *Multichannel News*. Retrieved from http://www.multichannel.com.

Besen, S. M., & Crandall, R.W. (1981). The deregulation of cable television. *Journal of Law and Contemporary Problems, 44*, 77-124.

Bray, H. (2014, April 28). Amid new technologies, TV is at a turning point. *Boston Globe*. Retrieved from http://www.bostonglobe.com.

Cable Communications Policy Act of 1984, 47 U.S.C. §551 (2011).

Cable Television Consumer Protection and Competition Act of 1992, 47 U.S.C. §§521-522 (2011).

Cablevision Systems Corp. (2014). *About multi-room DVR*. Retrieved from http://optimum.custhelp.com/app/answers/detail/a_id/2580/kw/dvr/related/1.

Carter, B. (2013, September 13). Bold play by CBS fortifies broadcasters. *The New York Times*. Retrieved from http://www.nytimes.com.

Corning, Inc. (2005). Broadband technology overview white paper: Optical fiber. Retrieved from http://www.corning.com/docs/opticalfiber/wp6321.pdf.

Crowley, S. J. (2013, October). *Capacity trends in direct broadcast satellite and cable television services.* Paper prepared for the National Association of Broadcasters. Retrieved from http://www.nab.org.

Diallo, A. (2013, October 14). Cable TV model not just unpopular but unsustainable. *Forbes.* Retrieved from http://www.forbes.com.

Dreier, T. (2014, March 10). SXSW '14: ESPN 'frustrated and disappointed' by TV Everywhere. *Streaming Media.* Retrieved from http://www.streamingmedia.com.

Dulac, S., & Godwin, J. (2006). Satellite direct-to-home. *Proceedings of the IEEE, 94,* 158-172. doi: 10.1109/JPROC.2006.861026.

Duverney, D. D. (1985). Implications of the 1983 regional administrative radio conference on direct broadcast satellite services: A building block for WARC-85. *Maryland Journal of International Law & Trade, 9,* 117-134.

EchoStar Communications Corp. (2014). *What is the Hopper?* Retrieved from http://www.dish.com/technology/hopper

Farrell, M. (2013, September 2). CBS, Time Warner Cable sign carriage agreement. *Multichannel News.* Retrieved from http://www.multichannel.com.

Farrell, M. (2014, April 22). Comcast adds 24K video subs in Q1. *Multichannel News.* Retrieved from http://www.multichannel.com.

Federal Communications Commission. (1965). Rules re microwave-served CATV (*First Report and Order*), 38 FCC 683.

Federal Communications Commission. (1966). CATV (*Second Report and Order*), 2 FCC2d 725.

Federal Communications Commission. (1969). Commission's rules and regulations relative to community antenna television systems (*First Report and Order*), 20 FCC2d 201.

Federal Communications Commission. (1970). CATV (*Memorandum Opinion and Order*), 23 FCC2d 825.

Federal Communications Commission. (1972a). Commission's rules and regulations relative to community antenna television systems (*Cable Television Report and Order*), 36 FCC2d 143.

Federal Communications Commission. (1972b). Establishment of domestic communications-satellite facilities by non-governmental entities (*Second Report and Order*), 35 FCC2d 844.

Federal Communications Commission. (1980). *Notice of Inquiry,* 45 F.R. 72719.

Federal Communications Commission. (1982). Inquiry into the development of regulatory policy in regard to direct broadcast satellites for the period following the 1983 Regional Administrative Radio Conference (*Report and Order*), 90 FCC2d 676.

Federal Communications Commission. (2010). Sky Angel U.S., LLC emergency petition for temporary standstill (*Order*), 25 FCCR 3879.

Federal Communications Commission. (2012). Media Bureau seeks comment on interpretation of the terms "multichannel video programming distributor" and "channel" as raised in pending program access complaint proceeding (*Public Notice*), 27 FCCR 3079.

Federal Communications Commission. (2013a). Annual assessment of the status of competition in the market for the delivery of video programming (*Fifteenth Report*), 28 FCCR 10496.

Federal Communications Commission. (2013b). *Definitions,* 47 CFR 76.1902.

Federal Communications Commission. (2013c). *Report on Cable Industry Prices,* 28 FCCR 9857.

Flint, J. (2012, July 16). Pay-TV industry not united on TV Everywhere. *Los Angeles Times.* Retrieved from http://www.latimes.com.

Holloway, D. (2014, March 19). Next TV: Hulu sale uncertainty swayed CBS' SVOD deals. *Broadcasting & Cable.* Retrieved from http://www.broadcastingcable.com.

Home Box Office v. FCC, 567 F.2d 9 (D.C. Cir. 1977).

Hoskins, C., McFadyen, S., & Finn, A. (2004). *Media economics: Applying economics to new and traditional media.* Thousand Oaks, CA: Sage.

James, M. (2013, December 4). A la carte TV pricing would cost industry billions, report says. *Los Angeles Times.* Retrieved from http://www.latimes.com.

Kafka, P. (2013, September 24). More cord-cutting drumbeats: "Very likely" cutters growing in number. *The Wall Street Journal.* Retrieved from http://allthingsd.com/20130924/more-cord-cutting-drumbeats-very-likely-cutters-growing-in-number.

Lafayette, J. (2014, April 28). Young viewers streaming more, pivot study says. *Broadcasting and Cable.* Retrieved from http://www.broadcastingcable.com.

Leichtman Research Group. (2013, 4Q). DVRs leveling off at about half of all TV households. *Research Notes.* Retrieved from http://www.leichtmanresearch.com.

Lenoir, T. (2014, January 14). High retrans stakes for multichannel operators in 2014. *Multichannel Market Trends.* Retrieved from http://www.snl.com.

Manjoo, F. (2014, February 15). Comcast vs. the cord cutters. *The New York Times*. Retrieved from http://www.nytimes.com.
McCord, L. (2014, April 29). Study: 61% of frequent binge-viewers millennials. *Broadcasting & Cable*. Retrieved from http://www.broadcastingcable.com.
National Cable and Telecommunications Association. (2014). Cable's story. Retrieved from. https://www.ncta.com.
National Conference of State Legislatures. (2013). *Statewide video franchising statutes*. Retrieved from http://www.ncsl.org/.
Nielsen. (2014, March). *An era of growth: The cross-platform report*. Retrieved from http://www.nielsen.com.
Nocera, J. (2007, November 24). Bland menu if cable goes a la carte. *The New York Times*. Retrieved from http://www.nytimes.com.
Office of Program Policy Analysis & Government Accountability. (2009). Benefits from statewide cable and video franchise reform remain uncertain. *OPPAGA Report*, No. 09-35. Tallahassee, FL: Author.
Olgeirson, I. (2013a, October 4). US multichannel faces stagnant subscribers, revenue upside. *Multichannel Market Trends*. Retrieved from http://www.snl.com.
Olgeirson, I. (2013b, December 20). Industry shows slow progress in TV Everywhere app expansion. *Multichannel Market Trends*. Retrieved from http://www.snl.com.
Olgeirson, I., Lenoir, T., & Young, C. (2014, March 13). Multichannel video subscription count drops in 2013. *Multichannel Market Trends*. Retrieved from http://www.snl.com.
Owen, D. (1985, June). Satellite television. *The Atlantic Monthly*, 45-62.
Parker, J. G. (2011). Statewide cable franchising: Expand nationwide or cut the cord? *Federal Communications Law Journal, 64*, 199-222.
Parsons, P. (2008). *Blue skies: A history of cable television*. Philadelphia: Temple University Press.
Parsons, P. R., & Frieden, R. M. (1998). *The cable and satellite television industries*. Boston: Allyn and Bacon.
Ramachandran, S., & Launder, W. (2013, February 28). Imagining a post-bundle TV world. *The Wall Street Journal*. Retrieved from http://online.wsj.com.
Robichaux, M. (2002). *Cable cowboy: John Malone and the rise of the modern cable business*. Hoboken, NJ: John Wiley & Sons.
Rubin, B. F. (2014, April 28). Charter buying subscribers that Comcast divests. *The Wall Street Journal*. Retrieved from http://online.wsj.com.
Satellite Home Viewer Act. (1988). Pub. L. No. 100-667, 102 Stat. 3949 (codified at.
scattered sections of 17 U.S.C.).
Satellite Home Viewer Improvement Act. (1999). Pub. L. No. 106-113, 113 Stat. 1501 (codified at scattered sections in 17 and 47 U.S.C.).
Sherman, A., & Lee, E. (2014, January 7). Billionaire Malone returns to empire building amid cord cutting. *Bloomberg*. Retrieved from http://www.bloomberg.com.
SNL Kagan. (2013). *Media trends*. Retrieved from http://www.snl.com.
SNL Kagan. (2014). *Economics of mobile programming*. Retrieved from http://www.snl.com.
Telecommunications Act. (1996). Pub. L. No. 104-104, 110 Stat. 56 (codified atscattered sections in 15 and 47 U.S.C.).
U.S. v. Southwestern Cable Co., 392 U.S. 157 (1968).
Wallenstein, A. (2013, June 3). Top Wall Street analyst: Pay TV "cord-cutting is real." *Variety*. Retrieved from http://variety.com.

8

IPTV: Video over the Internet

Jeffrey S. Wilkinson, Ph.D.*

Why Study Internet Protocol Television (IPTV)?

- IPTV continues to grow in popularity among consumers causing massive structural changes in the business of television, video, and mobile delivery
- Advertisers continue to scramble to take advantage of the phenomenal growth of Internet video as a means of reaching people
- IPTV is a global phenomenon that is not limited to American audiences.

Introduction

In 2012 it was predicted that 'TV Everywhere' (Dreier, 2012; Schaeffler, 2012) was upon us. Since then we have witnessed an explosion in new forms of video programs that has upended the television and film industries. The good news is that content keeps getting better; with more programs, more stories, in more genres than ever before, leading to more work, and more viewing overall. This phase of "TV distribution over the public Internet" (ibc, 2013) is causing massive changes across the industry, and the old system of launching and disseminating programs through the traditional networks has probably been disrupted forever. The once-hallowed 'pilot season' system is quickly eroding and the major networks now openly bypass the broadcast model with varying degrees of straight-to-Internet launches (Andreeva, 2014b).

But even with this exponential increase in viewer choices, there are some concerns in this new era of ubiquitous video. The best of the best still compete with the worst of the worst. When there are scores of zombie movies on-demand, how does one find and enjoy the best ones? In a time of limitless choice, viewer choices are still limited by the amount of time they have available to watch television.

The key to this transition is a fundamental change in how television signals are delivered, with IPTV replacing or supplementing the traditional television and video distribution systems discussed in the previous chapter. IPTV is defined as "multimedia services such as television/video/audio/text/graphics/ data delivered over IP-based networks managed to support the required level of quality of service (QoS)/quality of experience (QoE), security, interactivity, and reliability" (ITU-T Newslog, 2009). IPTV provides the complete range of personalized and interactive services desired by modern consumers. Globally, IPTV systems offer just about anything and everything on demand.

Getting IPTV into the home is varied and complicated. IPTV is the technology behind streaming, satellite delivery, wireline, and wireless delivery of audio and video (in a variety of formats). IPTV also extends all forms of content and services from the television to mobile devices such as smartphones and tablets.

* Professor and Chair, Department of Communication, University of Toledo (Toledo, Ohio).

From a practical perspective, IPTV offers a more efficient way to deliver almost any type of video signal over wired or wireless connections. As discussed in Chapter 7, traditional cable and satellite television providers are increasingly adopting IPTV-type protocols to deliver a greater range of programming and interactivity. At the same time, a new set of video service providers including Netflix, HuluPlus, and Amazon Prime are using IPTV to compete with traditional cable TV and satellite companies.

In addition to providing competitive programming with series such as Netflix's *House of Cards* and *Orange is the New Black* and Amazon Prime's *Beta House,* these services are positioning themselves to be the first to deliver the next generation of television signals knows as "UltraHD". This chapter explores the history of IPTV, the latest developments, and the factors, including UltraHD, that will help define the future of IPTV.

Background

In order to appreciate IPTV, it is necessary to examine the history of this technology. IPTV was created as a means of delivering individual video clips and, later, programs, to computers using both streaming and downloading of the content. As IPTV developed, playback devices included tablets, smartphones, and other computers, eventually including television receivers.

The current state of IPTV reflects the synergies of merging two distinct business models and industries. The term "streaming" came from computer-based networks and distribution, while cable and telco companies pushed IPTV through set-top boxes (STB).

Recent developments have rendered this distinction obsolete as living rooms merge with mobile devices to offer converged services for entertainment, information, and communication. SmartTVs are predicted to be in 70 million US households by 2017 (IPTV News, Sept 24, 2013). Even gaming consoles are used for far more than just the games and represent yet another means of delivering IPTV content.

Historically, cable has been around for generations, transitioning from analog to digital distribution with other industries in the 1990s. Streaming began when Progressive Networks launched Real Audio in 1995; video was offered the following year. The main issue for developers then was that most people had dial-up Internet connections, and files had to be configured for 14.4 and 28.8 kbps. Broadband was conceived as 150 kbps. Soon Microsoft launched Windows Media Player, followed by Apple with Quicktime, an MPEG-based architecture/codec. In 1999, Apple introduced QuickTime TV using Akamai Technologies' streaming media delivery service. A few years later came Flash, DIVX, and a number of MPEG players (for example, VLC remains popular worldwide because it plays a range of formats well).

As technology improved and bandwidth increased, new applications were created. Sports highlights, news, and music clips were most popular. Napster made (illegal) file-sharing a household phrase, and YouTube launched the modern age of ubiquitous user-generated content.

Types of Streaming:

One important distinction in IPTV is whether the content is live or on demand. Three leaders in on-demand streaming are Netflix, Hulu, and Roku. Providers need enough server power and space with enough capacity to hold content that can be configured so the maximum number of simultaneous users can be accommodated. For professional event streaming one also needs a dedicated broadcaster or encoding computer to create the streams from the live input (which may combine several sources mixed down through traditional broadcast equipment to a single digital feed).

It is also important to note that streaming is not downloading, but occurs when content has to buffer before playing. Early streaming on the Internet gave viewers a choice whether to store the video on the home computer. True streaming is often offered through special servers. According to Ellis (2012b), the most popular streaming servers continue to be Microsoft's Smooth Streaming, Apple's HTTP Live Streaming (HLS), and Adobe's Dynamic Streaming.

The advantages of hosting and streaming content are speed, control, and flexibility. Streamed material is played back almost immediately. Control is maximized for the host because the original content stays on the server and access is controlled using password, digital rights management (DRM), registration, or some other security feature.

Lastly, there is tremendous flexibility because the content can be configured in a variety of ways—

length, resolution, format, and more. Broadband connections allow delivery of high-definition content, which can be viewed on any of a number of players.

Consumer Streaming Platforms

There are a number of player-platforms for streaming video. The most popular are Windows Media Player (WMP) and Apple iTunes. The ubiquitous Flash plug-in used by YouTube is technically a browser plug-in rather than a streaming player, although the home user may not see the difference. And HTML5 includes video playback capability in the browser.

IPTV via P2P

While cable, telephony, and satellite are working to deliver the highest quality (and most expensive) service and programming into the home, there is one other form of IPTV that bears mention. Peer-to-peer networking (P2P) became famous a decade ago through the Napster file-sharing service and thus is often associated with questionable file sharing. But P2P is a cheap alternative to the STB offered by companies previously mentioned.

Well-known sources of P2P content include BitTorrent and Vuze. According to the annual update on Torrentfreak, the 2014 "Top 10" includes The Pirate Bay, Kickasstorrents, Torrentz, Extratorrent, Yify-Torrents, EZTV, 1337X, Bitsnoop, and RARBG (Ernesto, January 4, 2014).

Podcasting and Music Streaming

Given the emphasis and popularity of video, it is sometimes easy to forget that audio-only podcasting is also widely available. Podcasting is an important aspect of IPTV.

Podcasts are typically programs such as a talk show, hosted music program, or commentary. Podcasts are commonly pulled by the user as an automatic download via RSS through a service such as iTunes. According to Edison Research and Triton Digital, 39 million Americans report listening to a podcast within the past 30 days. This figure is a 25% increase over podcast listening in 2013 (Webster, 2014).

Music online also continues to be intricately involved with streaming. There are a number of services such as Pandora as well as radio stations that stream audio programs to listeners. Both live and on-demand programming is available. For more on digital music, see chapter 17.

Cable IPTV

While consumers are noticeably shifting services away from cable companies, the companies have remained competitive by offering IPTV services. One of the leaders in the U.S. has been Comcast which was in the process of acquiring Time Warner Cable for $45 billion in 2014.

Comcast not only has its IPTV Xfinity service, but inked a multi-year deal with Netflix (Baumgartner, February 23, 2014c) to deliver content to subscribers. The cable giant has a number of options from $29/month cable TV to a new customer bundled 'triple play' (Internet, television, telephone) for $89/month, one-year contract. (For more on cable, see Chapter 7).

Proprietary Telco Platforms

The biggest players from the telephony side are AT&T and Verizon. By going with a bundled package (combining cable, Internet, and smartphone) a single household can generate hundreds of dollars per month in revenue. In return, the company gets to use IPTV to supply all of the person's information, entertainment, and communication needs.

Information services now include news as well as banking, home security, weather, national, local, and even neighborhood political or social activities. In the same way, the definition of entertainment has expanded to include the range of live or recorded experiences involving music, sports, drama, comedy, and theater. Communication services now involve one-way and two-way interactions with family, friends, classmates, colleagues, and strangers.

The fight for who is chosen to deliver content and services is ongoing between modern media companies which have diversified from traditional roots and now delve into what we've known as broadcast, entertainment, telephone, cable, satellite, and Internet.

As IPTV becomes 'TV Everywhere' it includes HD, first-run films, older films, old TV shows, user-generated content, and an ever-increasing number of news, sports, and entertainment programming. It can include features such as selecting camera angles, language, and real-time commentary. It can synchronize multiple screens, so consumers are encouraged to use social media on their tablet or smartphone as they watch real-time video on their widescreen smart TV.

But in the middle of this viewing revolution, "TV Everywhere" has much in common with "TV the way it's always been," mostly passive viewing for entertainment and escape. Not everyone is ready to migrate to services such as one-touch surveillance and home security. While the myriad services are important, the foundation remains the home-TV film, or video watching experience.

IPTV Providers

Streaming Players

A series of in-between proprietary players have emerged in a highly-competitive niche market. At this time, the three most popular are Roku, Chromecast, and Apple TV. The three are remarkably similar and offer relatively low-cost streaming into the home (Baumgartner, 2014b). Each of the players also needs a subscription to either Netflix, Hulu, or some other service in order to actually stream shows and movies to the television (Alter, 2014).

Figure 8.1

Streaming Players: Roku, Chromecast & Apple TV

Source: A. Grant

Roku. In early 2014, Roku unveiled a high definition multimedia interface (HDMI)-connected Streaming Stick for $49.99 to give it an entry-level weapon to wield against its competition (Baumgartner, 2014e). Unlike the original $99 Roku Streaming Stick, the new HDMI Version is not limited to newer-generation, "Roku-Ready" TVs that use ports that support Mobile High-Definition Link (MHL) technology. The new Streaming Stick supports Roku's 1,200-plus apps/channels, including authenticated TV Everywhere apps such as HBO GO, Watch Disney, WatchESPN, and TWC TV from Time Warner Cable. In addition, the Streaming Stick comes with a remote. In early 2014, it was estimated that around 8 million Roku units had been sold in the U.S.

Google Chromecast. Google's entry in the market is a plug-in streaming device similar to Roku consisting of a small dongle that plugs into your television's HDMI port. The Chromecast is not a standalone device, requiring a device such as a smartphone, tablet, or a PC in order to watch streaming video (Flacy, 2013). The device streams high-definition programming from the Internet to the dongle via Wi-Fi. Google Chromecast was named by Time Magazine as '2013's best gadget' (McCracken, 2013). The $35 device enables you to stream music, movies, and TV shows from Netflix, YouTube, HuluPlus, HBO Go, Google Play Movies, and more. Chromecast can also be used to send videos from mobile devices and the computer to the TV.

Apple TV. Launched in 2007, the current third-generation Apple TV runs through a small square device that attaches to the HDMI port of your television. Like everything Apple, it works with iTunes to bring television, film, music, and gaming to the television. In terms of popularity, in mid-2013 Apple announced that iTunes users were purchasing over 800,000 TV episodes and over 350,000 movies per day (HBO GO, 2013). Critics note that Apple TV is still far more expensive than Roku or Chromecast, even for households that are already 'everything-Apple.'

Amazon Fire TV. Released in April 2014, Amazon's Fire TV adds a needed element to these streaming devices—voice activated search. No longer do users struggle to type out search terms using awkward remotes. Fire TV costs $99 and works with services such as Netflix, Watch ESPN, HuluPlus, and of course, Amazon Instant Video. Fire TV also has an optional gaming remote.

Streaming Content Providers

The engine driving the massive industry changes is new companies that bring movies and television programs into the home. The most well-known are Netflix, Hulu, and Amazon.

Netflix. Netflix has taken the number one position in subscribers, headlines, and criticisms from competitors. The company was launched in 1998 by Reed Hastings (see sidebar later in this chapter) as a DVD rental service by mail. After three years Netflix was still not making a profit (despite having 300,000 customers) and at that stage they almost sold out to Blockbuster to be the online version of the VHS video rental chain. But Netflix stayed independent, and as consumer preferences switched from tape to DVD, the company prospered and posted its first profitable year in 2003. By 2005 Netflix was shipping one million DVDs per day and reported success with its algorithm to recommend movies based on previous rentals. In February 2007, the company delivered its billionth DVD but began to shift from DVD mailings to Internet streaming video-on-demand. The timing was perfect.

As national DVD sales steadily dropped from 2006 to 2011 and digital IPTV capability grew, Netflix rapidly grew. One possible misstep occurred in 2011, when Netflix started a storm of criticism when it proposed splitting the streaming service from the DVD service and rebranding the disk-by-mail service as Qwikster. The move was roundly criticized by customers, and Netflix lost 800,000 customers that quarter (Pepitone, 2011). But since then, the company has rebounded and continued to grow, ending 2013 with its best quarter in three years. The company reported more than 2.3 million new households signed up for its streaming service, bringing it to over 33 million in the U.S. and 44 million global subscribers (Stelter, 2014).

Despite the phenomenal growth—from 2.7 to 29 million subscribers between 2005 and 2013—the most significant aspect of Netflix is that the company has transformed from being just a repository for content into being a curator, and now, a producer of original content (Warren, 2013).

Hulu Plus. Hulu is a joint venture launched in 2007 by NBCUniversal (Comcast), Disney-ABC, Fox, and other media companies to provide on-demand streaming programs through the Web. While actual numbers are not released, the company announced it had 3 million subscribers by the end of 2012 and 5 million paying customers by the end of 2013 (Lawler, 2013). Similar to Apple TV, Hulu says its revenues are over $1 billion for its Hulu Plus service (Lawler, 2013). Like Netflix, a yearly subscription costs less than $100. Hulu's greatest strength is that it gets the latest episodes from selected programs from major networks. Hulu offers current season TV shows many other services don't have. Like Netflix and other competitors, Hulu has plans to double its original program offerings.

Amazon Prime. The streaming video service from the online book giant is a little less expensive than Netflix or Hulu Plus ($79.99 per year versus $96 for the other two). Because of the Amazon connection, subscribers also get benefits like free two-day shipping on all products sold by Amazon and one free book a month from the Amazon Lending Library to use on a Kindle device.

Telco ISP

To stay competitive, telephone companies have also embraced IPTV. The two largest and most competitive comprehensive service providers are AT&T and Verizon.

AT&T's U-verse TV offers a variety of packages starting at $29 for television channels and include DVR, VOD, and HD. HD service generally costs an extra $10 per month. To compete with Google Fiber, AT&T launched GigaPower in selected markets, starting with symmetrical speeds of 300 Mbps with plans to further upgrade to 1 Gbps (Baumgartner, March 7, 2014g).

Verizon's FiOS TV offers a variety of packages to compete with other MVPDs. The FiOS Quantum triple-play offered subscriber discounts and continued adding subscribers through early 2014. Verizon ended 2013 with over 6.1 million FiOS Internet and 5.3 million FiOS video subscribers (Baumgartner, 2014g), compared to U-verse with 5.46 million (Leichtman Research Group, 2014). AT&T's U-verse generally offers slower speed but is also less costly than its competition.

Recent Developments

Merger Mania and Electronic Sell-Through

Verizon has indicated it is not planning to build FiOS beyond the current local franchise agreements, which today covers about 18.5 million homes passed (Baumgarner, 2014a). Instead, Verizon is moving toward a 'virtual MSO' to sell pay-TV service outside its FiOS territories. The company has purchased Intel

Media for its "OnCue" platform to prepare for next generation IP video service. Verizon's FiOS TV service relies on QAM/MPEG transport to deliver video services to the TV. The new plan with Intel media is for Verizon to offer service over IP to TVs and to mobile devices (Baumgartner, 2014a).

Verizon Digital Media Services (VDMS) is a cloud video unit that bought EdgeCast and upLynk to power apps like Watch Disney and Watch ABC. VDMS includes Redbox Instant, a direct competitor to Netflix. VDMS's aims to establish a cloud-based multiscreen video architecture similar to Comcast's X1 service. Comcast's VIPER system is a multiscreen video distributor that uses adaptive bit rate streaming to serve PCs, gaming consoles, tablets, smartphones, smart TVs, and video gateways and clients.

In early 2014, Comcast announced plans to acquire Time Warner Cable in a $45 billion deal to create a cable and technology giant covering one-third of the U.S. TV households. (Farrell, February 13, 2014). According to Comcast, the deal will accelerate rollout of its X1 operating system and cloud-based DVR and its XFinity TV mobile apps (Farrell, 2014a). Time Warner Cable has been a leader in community Wi-Fi, and the merger will combine its more than 30,000 hotspots, primarily in Los Angeles and New York City, and its in-home management system, IntelligentHome, with Comcast's offerings.

Comcast has also made a deal with Sony Pictures Home Entertainment to get into the home-rental movie business (Lieberman, March 10, 2014). Industry watchers predict that DVD and Blu-ray delivery of video programs and films will be replaced by online streaming and downloads. The new model is referred to as "Electronic Sell-through." By joining with Sony, Comcast can compete head-to-head with Apple and Amazon for direct sales of movie and TV shows such as *Breaking Bad* and *House of Cards*.

Verizon FiOS also moved from its "Flex view" brand to electronic sell-through and expects its numbers to vastly improve. According to NPD Group, Verizon Flex View held about 1% of the EST movies market in mid-2013. Apple iTunes held 61% of that market followed by Amazon Instant Video (15%), Sony PlayStation Store (7%), Vudu and Xbox Video (4% each) and Google Play (3%) (Baumgartner, 2014d).

A new IP-delivered video service billed as the 'next generation cable operator' called Layer3 TV has also entered the market (Baumgartner2014k). The platform will support live TV, on-demand content, and possibly DVR services. Details are few, but the startup uses a 'programmer-friendly business model' which is 'the antithesis of Aereo' (Baumgartner, 2014m).

Programmers are trying to loosen up on distribution rights that enable over-the-top delivery. Sony plans to test an over-the-top subscription video service in 2014, and Dish Network and Disney cut a deal giving Dish the rights to distribute OTT video subscription packages (Baumgartner, 2014m).

Bottlenecks and Crashes of Live Events

History was made when ABC streamed the live broadcast of the Academy Awards in 2014. The ABC.go.com app consistently crashed, however, due to unexpected high traffic (Grotticelli, 2014a).

A company owned by Verizon, upLynk, used its HD Adaptive Streaming Platform to reach various mobile platforms. During the Oscars, upLynk took a single high-definition feed from the event and uploaded it to its servers. But somewhere in the path, "a bottleneck ensued and thousands of requested live streams were not served" (Grotticelli, 2014a).

At around the same time, meanwhile, competitor Akamai worked with NBC to stream the Sochi Olympics and Fox broadcast of the NFL Super Bowl XLVIII. The NBC coverage of the U.S. hockey team against Russia generated 600,000 online viewers and the U.S. vs. Canada game recorded 2.1 million unique users (requiring 3.5 Tbps of bandwidth). Fox said they registered 1.1 million concurrent users in the third quarter of the 2014 Super Bowl stream.

Original Programming and Going Independent

In 2014, Discovery Communications and Google announced plans to offer a streaming service called Tapp that is, in essence, a 'celebrity video network' (Patel, 2014). The service is billed as a platform for celebrities, "the nation's top thought leaders," and other content creators. For example, Tapp has a channel devoted to former Alaska governor and vice presidential candidate Sarah Palin (Weprin, 2014). With over four million followers on Facebook, Palin is a natural draw for having her own channel. Internet-based distribution is much less expensive than broadcast or cable, and it's estimated that Tapp needs 25,000 subscribers to pay $9.95 each per month in order to be profitable (Palmeri & Erlichman, 2014). This project is remarkably similar to The Blaze, an online video subscription service and (later) TV channel started by former Fox News host Glenn Beck. As of early 2014, analysts estimated that The Blaze had 300,000 subscribers, each of whom pay $10 per month for the streaming service.

Meanwhile a number of Hollywood stars are trying to cut out the middleman altogether and creating original programs that get them out of the old system. Kevin Spacey, Rob Schneider, and others are launching new shows for online audiences where they own all aspects of the shows and are banking on their star power to attract viewers (Andreeva, 2014a; Lafayette, 2014). Mounting a series without the backing of a network is increasingly possible given the technology costs for production—shooting on location with inexpensive HD cameras, postproduction editing on affordable workstations using a variety of software, and then reducing actors' costs. The bottom line is that the dissemination of product—the programs—is now being done outside of the traditional Hollywood machine.

Current Status

According to research from LRG, there remains a fair amount of share-shifting as consumers experiment with video alternatives. As of early 2014, the 13 largest multi-channel video providers account for over 94.6 million subscribers. The top nine cable companies now report an aggregate of 49.6 million video subscribers while satellite TV has 34.3 million subscribers and the top two telcos have 10.7 million subscribers [see Table 8.1]

Overall, cable companies report subscriber losses across the board, continuing the trend from previous years. The main beneficiaries have been the telcos AT&T and Verizon, which have added substantial numbers of subscribers (1.46 million in 2013, 1.298 million in 2012) and to a lesser extent, satellite companies. DirecTV and Dish, which added 170,000 video subscribers in 2013 far below the 288,000 net additions reported in 2012.

Table 8.1

Comparison of Video Programming Services

Multi-Channel Video Provider	Subscribers at End of 2013	Net Adds in 2013
Cable Companies		
Comcast	21,690,000	(305,000)
Time Warner	11,393,000	(825,000)
Charter	4,342,000	(121,000)
Cablevision	2,813,000	(80,000)
Suddenlink	1,177,400	(33,800)
Mediacom	945,000	(55,000)
Cable ONE	538,894	(54,721)
Other Major Private Cable Companies*	6,675,000	(260,000)
Total Top Cable	49,574,294	(1,734,521)
Satellite TV Companies (DBS)		
DirecTV	20,253,000	169,000
DISH	14,057,000	1,000
Total Top DBS	34,310,000	170,000
Telephone Companies		
AT&T U-verse	5,460,000	924,000
Verizon FiOS	5,262,000	536,000
Total Top Telephone Companies	10,722,000	1,460,000
Total Multi-Channel Video	94,606,294	(104,521)

Source: Leichtman Research Group

Controversial Counter-Cable Cord-Cutting Continues

More than one million subscribers dropped cable TV in favor of streaming between 2008 and 2011 (Lang, 2012). This trend is continuing, albeit in drips and drops.

The top 13 multichannel video programming distributors, representing 94% of the market, reported losing just over 100,000 subscribers in 2013 (Baumgartner, 2014l). This is only 0.1% of all subscribers, but as noted in the research by Morgan Stanley, the more TV viewers watch, the more they plan on cutting the cord in favor of streaming services such as Netflix and Amazon Prime (Lang, 2014). The largest group of potential cord cutters are in the highly sought after 18-29 demographic.

The big winners continue to be Netflix, Hulu, HBO Go, Apple TV, and other video streaming platforms. A Morgan Stanley national survey of 2,500 adults noted that the average total number of hours of TV content consumed each week climbed to nearly 21 hours, from 19 hours in 2012 and 16.7 hours in 2011 (Lang, 2014). Viewing on tablets, computers, and portable devices continues to increase while watching on television remains relatively stable.

UltraHD TVs

UltraHD is another name for the 4K revolution discussed in Chapters 4 and 6, and it has the potential to change the way video is delivered and consumed. According to research from Parks Associates, there will be 70 million US households with connected TVs by 2017, and 191 million online video users in the US by 2017, rising from 175 million at the end of 2014. The research adds that 25% of smart TV app users recall seeing an in-app ad, with 84% responding to that ad—the highest response rates to app advertising among current connected consumer electronics platforms.

Gaming Integrates, Merges with IPTV

Sony announced it is integrating original TV programming into its PlayStation gaming platform in order to broaden the platform's potential appeal (Sharma, 2014). The company plans to offer original series through the console in addition to an online pay-TV service. Users will be able to stream shows through the 30-million connected PlayStation Network. Similarly, Microsoft has similar plans for the Xbox and was working with Steven Spielberg to develop original TV series for the gaming console.

It has also been a banner year for Twitch, a streaming platform dedicated to video games (Siegal, 2014). Launched in 2011, Twitch reaches over 45 million viewers each month, and more in prime time than MTV, TNT, and AMC. In 2013 Twitch integrated into Sony's PlayStation 4 and Microsoft's Xbox. It is anticipated that streaming games is capturing the attention of the vital 18-49 demographic, "tearing teenagers and adults alike away from the television to watch professional gamers." (Siegal, 2014).

Factors to Watch

Smart TV versus Standalone Streamers

According to Anthony (2014), the rise in smart TV adoption may result in the decline of the standalone media streamer. The last few years, services like Netflix, iTunes, Spotify, and BBC iPlayer improved their libraries and quality of service to the point they were the easiest way to consume media. But the new generation of TVs has streaming software built in.

TV Migrates to the Automobile

Look for a proliferation of in-vehicle video services by the end of the decade. QuickPlay Media and AT&T jointly launched an in-vehicle video service made possible by what AT&T said is the first connected car innovation center in the U.S., a 5,000 square foot facility in Atlanta (Grotticelli, 2014d).

The new AT&T Drive Studio focuses on connected car technologies. The partnership aims to securely stream hundreds of live linear TV channels and hours of premium VOD content. By the end of the decade, 70 to 80 percent of new vehicles are expected to include an infotainment system equipped with Internet access and downloadable apps. (For more on connected automobiles, see Chapter 13.)

The Second Screen Experience

As the amount of IPTV content increases, marketers continue to seek creative ways to engage viewers. Mobile devices and social media have inspired further experiments to reach consumers. For example, just as Google is used as a verb to mean searching for something, an app exists to help you find aural content—a song, a speech, or a soundbite.

A company called Shazam markets itself as a second screen experience for viewers to call up sound-related content for whatever they are watching (Baer, 2014). The company announced "this year, 700,000 people Shazammed the Super Bowl and more than a million Shazammed the Grammies." On the show *X Factor*, viewers can Shazam to vote for singing contestants. While admittedly not a good fit for dramas or sitcoms, this service does lend itself to the ever-expanding universe of talk and reality shows.

Conclusion

Of all of the technologies discussed in this book, IPTV may have the greatest potential to affect common media use patterns. The capability of choosing almost any content, any time or anywhere could make viewing more selective, or it could lead to binging or other, hyper-focused, behaviors. Ironically, the companies that own the content may discover that limiting the availability of certain types of content could lead to greater popularity, forcing viewers to "use or lose" content.

IPTV: Video Over the Internet Visionary: Reed Hastings

Netflix CEO and founder **Reed Hastings** has helped usher in the age of TV everywhere. The California entrepreneur dabbled with software development companies and started Netflix in 1997. The idea was said to have come from being fined for an overdue video rental. "I had a big late fee for 'Apollo 13.' It was six weeks late and I owed the video store $40. I had misplaced the cassette. It was all my fault. I didn't want to tell my wife about it. And I said to myself, 'I'm going to compromise the integrity of my marriage over a late fee?' (Zipkin, 2006).

As the story goes, he had a health club membership and wondered if that model would transfer to video and entertainment. In 2014 Hastings was forced to sign deals with ISPs to assure the quality of service for Netflix remained high. He has been an outspoken proponent of net neutrality because he feels his company would otherwise be at a disadvantage (Netflix company blog, 2006).

Hastings is also an active educational philanthropist and is currently on the board of several organizations including CCSA, Dreambox Learning, KIPP, and Pahara (Netflix, n.d.). According to the Netflix website, Hasting served in the Peace Corps as a high school math teacher in Swaziland. He is married with two children.

Bibliography

Alter, Maxim (March 26, 2014). Breakdown: What is the best TV, movie streaming option? WCPO.com. Retrieved from http://www.wcpo.com/news/technology/netflix-vs-hulu-vs-redbox-vs-amazon-prime-whats-the-best-streaming-option-on-the-market.

Andreeva, Nellie (January 20, 2014a). Rob Schneider challenges TV biz model with independently produced comedy series he co-created, financed, & stars in. Deadline Hollywood. Retrieved from http://www.deadline.com/2014/01/rob-schneider-self-made-sitcom-real-rob/.

Andreeva, Nellie (January 22, 2014b). The challenges of bypassing pilot season. Deadline Hollywood. Retrieved from http://www.deadline.com/2014/01/the-challenges-of-bypassing-pilot-season/.

Anthony, Sebastian (January 6, 2014). CES 2014: The death of Roku, and can UHDTV succeed where 3D TV failed? Extremetech.com. Retrieved from http://www.extremetech.com/computing/173992-ces-2014-the-death-of-roku-and-can-uhdtv-succeed-where-3d-TV-failed/.

Baer, Drake (March 07, 2014). Now that Shazam has your ear, it wants your eyes, too. FastCompany.com. Retrieved from http://www.fastcompany.com/3026906/lessons-learned/now-that-shazam-has-your-ear-it-wants-your-eyes-too.

Baumgartner, Jeff (January 21, 2014a). Verizon CFO: Intel Media buy speeds up FiOS's next-gen IPTV plan. Multichannel News. Retrieved from http://www.multichannel.com/distribution/verizon-cfo-intel-media-buy-speeds-fios%E2%80%99s-next-gen-iptv-plan/147787.

Baumgartner, Jeff (February 22, 2014b). Roku flirting with an IPO: Report. Multichannel News. Retrieved from http://www.multichannel.com/technology/roku-flirting-ipo-report/148468.

Baumgartner, February 23, 2014c). Multichannel News. Retrieved from http://multichannel.com/news/content/comcast-netflix-forge-stronger-streaming-connection/356109.

Baumgartner, Jeff (March 03, 2014d). Verizon FiOS starts selling full TV seasons: Telco phasing out 'Flex View' brand as it expands its electronic sell-through offering. Multichannel News. Retrieved from http://www.multichannel.com/distribution/verizon-fios-starts-selling-full-tv-seasons/148639.

Baumgartner, Jeff (March 04, 2014e). Roku Unleashes Its Chromecast-Killer: $49.95 Streaming Stick Is Compatible With Any TV With An HDMI Port. Multichannel News. Retrieved from http://www.multichannel.com/distribution/roku-unleashes-its-chromecast-killer/148669.

Baumgartner, Jeff (March 05, 2014f). Dishing out some OTT. Multichannel News. Retrieved from http://www.multichannel.com/blog/bauminator/dishing-out-some-ott/373097.

Baumgartner, Jeff (March 07 2014g). Most Consumers Haven't Heard Of 4K: Study. Multichannel News. Retrieved from http://www.multichannel.com/technology/most-consumers-haven%E2%80%99t-heard-4k-study/148724.

Baumgartner, Jeff (March 7, 2014h). Verizon Pitches LTE/FiOS Service Bundle Discounts. Multichannel News. Retrieved from http://www.multichannel.com/distribution/verizon-pitches-ltefios-service-bundle-discounts/148740.

Baumgartner, Jeff (March 7, 2014i). AT&T's 'GigaPower' Coming To Dallas. Multichannel News. Retrieved from http://www.multichannel.com/distribution/att%E2%80%99s-%E2%80%98gigapower%E2%80%99-coming-dallas/148733.

Baumgartner, Jeff (March 9, 2014j). Aereo Shuts Down in Denver and Salt Lake City. Multichannel News. Retrieved from http://www.multichannel.com/distribution/aereo-shuts-down-denver-and-salt-lake-city/148762.

Baumgartner, Jeff (March 13, 2014k). Industry Vets Start Up 'Next-Generation' Cable Operator. Multichannel News. Retrieved from http://www.multichannel.com/distribution/industry-vets-start-%E2%80%98next-generation%E2%80%99-cable-operator/148851.

Baumgartner, Jeff (March 13, 2014l). Dish Unleashes 'Super Joey': Client Device Expands Recording Capabilities Of Hopper With Sling HD-DVR, Costs $10 Per Month. Multichannel News. Retrieved from http://www.multichannel.com/distribution/dish-unleashes-%E2%80%98super-joey%E2%80%99/148849.

Baumgartner, Jeff (March 14, 2014m). Top U.S. pay-TV providers lost 105,000 subs in 2013. Multichannel News. Retrieved from http://www.multichannel.com/distribution/top-us-pay-tv-providers-lost-105000-subs-2013/148867.

Baumgartner, Jeff (March 14, 2014n). Industry vets start up 'Next-Generation' cable operator. Multichannel News. Retrieved from http://www.multichannel.com/distribution/industry-vest-start-up-next-generation-cable-operator/.

Baumgartner, Jeff (March 19, 2014o). Next TV: Comcast Adds 18 Live Channels To Its Xfinity TV Go App Lineup. Broadcasting & Cable.com. Retrieved from http://www.broadcastingcable.com/news/technology/next-tv-comcast-adds-18-live-channels-its-xfinity-tv-go-app-lineup/129880.

Baumgartner, Jeff (March 20, 2014p). Netflix CEO: Some Big ISPs 'Extracting A Toll Because They Can'. Multichannel News. Retrieved from http://www.multichannel.com/distribution/netflix-ceo-some-big-isps-%E2%80%98extracting-toll-because-they-can%E2%80%99/149000.

Dreier, T. (2012, February 28). The state of media and entertainment video 2012. *Streamingmedia.com*. Retrieved from http://streamingmedia.com/Articles/Editorial/Featured-Articles/The-State-of-Media-and-Entertainment-Video-2012-80946.aspx.

Ellis, L. (2012, February 21). For the love of broadband. *Multichannel News*. Retrieved from http://www.multichannel.com/blog/Translation_Please/33116-For_The_Love_of_Broadband.php.

Ernesto (January 4, 2014). Top ten most popular torrent sites of 2014. Torrentfreak.com. Retrieved from http://torrentfreak.com/top-10-popular-torrent-sites-2014-140104/.

Farrell, Mike (February 13, 2014a). It's Official: Comcast to Merge With Time Warner Cable Transaction to Create 'World Class Technology and Media Company'. Multichannel News. Retrieved from http://www.multichannel.com/cable-operators/its-official-comcast-merge-time-warner-cable/148282.

Farrell, Mike, (March 19, 2014b). Kagan: Pay-TV Shed 251K Subscribers in 2013. Multichannel News. Retrieved from http://www.multichannel.com/distribution/kagan-pay-tv-shed-251k-subscribers-2013/148966.

Flacy, Mike (December 21, 2013). Streaming Smackdown: Google Chromecast vs. Roku 3 vs. Apple TV. Digitaltrends.com. Retrieved from http://www.digitaltrends.com/home-theater/google-chromecast-versus-roku-3-versus-apple-tv/#ixzz2x6NKQP2w.

Grotticelli, Michael (March 7, 2014a). Is IP Streaming Up to the Task for Live Large Scale TV Events? CreativePlanetnetwork.com. Retrieved from http://www.creativeplanetnetwork.com/dcp/news/ip-streaming-task-live-large-scale-tv-events/65708.

Grotticelli, Michael (March 13, 2014b). IPTV Battle Lines Forming. CreativePlanetnetwork.com. Retrieved from http://www.creativeplanetnetwork.com/dcp/news/iptv-battle-lines-forming/65764.

Grotticelli, Michael (March 13, 2014c). Verizon Research: Millennials Want It Now! CreativePlanetnetwork.com. Retrieved from http://www.creativeplanetnetwork.com/dcp/news/verizon-research-millennials-want-it-now/65766.

Grotticelli, Michael (March 20, 2014d). New In-Car IP 'Infotainment' Service Introduced by AT&T and QuickPlay Media. CreativePlanetnetwork.com. Retrieved from http://creativeplanetnetwork.com/dcp/news/new-car-ip-infotainment-service-introduced-att-and-quickplay-media/65842.

HBO GO & WatchESPN Come to Apple TV (June 13, 2013). Apple Press Info. Apple.com. Retrieved from https://www.apple.com/pr/library/2013/06/19HBO-GO-WatchESPN-Come-to-Apple-TV.html.

Ibc.org (December 11, 2013). Pay TV operators cash in on over the top demand. Ibc.org. Retrieved from http://ibc.org/page.cfm/action=library/libID=2/libEntryID=223/listID=3.

IPTV news (September 24, 2013). Retrieved from http://www.iptv-news.com/2013/09/70-million-us-households-to-have-smart-tv-by-2017/.

ITU-T Newslog (2009, February 3). New IPTV standard supports global rollout. Retrieved from http://www.itu.int/ITU-T/newslog/CategoryView,category,QoS.aspx.

Lafayette, Jon (January 20, 2014). Original programming costs raise concern. Broadcasting & Cable. Retrieved from http://www.broadcastingcable.com/news/currency/original-programming-costs-raise-concerns.

Lang, Brent (April 3, 2012). Cord cutting is real: 1 million TV subscribers lost ot streaming services (update). The Wrap.com. Retrieved from http://www.thewrap.com/media/column-post/cord-cutting-real-1-million-subscribers-ditch-cable-streaming-services-36769.

Lang, Brent (March 13, 2014). Watch Out, Cable Companies: More Avid TV Viewers Planning on Cutting the Cord. The Wrap.com. Retrieved from http://www.thewrap.com/cord-cutting-morgan-stanley-streaming-netflix-amazon-prime.

Lawler, Richard (December 22nd 2013). Hulu Plus passes 5 million subscribers, plans to double its original content. Engadget.com. Retrieved from http://www.engadget.com/2013/12/22/hulu-plus-passes-5-million-subscribers/.

Leichtman Research Group (LRG) (March 14, 2014). Major Multi-channel video providers lost about 105,000 subscribers in 2013. Leichtmanresearch.com. Retrieved from http://www.leichtmanresearch.com/press/031414release.html.

Lieberman, David (March 10, 2014). Sony agrees to sell movies and TV shows via Comcast. Deadline.com. Retrieved from http://www.deadline.com/2014/03/sony-agrees-to-sell-movies-and-tv-shows-via-comcast/.

McCracken, Harry (December 03, 2013). Top Ten Gadgets. Time.com. Retrieved from http://techland.time.com/2013/12/04/technology/slide/top-10-gadgets/.

McQuivey, James L., Cooperstein, David M., Hayes, Alexandra (March 7, 2014). The Clash Of The Digital Platforms: Why 2014 Will See A Flood Of Innovation From Platform Players And How You Can Profit From It. Forrester.com. Retrieved from http://www.forrester.com/The+Clash+Of+The+Digital+Platforms/fulltext/-/E-RES115373.

Netflix company website, n.d., retrieved from https://pr.netflix.com/WebClient/loginPageSalesNetWorksAction. do?contentGroupId= 10478&contentGroup=Management+Team.

Netflix company blog, March 20, 2014 retrieved from http://blog.netflix.com/.

Palmeri, Christopher, and Erlichman, Jonathan (March 11, 2014). Discovery backing web-video service along with Schmidt. Bloomberg.com. Retrieved from http://www.bloomberg.com/news/2014-03-11/discovery-backing-web-video-service-along-with-schmidt.html.

Patel, Sahil (March 11, 2014). Discovery and Eric Schmidt back new subscription online video platform. VideoInk.com. Retrieved from http://www.thevideoink.com/news/discovery-eric-schmidt-back-new-svod-network-former-nbc-cnn-execs/.

PayTV operators cash in on over the top demand. (December 11, 2013). IBC. Ibc.org. Retrieved from http://www.ibc.org/page.cfm/action=library/libID=2/libEntryID=223/listID=3.

Palmieri, Christopher, and Erlichman, Jonathan (March 11, 2014). Discovery backing web-video service along with Schmidt. Bloomberg.com. Retrieved from http://www.bloomberg.com/news/print/2014-03-11/discovery-backing-web-video-service-along-with-schmidt.html.

Pepitone, Julianne (October 24, 2011). Netflix loses 800,000 subscribers. CNN.com. Retrieved from http://money.cnn.com/2011/10/24/technology/netflix_earnings/.

Schaeffler, J. (2012, February 20). Ubiquitous video: Everything to everyone, always. *Multichannel News*. Retrieved from http://www.multichannel.com/blog/mixed_signals/33113-Ubiquitous_Video_Everything_to_Everyone_Always.php.

Sharma, Amol, (March 19, 2014). Sony to Add Original TV Shows for PlayStation Move Is Part of Effort to Broaden Console's Appeal Beyond Videogames. WallStreetJournal.com. Retrieved from http://online.wsj.com/news/articles/SB10001424052702304026304579449751926523032?mg=reno64-wsj&url=http%3A%2F%2Fonline.wsj.com%2Farticle%2FSB10001424052702304026304579449751926523032.html.

Siegal, Jacob (January 16, 2014). Twitch: The streaming platform that's knocking off TV stations. Retrieved from http://bgr.com/2014/01/16/twitch-streaming-statistics-2013/.

Stelter, Brian (January 23, 2014). Netflix stock soars 16% on huge subscriber growth. CNN.com. Retrieved from http://money.cnn.com/2014/01/22/technology/netflix-earnings/.

Warren, Christina (May 9, 2013). Netflix's Amazing Growth and AOL's Dismal Decline. Mashable.com. Retrieved from http://mashable.com/2013/05/09/netflix-aol-chart/.

Webster, Tom (April 14, 2014). A Major Shift in Podcast Consumption. Edison Research. Retrieved from http://www.edisonresearch.com/home/archives/2014/04/a-major-shift-in-podcast-consumption.php.

Weprin, Alex (March 14, 2014). Sarah Palin plans 'Rogue TV'. Capitalnewyork.com. Retrieved from http://www.capitalnewyork.com/article/media/2014/03/8541966/sarah-palin-plans-%E2%80%98 rogue-tv%E2%80%99#.

Zipkin, A. (December 17, 2006) The Boss: Out of Africa, onto the web. Retrieved from http://www.nytimes.com/2006/12/17/jobs/17boss.html.

9

Radio Broadcasting

Gregory Pitts, Ph.D.*

> *(While) radio is valued, radio is also taken for granted. Because it is so pervasive, radio is sometimes overlooked, just like water or electricity.*
>
> —National Association of Broadcasters, (n.d.)

Why Study Radio Broadcasting?

- Broadcast radio remains important because it's free and easy to use—there's no subscription or expensive receiver.
- Around the world, there are billions of AM or FM receivers. In the U.S. alone, there are 800 million radios.
- Radio technology plays a major communications role around the world for populations in political, economic, and social transition.

Introduction

Radio experienced its golden age in the 1930s and 1940s. It was *the* medium for live entertainment and news broadcasts, but the enthusiastic embrace of television after World War II put radio on a technology path for the next 60 years that has been dominated by a music box role and efforts to remain relevant to consumers, as newer technologies have emerged. The most exotic innovation—beyond improvements in the radio tuner itself—was the arrival of FM stereo broadcasting in 1961.

Today, audiences are redefining the term "radio." Satellite-delivered audio services from SiriusXM; streaming audio options such as Pandora or other online options; and audio downloads (both music and full-length programming); are allowing consumers to think anew about the meaning of the word *radio*. Radio is just as likely to refer to personal audio media—multiple selections and formats of audio content provided by a variety of sources—beyond FM and AM radio.

Radio has become a generic term for audio entertainment supplied by terrestrial broadcast frequency, satellite, Internet streaming, cellphones, and portable digital media players—via podcasts, some of which come from traditional radio companies and still others that are the product of technology innovators (Green, et al., 2005). Where the technology of over-the-air radio remains most helpful is through the ability to build an audio brand receivable by a large, mainstream audience in a free and easy to consume product.

Radio remains relevant because it is easy to use and it reaches millions (billions worldwide) of listeners each week. Radio doesn't require a smart device or streaming bandwidth. Receivers are designed to be easy to use. The public reports listening to radio each week—broadcast radio reaches 93% of the U.S. population—but the growth in the number of portable digital devices from smartphones to media players is changing the access pattern. Arbitron and Edison Research describe today's media environment as "The Infinite Dial" where broadband or high speed Internet access is changing the media consumption experience (Arbitron, 2012a).

The next frontier for smartphones, tablets, and other audio devices is the automobile, as car manufacturers add external inputs to enable these devices to work with automobile audio systems and some automobile manufacturers are adding Internet connectivity (Arbitron, 2012b)

* Professor, Department of Communications, University of North Alabama (Florence, Alabama).

As you read about radio technology, it is important to remember that technology is disbursed and adopted around the world at different rates. Much of the discussion in this chapter will take a North American or Western European perspective. The economic, regulatory, and political environments in the United States and other countries in these regions are different than in parts of the world with a different development agenda. In these places, radio likely occupies a more relevant position in society.

Countries that are transitioning to market economies and fair elections are just now beginning to license commercial or community radio stations. Where per capita income is low and literacy levels are low, radio often is more important than newspapers or television. The low cost of radio program production, the ease of translating content into multiple languages, and the low cost of a radio receiver make radio a very important information source. Free over-the-air programming entertains audiences but also provides civic and development news and information.

An effort by the U.S. radio industry to add technological sizzle through digital over-the-air radio broadcasting (called HD Radio) has largely fallen flat with only 8% of adults surveyed by the Pew project saying they'd ever listened to the service (Pew, 2012). However, the National Association of Broadcasters (NAB) has not ignored radio innovation. NAB identifies five initiatives that promote radio technology and program availability. These include:

- *NAB Labs* to provide for over-the-air radio innovation and public education
- *Radio-enabled mobile phones.* Mobile phones with built in FM radio tuners.
- *HD Radio* utilizing digital subcarriers to deliver information.
- *The National Radio Systems Committee* (NRSC) a jointly sponsored NAB and Consumer Electronics Association (CEA) committee to recommend technical standards for radio receivers.
- *The Radio Data System* (RDS) that allows analog FM stations to provide digital data services, such as song title and artist or traffic information. (NAB, n.d.).

Radio remains an important part of the daily lives of millions of people. Personality and promotion driven radio formats thrust themselves into the daily routines of listeners. Radio station ownership consolidation has led to greater emphasis on formats, format branding, and promotional efforts designed to appeal to listener groups and, at the same time, yield steady returns on investments for owners and shareholders through the sale of advertising time. But, advertising loads and the availability of personal choice technology, such as portable digital media players and personalized online audio, are driving listeners to look for other options.

Critics have been quick to point to the malaise brought upon the radio industry by consolidated ownership and cost-cutting but these impacts have not been entirely detrimental. The previous ownership fragmentation may never have allowed today's radio to reach the point of offering HD radio as a competitor to satellite and streaming technologies. Through consolidation, the largest owner groups have focused their attention on new product development to position radio to respond to new technological competition.

There have been technology stumbles in the past, such as the following:

- *FM broadcasting*, which almost died because of lack of support from AM station owners and ultimately took more than 35 years to achieve 50% of all radio listening in 1978.
- *Quad-FM* (quadraphonic sound) never gained market momentum.
- *AM stereo*, touted in the early 1980s as a savior in AM's competitive battle with FM.

These technologies did not fail exclusively for want of station owner support, but it was an important part of their failure. Larger, more consolidated ownership groups have greater economic incentive to pursue new technology.

HD radio best exemplifies the economy of scale needed to introduce a new radio technology but HD radio hasn't been seen by consumers as technology worthy of adoption in a marketplace with digital downloads, streaming audio, and satellite services.

This chapter examines the factors that have redirected the technological path of radio broadcasting.

Background

The history of radio is rooted in the earliest wired communications—the telegraph and the telephone—although no single person can be credited with inventing radio. Most of radio's "inventors" refined an idea put forth by someone else (Lewis, 1991).

Although the technology may seem mundane today, until radio was invented, it was impossible to simultaneously transmit entertainment or information to millions of people. The radio experimenters of 1900 or 1910 were as enthused about their technology as are the employees of the latest tech startup. Today, the Internet allows us to travel around the world without leaving our seats. For the listener in the 1920s, 1930s, or 1940s, radio was the only way to hear live reports from around the world.

Probably the most widely known radio inventor/innovator was Italian Guglielmo Marconi, who recognized its commercial value and improved the operation of early wireless equipment.

The one person who made the most lasting contributions to radio and electronics technology was Edwin Howard Armstrong. He discovered regeneration, the principle behind signal amplification, and invented the superheterodyne tuner that led to a high-performance receiver that could be sold at a moderate price, thus increasing home penetration of radios. In 1933, Armstrong was awarded five patents for frequency modulation (FM) technology (Albarran & Pitts, 2000).

The two traditional radio transmission technologies are amplitude modulation and frequency modulation. AM varies (modulates) signal strength (amplitude), and FM varies the frequency of the signal. It's also worth noting that the patent fights today between the three biggest technology companies—Google, Apple and Microsoft—reflect similar struggles in the evolution of radio between 1900 and 1940.

The oldest commercial radio station began broadcasting in AM in 1920, with the technology advantage of being able to broadcast over a wide coverage area. AM signals are low fidelity and subject to electrical interference.

FM, which provides superior sound, is of limited range. Commercial FM took nearly 20 years from the first Armstrong patents in the 1930s to begin significant service and did not reach listener parity with AM until 1978 when FM listenership finally exceeded AM listenership.

FM radio's technological add-on of stereo broadcasting, authorized by the Federal Communications Commission (FCC) in 1961, along with an end to program simulcasting (airing the same program on both AM and FM stations) in 1964, expanded FM listenership (Sterling & Kittross, 1990).

Other attempts, such as Quad-FM (quadraphonic sound), ended with disappointing results. AM stereo, touted in the early 1980s as the savior in AM's competitive battle with FM, languished for lack of a technical standard because of the inability of station owners and the FCC to adopt an AM stereo system (FCC, n.d.-a; Huff, 1992). Ultimately, consumers expressed minimal interest in AM stereo.

Why have technological improvements in radio been slow in coming? One obvious answer is that the marketplace did not want the improvements. Station owners invested in modest technological changes; they shifted music programming from the AM to the FM band. AM attracted listeners by becoming the home of low-cost talk programming.

Another barrier was the question of what would happen to AM and FM stations if a new radio service were created? Would existing stations automatically be given the first opportunity to occupy new digital space, as an automatic grant, eliminating the possibility that new competition would be created? What portion of the spectrum would a new digital service occupy? Were listeners ready to migrate to a new part of the spectrum? New radio services would mean more competitors for the limited pool of advertising revenue.

Radio listeners expressed limited interest in new radio services at a time when personal music choice blossomed through the offering of 45-RPM records, followed by 8-track, and then cassette tapes, and then CDs and portable digital media players. As radio owners were reluctant to embrace new technology offerings, consumers were not.

Radio is important because it was free and familiar but new devices throughout the years and today's unlimited digital dial appear to have captured the attention of the public, policy makers, and electronics manufacturers (Pew, 2012; Pew, 2011).

The Changing Radio Marketplace

The FCC elimination of ownership caps mandated by the *Telecommunications Act of 1996* set the stage for many of the changes that have taken place in radio broadcasting in the last decade. Before the ownership limits were eliminated, there were few incentives for broadcasters, equipment manufacturers, or consumer electronics manufacturers to upgrade the technology.

Outside of the largest markets, radio stations were individual small businesses. (At one time, station owners were limited to seven stations of each service. Later, the limit was increased to 18 stations of each service, before deregulation eventually removed ownership limits.) Analog radio, within the technical limits of a system developed nearly 100 years ago, worked just fine. The fractured station ownership system ensured station owner opposition to FCC initiatives and manufacturer efforts to pursue technological innovation.

The compelling question today is whether anyone cares about the technological changes in terrestrial radio broadcasting. Or, are the newest changes in radio coming at a time when public attention has turned to other sources and forms of audio entertainment? The personal audio medium concept suggests that *local* personality radio may not be relevant when listeners can access both mega-talent personalities and local stars through satellite services, online, and with podcasting.

Recent Developments

There are four areas where technology is affecting radio broadcasting:

1) New technologies that offer substitutes for radio.
2) Delivery competition from satellite digital audio radio services (SDARS).
3) New voices for communities: low power FM service.
4) New digital audio broadcasting transmission modes that are compatible with existing FM and AM radio.

New Competition—Internet Radio, Digital Audio Files, and Podcasts

Listening to portable audio devices and streaming audio are becoming mainstream practices. Online listening may include content originating from over-the-air stations, including HD Radio or FM stations, begging the question of which distribution venue will be more valuable to consumers in the future—broadband or over-the-air. Certainly over-the-air is the cheapest option; it is free, and perhaps that is the most compelling advantage to radio broadcasts. Where Wi-Fi is not available for free, consumers must rely on broadband connections and potential data costs. As access to office networks comes under greater scrutiny, the use of a digital connection to stream music (or video) may be prohibited.

Fifty-six percent of all Americans are now smartphone owners. (Smith, 2013). Six in ten (61%) own a digital device, including smartphones, MP3 players or tablets (Arbitron, 2012b). The Pew Research Center's Project Excellence in Journalism asked survey respondents which technology or device had the greatest impact on their lives. Only one person in five (22%) indicated radio had a big impact on the user's life.

The percentage of people who use their cell phone to listen to radio in their car has increased from 6% in 2010 to 21% in 2013 (Pew, 2013). About 40% of Americans listen to audio on digital devices each week, and this number is expected to double by 2015 (Pew, 2012). As this number rises, the number of hours and minutes per day spent listening to radio will likely drop.

High-speed Internet connectivity—wired or wireless—and the accompanying smart devices, phones, tablets, and digital media players, are technological challengers for terrestrial radio. These sexier technologies offer more listener choice, and they lure consumer dollars to the purchase of these technologies. (For more information on digital audio and Internet connectivity, see Chapters 17 and 20.) The opportunity to control music listening options and to catalog those options in digital form presents not just a technological threat to radio listening, but also a lifestyle threat of greater magnitude than listening to tapes or CDs. Listeners have thousands of songs that can be programmed for playback according to listener mood, and the playback will always be commercial-free and in high fidelity.

As digital audio file playback technology migrates from portable players to automotive units, the threat to radio will increase. Radio stations are responding by offering streaming versions of their program content. The iHeartRadio application from Clear Channel Digital is one example of a radio station owner's response to alternative delivery. TuneIn Radio and similar apps offer smart device access to local radio streams, plus national and international streams and online only services.

Consumers around the world use cell phones to receive music files, video clips, and video games. Missing from most mobile phones is an FM or AM tuner, once again fostering an environment where consumers grow up without the choice of broadcast radio listening. Apple, after substantial lobbying by the broadcast industry, now offers a conventional FM tuner on the iPod nano model.

The NAB's efforts to equip music players and mobile phones with radio tuners, through its *Radio Rocks My Phone* initiative has largely failed (NAB, n.d.). Neither of the two biggest smartphone makers, Apple and Samsung, offers an FM tuner. Ostensibly including the FM tuner will give listeners greater access to local information but this will only be of value if local stations commit to hire personnel to program local information—something more compelling than just music—rather than rely on technology to reduce personnel costs.

Nearly three-fourths of radio listening takes place in cars. Arbitron, Edison Research, and Scarborough Research report that more than half (55%) of 18-24 year olds use their iPod/MP3 player to listen to audio in the car (Arbitron, 2012b). The sophisticated audio systems in newer cars allow for the integration of consumer smartphones and other devices, bringing a host of new capabilities to the car (Cheng, 2010). Included in the features is the ability to run mobile programs such as Internet audio programs from Pandora to other Internet audio service as well as providing consumers with a seamless access to their own content from personal audio files.

Internet radios and mobile devices offer access to thousands more online audio/radio services than any FM tuner ever could (Taub, 2009). Streamingradioguide.com is a Web guide listing nearly 16,000 online U.S. stations—almost all of the commercial, noncommercial, and low-power radio stations in the U.S. By comparison, local over-the-air reception usually provides access to fewer than 100 stations even in a major metropolitan area.

Competition from Satellite Radio

Subscriber-based satellite radio service, a form of out-of-band digital "radio," was launched in the U.S. in 2001 and 2002 by XM Satellite Radio and Sirius Satellite Radio. The competitors received regulatory approval to merge in July 2008 to become SiriusXM. Satellite service was authorized by the FCC in 1995 and, strictly speaking, is not a radio service. Rather than delivering programming primarily through terrestrial (land-based) transmission systems, each service uses geosynchronous satellites to deliver its programming (see Figure 9.1). (Terrestrial repeaters enhance reception in some fringe areas, such as tunnels.)

Figure 9.1
Satellite Radio

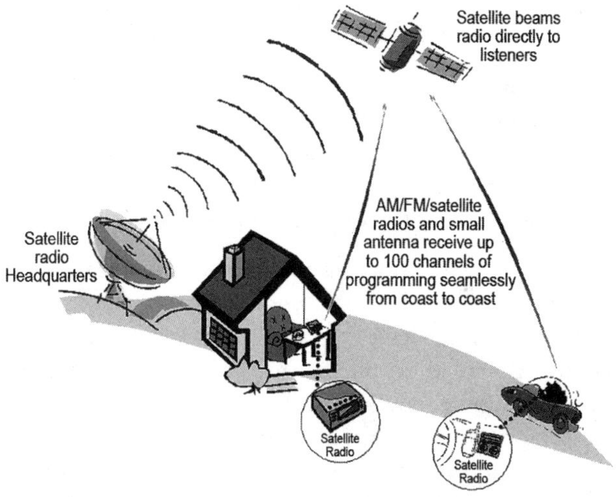

Source: J. Meadows & Technology Futures, Inc.

About 25.6 million people subscribe to the 135 music and news, sports and talk channels. SiriusXM concluded 2013 with the addition of 1.66 million net subscribers (SiriusXM, 2014). Users pay a monthly subscription fee of between $14 and $18 and must have a proprietary receiver to decode the transmissions.

SiriusXM depends on receiver placement in new cars. When automobile sales improve, subscription rates improve. There is a willingness among consumers to pay for audio service but the cost to SiriusXM to attract and retain subscribers remains high; the subscriber acquisition cost was $44 in 2013, a record low (SiriusXM, 2014).

The conversion rate from trial subscriptions is between 45-46% (Team, 2013). The company has also modified its business model to offer online streaming through Internet connectivity, though online streaming is limited only to existing subscribers and requires an additional fee.

Helping the growth of satellite radio has been an array of savvy partnerships and investments, including alliances with various automobile manufacturers and content relationships with personalities and artists, including Howard Stern, Martha Stewart, Oprah Winfrey, Jamie Foxx, Barbara Walters, and Opie & Anthony. SiriusXM Radio is the "Official Satellite Radio Partner" of the NFL, Major League Baseball, NASCAR, NBA, NHL, PGA Tour, and major college sports.

New Voices for Communities: Low-Power FM Service

The FCC approved the creation of a controversial new classification of noncommercial FM station in January 2000 (Chen, 2000). LPFM, or low-power FM, service limits stations to a power level of either 100 watts or 10 watts (FCC, n.d.-b).

The classification was controversial because existing full-power stations were afraid of signal interference. The service range of a 100-watt LPFM station is considered to be about a 3.5-mile radius. In practice, the signals cover a radius of 10–12 miles. LPFM stations do not have the option of adopting HD Radio but they do provide free analog programming to supplement existing FM broadcasts. (FCC, 2012).

After an initial growth spurt of licensing in the early 2000s, LPFM growth stagnated. President Obama signed the Local Community Radio Act of 2010 into law on January 4, 2011. The LCRA seeks to expand the licensing opportunities for LPFM stations by eliminating the third adjacent channel separation requirement enacted by Congress at the behest of commercial broadcasters in 2000 when LPFM service was created (FCC, 2012).

The change increases the likelihood of finding a frequency for an LPFM station. Under the old rules, a full power FM on 102.9 MHz would be protected from any LPFM closer than 102.1 or 103.7. This didn't just protect the full power station; it decreased the likelihood of finding a frequency for a station operation. Now an LPFM could operate on 102.3 or 103.5.

A second wave of LPFM interest occurred between October 17 and November 14, 2013 when 2,819 LPFM station applications were submitted. As of mid-2014, the FCC had approved 1,149 construction permits—authorizations to build a station (REC Networks, 2014). These permits allow 18 months for the stations to begin operation.

Getting the stations built and operational will demonstrate public interest in the service and its ability to offer community programming mostly abandoned by mainstream commercial radio stations.

New Digital Audio Broadcasting Transmission Modes

Most free, over-the-air AM and FM radio stations broadcast in analog but their on-air and production capabilities, and the audio chain, from music and commercials to the final program signal delivered to the station's transmitter, is digitally processed and travels to a digital exciter in the station's transmitter where the audio is added to the carrier wave. The only part of the process that remains analog is the final transmission of the over-the-air FM or AM signal.

AM and FM radio made the critical step toward digital transmission in 2002, when the FCC approved the digital broadcasting system proposed by iBiquity Digital (marketed as HD radio). The FCC calls stations providing the new digital service hybrid broadcasters because they continue their analog broadcasts (FCC, 2004). iBiquity's identification of "HD" Radio is a marketing label not a reference to an FCC station classification.

The HD Radio digital signal eliminates noise and multipath distortion, provides better audio quality—including surround sound—and provides digital information delivery, including traffic information, song tagging, and other forms of consumer information.

IBOC technology consists of an audio compression technology called perceptual audio coder (PAC) that allows the analog and digital content to be combined on existing radio bands, and digital broadcast technology that allows transmission of music and text while reducing the noise and static associated with current reception. The system does not require any new spectrum space, as stations continue to broadcast on the existing analog channel and use the new digital system to broadcast on the same frequency.

Figure 9.2

Hybrid and All-Digital AM & FM IBOC Modes

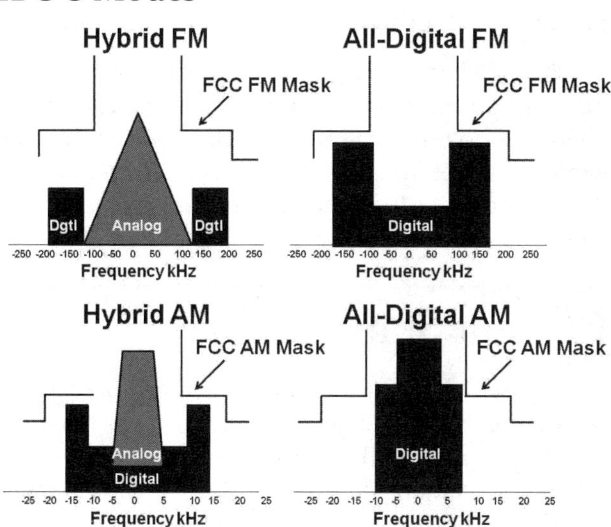

Source: iBiquity

As illustrated in Figure 9.2, this digital audio broadcasting (DAB) system uses a hybrid in-band, on-channel system (IBOC) that allows simultaneous broadcast of analog and digital signals by existing FM stations through the use of compression technology, without disrupting the existing analog coverage.

The FM IBOC system is capable of delivering near-CD-quality audio and new data services including song titles, and traffic and weather bulletins. The so-called killer application to attract consumers to HD radio is the ability to offer a second, third or fourth audio channel. For example, many affiliates of National Public Radio offer classical music or jazz on their main channel and talk or other programming on the side channels.

For broadcasters, HD radio will allow one-way wireless data transmission similar to the radio broadcast data system (RBDS or RDS) technology that allows analog FM stations to send traffic and weather information, programming, and promotional material from the station for delivery to smart receivers. HD radio utilizes multichannel broadcasting by scaling the digital portion of the hybrid FM broadcast. IBOC provides for a 96 Kb/s (kilobits per second) digital data rate, but this can be scaled to 84 Kb/s or 64 Kb/s to allow 12 Kb/s or 32 Kb/s for other services, including non-broadcast services such as subscription services.

Terrestrial digital audio broadcasting involves not only regulatory procedures but also marketing hurdles to convince radio station owners, broadcast equipment manufacturers, consumer and automotive electronics manufacturers and retailers, and most important, the public, to support the technology. You can't listen to an HD signal unless you have a radio receiver equipped to decode HD signals. As with satellite radio's earliest efforts to attract subscribers, receiver availability has been a significant barrier to consumer acceptance. New car models are including HD radio receivers but they are also including satellite radio, Bluetooth, or USB ports for external devices, and the next push appears to be for Internet equipped cars that will allow for direct audio streaming to vehicles. There is no current plan to eliminate analog FM and AM broadcasts, and the HD radio transmissions will not return any spectrum to the FCC for new uses. So far, consumers don't seem to want HD radio when other technologies are available.

iBiquity Digital's HD radio gives listeners, who are used to instant access, one odd technology quirk to get used to. Whenever an HD radio signal is detected, it takes the receiver approximately 8.5 seconds to lock onto the signal. The first four seconds are needed for the receiver to process the digitally compressed information; the next 4.5 seconds ensure robustness of the signal (iBiquity, 2003).

The hybrid (analog and HD) operation allows receivers to switch between digital and analog signals, if the digital signal is lost. Receivers compensate for part of the lost signal by incorporating audio buffering technology into their electronics that can fill in the missing signal with analog audio. For this approach to be effective, iBiquity Digital recommends that analog station signals operate with a delay, rather than as a live signal. Effectively, the signal of an analog FM receiver, airing the same programming as an HD receiver, would be delayed at least 8.5 seconds. As a practical matter, iBiquity Digital notes, "Processing and buffer delay will produce off-air cueing challenges for remote broadcasts…" (iBiquity, 2003, p. 55).

A different form of DAB service is in operation in a number of countries. The Eureka 147 system broadcasts digital signals on the L-band (1452-1492 MHz) or a part of the spectrum known as Band III (around 221 MHz). The service is in operation in fourteen countries around the world, including in the

UK, with 85% coverage and 14 million receiving devices in the market. Germany expects 99% population coverage by 2014. China operates stations reaching 8% of the country's population. China has the potential to establish DAB as a global format. With a state controlled economy and technology marketplace, the government of China could direct a country-wide rollout of the technology, creating a marketplace with 1.3 billion potential listeners. Eighteen additional countries are conducting experimental tests and 13 more have expressed interest in the system (World DAB, n.d.).

Eureka technology is not designed to work with existing AM and FM frequencies. That is one reason why broadcasters in the United States rejected the Eureka 147 system in favor of the "backward and forward" compatible digital technology of iBiquity Digital's IBOC that allows listeners to receive analog signals without having to purchase a new receiver for the DAB system (FCC, 2004b).

The World DAB Forum, an international, non-government organization to promote the Eureka 147 DAB system, reports that more than 20 countries around the world have regular DAB services, another 15 countries are conducting trials or establishing regulations and another 14 countries have shown interest in the service (World DAB, 2013). Missing from the list of country participants is the United States. Over 500 million people around the world can receive the 1,400 different DAB/DAB+/DMB services (World DAB, 2013). As with digital broadcasting in the United States, proponents of Eureka 147 cite the service's ability to deliver data as well as audio. Examples of data applications include weather maps or directional information that might be helpful to drivers or emergency personnel.

Unlike iBiquity Digital, the Eureka system is gaining market strength in a number of countries. Receivers now cost about 20 Euros (equal to about $26 U.S.), and more than 80 companies offer a receiver or compatible product (World DAB, 2013). What makes DAB enticing to governments is that the service would eliminate the current spectrum used for AM/FM broadcasts. Moreover, the DAB service has the potential to allow improved services for disaster communication, mobile TV, and even something as basic as controlling street lights.

DAB is attractive because multiple information platforms could use the services, and the DAB system is energy efficient. Receivers in mobile locations could deliver data and text or video but rely only on a simple solar panel for power.

An early effort by iBiquity Digital and its broadcast equipment manufacturing partners introduced HD Radio to other countries, particularly in Europe. While the technology attracted some attention, it does not appear to be gaining station or regulatory converts.

Current Status

Terrestrial broadcast radio remains a big part of the lives of most people. Each week, more than 241 million people hear the nearly 15,859 FM, LPFM, and AM radio stations in the U. S. Advertisers spent $17.65 billion on radio advertising in 2013—a substantial amount but still less than the industry's $21.3 billion in 2007. The decline resulted from the soft economy as well as advertising shifts to other media, including online sources (FCC, 2012; RAB, n.d. & U.S. Bureau of the Census, 2011).

In fairness, ad expenditures for radio are trending up since sinking to about $16 billion in 2009 but overall ad spending is projected to rise about 20% between 2013 and 2017. Radio's revenue increase will only be about 4%. Radio listening statistics are impressive but a closer examination shows persistent radio industry declines, ranging from declining ad revenue to a daily drop in listeners.

The FCC estimates there are more than 800 million radios in use in U.S. households, private and commercial vehicles, and commercial establishments. All of these radios will continue to receive analog signals from radio stations. Radio listening is steady at 93% of the U.S. population reached each week, but Internet access at any location is now 85%. Time spent listening to radio was about 3 hours per day in 2003; it was slightly under 2 hours in 2013 (RAB, 2003; RAB, n.d.). It's not that radio doesn't have something to offer; it's just that other services have more ways to offer content. Still, the radio content is free—except for the opportunity cost of listening to commercials—and some sort of signal is available nearly anywhere in the world you travel.

Table 9.1

Radio in the United States at a Glance

Number of Radios

Households with radios	99%
Average number of radios per household	8
Number of radios in U.S. homes, autos, commercial vehicles and commercial establishments	800 million

Source: U.S. Bureau of the Census (2012) and FCC (2004)

Radio Station Totals

AM Stations	4,725
FM Commercial Stations	6,624
FM Educational Stations	4,057
Total 15,406	
FM Translators and Boosters	6,082
LPFM Stations	774

Source: FCC (2014)

Radio Audiences

Persons Age 12 and Older Reached by Radio:	
Each week:	92.3% (About 241.5 million people)
Each day:	77% (About 187 million people)
Persons Age 12 and Older Radio Reach by Daypart:	
6–10am	74.9%
10–3pm	82.3%
3–7pm	81.5%
7–12 Mid.	59.4%
12 Mid–6am	25.7%
Where Adults Age 18 and Older Listen to the Radio:	
At home:	17.8% of their listening time
In car:	60.4% of their listening time
At work or other places:	10.4% of their listening time
Daily Share of Time Spent With Various Media:	
Broadcast Radio	77%
TV/Cable	95%
Newspapers	35%
Internet/Web	64%
Magazines	27%

Source: Radio Advertising Bureau Why Radio Fact Sheet(n.d.-b)

Satellite Subscribers

SiriusXM Satellite Radio	25.6 million

Source: SiriusXM Satellite Radio (2014)

Factors to Watch

Radio stations have been in the business of delivering music and information to listeners for nearly a century. Public acceptance of radio, measured through listenership more than any particular technological aspect, has enabled radio to succeed. Stations have been able to sell advertising time based on the number and perceived value of their audience to advertising clients. Technology, when utilized by radio stations, focused on improving the sound of the existing AM or FM signal or reducing operating costs.

Digital radio technology has modest potential to return the radio industry to a more relevant status among consumers. An FCC initiative, championed by Commissioner Ajit Pai, to overhaul AM radio service probably won't make much difference. The initiative proposes to examine HD radio for AM and to allow AM stations to simulcast on FM. Even if implemented, these actions would depend on public acceptance to be meaningful (Wyatt, 2013).

The plethora of alternative delivery technology is changing consumer perceptions about radio. The increased number of people listening to audio through a smartphone or other online device is increasingly marginalizing traditional broadcast radio.

The distinction however is that people still listen to something they define as radio. The programming is important; whether delivered by new technology or old, listeners value the content. It is up to radio programmers to respond with meaningful programming delivered through an appropriate technological pipeline.

A decade ago, Chris Andersen (2006), writing in *Wired*, promoted the notion of *The Long Tail*, where consumers, bored with mainstream media offerings, regularly pursue the digital music road less traveled. As *Business Week* noted, "Listeners, increasingly bored by the homogeneous programming and ever-more-intrusive advertising on commercial airwaves, are simply tuning out and finding alternatives" (Green, et al., 2005).

Consumer ability to easily store and transfer digital audio files to and from a variety of small personal players that have a convenience advantage as the technology gets easier to interface with car audio systems. The competitive nature of radio suggests that the battle for listeners will lead to fewer format options and more mass appeal formats, as stations attempt to pursue an ever-shrinking radio advertising stream.

Localism—the ability of stations to market not only near-CD-quality audio content but also valuable local news, weather, and sports information—has been cited as the touchstone for the terrestrial radio industry. In his farewell address in 2005, retiring NAB President Eddie Fritts presented this challenge: "Our future is in combining the domestic with the digital. Localism and public service are our franchises, and ours alone" (Fritts farewell, 2005, p. 44).

But even radio localism is under attack by newspapers and television stations, as well as bloggers and Twitter feeds that offer micro-local content through Internet and smart device delivery. Why wait for the radio station to provide weather, traffic or sports scores when the content is available on your mobile phone as text, audio or video? But, ease of access and use of radio technology that requires only a receiver and no downloading or monthly fees may keep radio relevant for a core group of listeners unwilling to pay for new services and new technologies.

Radio Broadcasting Visionaries: Robert W. Pittman & Pete Tridish

Robert W. Pittman is Chairman and Chief Executive Officer of Clear Channel, the largest group owner of radio stations in the United States. Clear Channel was, historically speaking, a radio company. The company traces its founding in 1972 to ownership of a single clear channel AM station, WOAI, in San Antonio, Texas. Pittman wasn't the company founder but he is notable today for an earlier role as the co-founder and programmer who led the team that created MTV; he later became the CEO of MTV Networks. MTV changed how people heard and *saw* music. The awareness that music didn't just come from the radio but through a video screen emerged with the introduction of MTV.

In his current role, he is also changing radio by no longer being just a radio company. The company owns radio stations, billboards and a mixture of online and mobile services. Through these platforms, Clear Channel Media + Entertainment has a greater reach than any other radio or television company in the United States. Over 243 million people listen every month but they aren't listening only to over-the-air broadcasting. Listeners stream music through Clear Channel's i-Heart-Radio app and bypass the traditional technology.

Clear Channel is not the technology leader pursuing media changes but it is making savvy business changes to capture eyes and ears of listeners and viewers. These alternate routes to audio consumption are capable of altering traditional radio listenership approaches. As CEO, Bob Pittman is the leader of a significant radio company that is following a new path.

Pete Tridish describes himself as a radio engineer, policy advocate and trouble maker. (Who?, n.d.). The Federal Communications Commission would rightfully call Tridish a pirate because he was part of a group operating an unlicensed (and therefore illegal) FM radio station called Radio Mutiny, 91.3 FM Pirate Radio, in Philadelphia. Tridish is also a founding member of the Prometheus Radio Group, a social justice advocacy group stressing the need for and importance of community radio.

Tridish has provided engineering consultation and guidance for groups seeking to operate community radio stations. He has been an outspoken critic of the FCC, as well as speaking out against media consolidation and in favor of community or grass-roots organizations seeking to operate a station. Though Tridish has left Prometheus, in the recent LPFM filing window Prometheus provided engineering and support to more than 1,000 LPFM applicants, through free online information or actual applicant consultation (Stine, 2014). More than 2,800 groups or individuals applied for LPFM stations. These stations hope to use traditional FM technology to deliver programming that is at least atypical of the programming provided by most commercial and noncommercial radio stations.

About 3,258 applicants requested LPFM stations in the first filing window in 2000. Only about 800 of those stations are on the air today. It may be that LPFM will not have much of a community impact but LPFM represents the only means available to community groups wanting to at least try to return radio to a local voice. Pete Tridish was a visionary leader who evangelized on behalf of community radio. If LPFM becomes a meaningful community service, both Tridish and Prometheus Radio deserve some of the credit.

Bibliography

Albarran, A. & Pitts, G. (2000). *The radio broadcasting industry*. Boston: Allyn and Bacon.
Anderson, C. (2006). *The long tail: Why the future of business is selling less of more*. New York: Hyperion.
Arbitron (2012a). The infinite dial 2012: Navigating digital platforms. Retrieved April 27, 2012 from http://www.arbitron.com/study/digital_radio_study.asp.
Arbitron (2012b). Weekly online radio audience jumps more than 30 percent in past year says new Arbitron/Edison research study. Retrieved April 28, 2012 from http://arbitron.mediaroom.com/index.php?s=43&item=813.
Boehret, K. (2010, January 27). Reaching for the height of radio. *The Wall Street Journal*, p. D8.
Chen, K. (2000, January 17). FCC is set to open airwaves to low-power radio. *Wall Street Journal*, B12.
Cheng, R. (2010, April 28). Car phones getting smarter. *The Wall Street Journal*, p. B5.
Federal Communications Commission. (n.d.-a). *AM stereo broadcasting*. Retrieved May 1, 2010 from http://www.fcc.gov/mb/audio/bickel/amstereo.html.
Federal Communications Commission. (n.d.-b). *Low-power FM broadcast radio stations*. Retrieved May 1, 2010 from http://www.fcc.gov/mb/audio/lpfm.
Federal Communications Commission. (2004). In the matter of digital audio broadcasting systems and their impact on the terrestrial radio broadcast services. *Notice of proposed rulemaking*. MM Docket No. 99-325. Retrieved April 20, 2010 from http://hraunfoss.fcc.gov/edocs_public/attachmatch/FCC-04-99A4.pdf.
Federal Communications Commission. (2012). In the matter of economic impact of low-power FM stations on Commercial FM radio: Report to Congress pursuant to section 8 of the Local Community Radio Act of 2010. Retrieved April 21, 2012 from http://transition.fcc.gov/Daily_Releases/Daily_Business/2012/db0105/DA-12-2A1.pdf.
Federal Communications Commission. (2014, March 31)). *Broadcast station totals as of March 31, 2014*. Retrieved April 22, 2014 from http://www.fcc.gov/document/broadcast-station-totals-march-31-2014.
Fritts' farewell: Stay vigilant, stay local. (2005, April 25). *Broadcasting & Cable*, 44.
Green, H., Lowry, T., & Yang, C. (2005, March 3). The new radio revolution. *Business Week Online*. Retrieved February 15, 2006 from http://yahoo.businessweek.com/technology/content/mar2005/ tc2005033_0336_tc024.htm.
Huff, K. (1992). AM stereo in the marketplace: The solution still eludes. *Journal of Radio Studies*, *1*, 15-30.
iBiquity Digital Corporation. (2003). *Broadcasters marketing guide, version 1.0*. Retrieved March 10, 2006 from http://www.ibiquity.com/hdradio/documents/BroadcastersMarketingGuide.pdf.
Lewis, T. (1991). *Empire of the air: The men who made radio*. New York: Harper Collins.
National Association of Broadcasters. (n.d.). *Innovation in Radio*. Retrieved April 23, 2012 from http://www.nab.org/radio/innovation.asp.
Pew Research Center. (2013). Percent of Cellphone Owners Who Listen to Online Radio Through Phone in the Car. Retrieved April 1, 2014 from http://www.journalism.org/media-indicators/percent-of-cellphone-owners-who-listen-to-online-radio-through-phone-in-the-car/.
Pew Research Center. (2012). The state of the news media 2012: An annual report on American journalism. Retrieved April 22, 2012 from http://stateofthemedia.org/.
Pew Research Center. (2011). The state of the news media 2011: An annual report on American journalism. Retrieved April 22, 2012 from http://pewresearch.org/pubs/1924/state-of-the-news-media-2011.
Radio Advertising Bureau. (n.d.). *Why Radio Factsheets*. Retrieved April 9, 2014 from http://www.rab.com/whyradio/WRFacts.cfm.

Radio Advertising Bureau. (2003). Radio marketing guide and factbook for advertisers 2003-2004 edition. New York: Radio Advertising Bureau.

REC Networks (2014, April 14). Granted application order of finish. Retrieved April 14, 2014 from http://recnet.net/grant/?f=1101.

Riismandel, P. (2013, December 3) With 241 million users radio kicks Facebook and Twitter ass. Re trieved March 29, 2013 from http://www.radiosurvivor.com/2013/12/03/with-241-million-users-radio-kicks-facebook-and-twitter-ass/.

SiriusXM Radio. (2014, February 4). SIRIUS XM Reports Fourth-Quarter and Full Year 2013 Results. Retrieved April 3, 2014 from http://investor.siriusxm.com/releasedetail.cfm?ReleaseID=823023.

Sterling, C. & Kittross, J. (1990). *Stay tuned: A concise history of American broadcasting*. Belmont, CA: Wadsworth Publishing.

Stine, R. J. (2014, January 2). LPFM Application Tally Surprises. Retrieved April 2, 2014 from http://www.radioworld.com/article/lpfm-application-tally-surprises-/223002#sthash.Q8OPKJkZ.dpuf.

Taub, E. A. (2009, December 30). Internet radio stations are the new wave. *The New York Times*. Retrieved May 2, 2010 from http://www.nytimes.com/2009/12/31/technology/personaltech/31basics.html?_r=1&scp=3&sq=internet%20radio%20receivers&st=Search.

Team, T. (2013, April 12). Can SiriusXM Tune in big subscriber growth this year? *Forbes,* Retrieved April 3, 2014 from http://www.forbes.com/sites/greatspeculations/2013/04/12/can-sirius-xm-tune-in-big-subscriber-growth-this-year/.

U.S. Bureau of the Census. (2012). *Statistical abstract of the United States*. Washington, DC: U.S. Government Printing Office.

World DAB: The World Forum for Digital Audio Broadcasting. (n.d.). Country information. Retrieved April 22, 2012 from http://www.worlddab.org/country_information_\.

World DAB (2013). World DMB Global Update. Retrieved April 10, 2014 from http://www.worlddab.org/ public_document/file/387/WorldDMB-Global_Update-Sept_2013_-LoRes.pdf?1379926526.

Who? (n.d.) Pete Tridish . Net: Radio engineer. Policy advocate. Troublemaker… Retrieved April 3, 2014 from http://www.petetridish.net/?page_id=7.

Wyatt, E. (2013, September 9). A quest to save AM before it's lost in static. *The New York Times*, p. A1.

10

Digital Signage

Jennifer Meadows, Ph.D.*

Introduction

Chances are you have probably never heard of digital signage or considered it an important communication technology. This type of signage is everywhere although many of us go through the day without ever really noticing it. When you hear the term digital signage perhaps the huge outdoor signs that line the Las Vegas Strip or Times Square in New York come to mind. But, there is more to digital signage.

Consider the menu board at a local fast food restaurant, the AMBER Alert signs on the highway, the sign announcing an exercise class schedule at a health club, even the signs promoting interest rates while you wait in line at the bank. These are all examples of digital signage.

The Digital Screenmedia Association defines digital signage as "the use of electronic displays or screens (such as LCD, LED, plasma or projection, discussed in Chapter 6) to deliver entertainment, information and/or advertisement in public or private spaces, outside of home" (Digital Screenmedia, n.d.). Digital signage is frequently referred to as DOOH or digital outside of the home.

The Digital Signage Federation expands the definition a little to include the way digital signage is networked and defines digital signage as "a network of digital displays that is centrally managed and addressable for targeted information, entertainment, merchandising and advertising" (Digital Signage, n.d.). Either way, the key components to a digital signage system include the display/screen, the content for the display, and some kind of content management system.

The displays or screens for digital signage can take many forms. LCD and LED LCDs are the most common type of screen but you will also see plasma, OLED and projection systems. Screens can be flat or curved and can include touch and gesture interactivity. Screens can range in size from a tiny digital price tag to a huge stadium scoreboard.

Content for digital signage can also take many forms. Similar to a traditional paper sign, digital signs can contain text and images. This is where the similarity ends, though. Digital signage content can also include video (live and stored, standard, high definition and 4K), animation, RSS feeds, and social networking.

Unlike a traditional sign, digital signage content can be changed quickly. If the hours at a university library were changed for spring break, a new traditional sign would have to be created and posted while the old sign was taken down. With a digital sign, the hours can be updated almost instantly.

Another advantage of digital signage is that content can also be delivered in multiple ways beside a static image. The sign can be interactive with a touch screen or gesture control. For example, the Coke Happiness Machine allows users to pay for a drink with a song or a hug (Coca-Cola Company, 2012) while the Pepsi Like Machine trades drinks for "Likes" on Facebook (Hall, 2013a). Wayfinding signs in the Venetian Hotel and Casino allow users to input the desired location while the sign delivers a visual route. Digital signs can also include technologies that allow the system to recognize the viewer's age and sex. Some even include facial recognition technology.

* Professor and Chair, Department of Communication Design, California State University, Chico (Chico, California).

Table 10.1
Paper versus Digital Signage

Traditional/Paper Signs	Digital Signage
• Displays a single message over several weeks	• Displays multiple messages for desired time period
• No audience response available	• Allows audience interactivity
• Changing content requires new sign creation	• Content changed quickly and easily
• Two-dimensional presentation	• Mixed media = text, graphics, video, pictures
• Lower initial costs	• High upfront technology investment costs

Source: Janet Kolodzy

All of these technologies allow digital signage to deliver more targeted messages in a more efficient way. For example, a traditional sign must be reprinted when the message changes, and it can only deliver one message at a time. Digital signage can deliver multiple messages over time, and those messages can be tailored to the audience using techniques such as dayparting.

Dayparting is when the day is divided up into segments so a targeted message can be delivered to a target audience at the right time (Dell, 2013). For example a digital billboard may deliver advertising for coffee during the morning rush hour and messages about prime time television shows in the evening. Those ads could be longer because drivers are more likely to be stopped or slowed because of traffic. An advertisement for a mobile phone during mid-day could be very short because traffic will be flowing.

Digital signage also allows viewers to interact with signs in new ways such as the touch and gesture examples above and through mobile devices. In some cases, viewers can move content from a sign to their mobile device using near field communication (NFC) technology or Wi-Fi. For example, a bus stop could advertise a movie, and just by touching an NFC enabled phone to the sign, a viewer can receive the trailer and local theater screening times.

All of this interactivity brings great benefits to those using digital signs. Audience metrics can be captured such as dwell time (how long a person looks at the sign), age and sex of the user, what messages were used interactively, and more.

Considering all of the advantages of digital signage, why are people still using traditional signs? Although digital signage has many advantages such as being able to deliver multiple forms of content; allowing interactivity; multiple messages and dayparting; easily changeable content; and viewer metrics; the big disadvantage is cost. Upfront costs for digital signage are much higher than for traditional signs. Over time, the ROI (return on investment) is generally high for digital signage but for many businesses and organizations this upfront cost is prohibitive. See Table 10,1 for a comparison traditional and digital signs.

Cost is often a factor in choosing a digital signage system. There are two basic types of systems: premise based and Software as a Service (SaaS). A premise based system means that the entire digital signage system is in-house. Content creation, delivery, and management are hosted on location. Advantages of premise based systems include control, customization, and security. In addition, there isn't an ongoing cost for service. Although a doctor's office with one digital sign with a flash drive providing content is an inexpensive option, multiple screen deployments with a premise based system are generally expensive and are increasingly less popular.

SaaS systems use the Internet to manage and deliver content. Customers subscribe to the service, and the service usually provides a content management system that can be accessed over the Internet. Templates provide customers an easy interface to create sign content, or customers can upload their own content in the many forms described earlier. Users then pay the SaaS provider a regular fee to continue using the service. Both Premises and SaaS systems have advantages and disadvantages and many organizations deploy a mix of both types.

So, to review the technology that makes digital signage work, first there needs to be a screen or display. Next, there should be some kind of media player, which connects to the screen. Content can be added to the media player using devices such as flash drives and DVDs, but more commonly the media player resides on a server and content is created, managed, and delivered using a server on site or accessed over the Internet.

Digital signage, then, is a growing force in out of home signage. Kelsen (2010) describes three major types of digital signage: point of sale, point of wait, and point of transit. These categories are not mutually exclusive or exhaustive but they do provide a good framework for understanding general categories of digital signage use.

Point of sale digital signage is just what it sounds like, digital signage that sells. The menu board at McDonald's is an example of a point of sale digital sign. A digital sign at a Foot Locker store in New York City allowed customers to design and order their own New Balance shoe right in the store (Foot Locker, 2013).

Point of Wait digital signage is placed where people are waiting and there is dwell time, such as a bank, elevator, or doctor's office. One example is the digital signage deployed inside taxis.

Finally, **Point of Transit** digital signs target people on the move. This includes digital signs in transit hubs such as airports and train stations, and signage that captures the attention of drivers like digital billboards and walkers, such as street facing retail digital signs. Figure 10.1 gives examples of where you might find the three different types of digital signage.

Figure 10.1
Types of Digital Signage

POINT OF WAIT
Taxi
Doctor's Office
Elevator
Bank Line

POINT OF SALE
Restaurant Menu Board
Price Tags
Point of Payment
Mall

POINT OF TRANSIT
Airport
Billboard
Bus Shelter
Store Window

Source: J. Meadows

Background

The obvious place to begin tracing the development of digital signage is traditional signage. The first illuminated signs used gas, and the P.T. Barnum Museum was the first to use illuminated signs in 1840 (National Park Service, n.d.). The development of electricity and the light bulb allowed signs to be more easily illuminated, and then neon signs emerged in the 1920's (A Brief History, 1976). Electronic scoreboards such as the Houston Astros scoreboard in the new Astrodome in 1965 used 50,000 lights to create large messages and scores (Brannon, 2009).

Billboards have been used in the United States since the early 1800s. Lighting was later added to allow the signs to be visible at night. In the 1900s billboard structure and sizes were standardized to the sizes and types we see today (History of OOH, n.d.).

The move from illuminated signs to digital signs developed along the same path as the important technologies that make up digital signage discussed earlier: screen, content and networks/content management. The development of the VCR in the 1970's contributed to digital signage as the technology provided a means to deliver custom content to television screens. These would usually be found indoors because of weather concerns.

A traditional CRT television screen was limited by size and weight. The development of projection and flat screen displays advanced digital signage. James Fergason developed LCD display technology in the early 1970's that led to the first LCD watch screen (Bellis, n.d.).

Plasma display monitors were developed in the mid 1960's but were also limited in size, resolution and color, like the LCD. It wasn't until the mid 1990's that high definition flat screen monitors were developed and used in digital signage. Image quality and resolution are particularly important for viewers to be able to read text, which is an important capability for most signage.

As screen technologies developed, so did the content delivery systems. DVDs replaced VCRs, eventually followed by media servers and flash memory devices. Compressed digital content forms such as jpeg, gif, flv, mov, and MP3 allowed signage to be quickly updated over a network.

Interactivity with screens became popular on ATM machines in the 1980's. Touch screen technology was developed in the 1960's but wasn't developed into widely used products until the 1990's when the technology was incorporated in personal digital assistant devices (PDAs) like the Palm Pilot.

The Portfoliowall, developed by Alias/Wavefront and released in 2001, was a major step in touch and gestural displays. "The Portfoliowall used a simple, easy-to-use, gesture-based interface. It allowed users to inspect and maneuver images, animations, and 3D files with just their fingers. It was also easy to scale images, fetch 3D models, and play back video" (Ion, 2013).

Digital billboards first appeared in 2005 and immediately made an impact, but not necessarily a good one. Billboards have always been somewhat contentious. Federal legislation, beginning with the Bonus Act in 1958 and the Highway Beautification Act in 1965, tie Federal highway funding to the placement and regulation of billboards (Federal Highway Administration, n.d.). States also have their own regulations regarding outdoor advertising including billboards.

The development of digital billboards that provide multiple messages and moving images provoked a new look at billboards as driver distractors. Although Federal Transportation Agency rules state billboards that have flashing, intermittent or moving light or lights are prohibited, the agency determined that digital billboards didn't violate that rule as long as the images are on for at least 4 seconds and are not too bright (Richtel, 2010).

States also have their own regulations, and there has been a recent push for more regulation on the state level. For example, a law passed in Michigan in early 2014 limits the number of digital billboards and their brightness, how often the messages can change, and how far apart they can be erected. Companies can erect a digital billboard if they give up three traditional billboards or billboard permits (Eggert, 2014).

While distraction is an issue with billboards, privacy is an overall concern with digital signage, especially as new technologies allow digital signs to collect data from viewers. The Digital Signage Federation adopted its Digital Signage Privacy Standards in 2011. These voluntary standards were developed to "help preserve public trust in digital signage and set the stage for a new era of consumer-friendly interactive marketing (Digital Signage Federation, 2011)."

The standards cover both directly identifiable data such as name, address, date of birth, and images of individuals, as well as pseudonymous data. This type of data refers to that which can be reasonably associated with a particular person or his or her property. Examples include IP addresses, Internet usernames, social networking friend lists, and posts in discussion forums. The standards also state that data should not be knowingly collected on minors under 13.

The standards also recommend that digital signage companies use other standards for complementary technologies, such as the Mobile Marketing Association's Global Code of Conduct and the Code of Conduct from the Point of Purchase Association International.

Other aspects of the standard include fair information practices such as transparency and consent. Companies should have privacy policies and should give viewers notice.

Audience measurement is divided into three levels. Level I is audience counting such as technologies that record dwell time but no facial recognition. Level II is audience targeting. This is information that is collected and aggregated and used to tailor advertisements in real time. Finally Level III is audience identification and/or profiling. This is information collected on an individual basis. The information is retained and has links to individual identity such as a credit card number.

The standards recommend that Level I and II measurement should have opt out consent while Level III should have opt in consent.

As the use of digital signage has grown, so too has the number of companies offering digital signage services and technologies. Some of the long standing companies in digital signage include Scala, BrightSign, Dynamax, Broadsign, and Four Winds Interactive. Other more traditional computer and networking companies including IBM and Cisco have also begun to offer full service digital signage solutions.

Recent Developments

A fast growing technology, digital signage continues along a path of fast innovation as companies race to develop more interactive and engaging signage. Touch and gesture control, facial recognition, and augmented reality technologies are being incorporated into signage. Mobile and social networking technology is also being increasingly incorporated into digital signage. And, as the signage becomes more engaging and perhaps distracting, there are increasing attempts to regulate it, especially roadside billboards.

Motion control video games systems such as the Xbox Kinect moved this technology from science fiction into the home. Now millions of people are swiping and kicking the air to make things happen on a screen. That same motion detection technology is being used in digital signage.

For example, Lifetime Television advertised their new show *The Witches of East End* in a New York City subway station using two 9-screen digital video walls with gesture technology. As people walked by the sign, a giant eyeball followed them. If a person stopped to look at the eyeball, the sign would change into a wooded scene where the viewer could create fire with arm movements. The sign was successful in engaging the audience. Josh Cohen, of sign creator Pearl Media, was enthusiastic about the potential of interactive digital signage. "The digital screen now, which is replacing the static image, when it's just flipping and showing normal content, it doesn't really do much more than show a digital static image," he said. "But when we're able to allow people to interact—whether it's gesture, mobile, social, touch, photo-capture, augmented reality—it allows consumers to have a personal connection to the advertisement, to the client, to create what we call a lasting impression and increase the opportunity for them to share via mobile and social, which I think is a big part of the digital signage interactive opportunities, which are sometimes missed in just your traditional digital signage" (Hall, 2013b).

Gesture based digital signage isn't just to capture audiences for advertising. GestureTek, one company specializing in gesture based digital signage, has products designed for medical rehabilitation. Much like a video game, these systems allow physical and occupational therapists to use digital signage to help stroke victims, and children with cerebral palsy to improve balance, range of movement, and hand eye coordination (Rehabilitation, 2014).

Facial recognition and demographic identification is improving. For example, Intel offers its Audience Impression Metric Suite for digital signage to deliver audience analytics and metrics. This system uses "audience identification" technology rather than facial recognition in order to ensure privacy. No personal information is captured nor are images recorded. Once real time video is scanned and analyzed, it is destroyed. The system collects age, sex, and dwell time so messages can be targeted.

One installation with the system was used with a Jello promotion. A vending machine was set up to deliver samples of an "adults only" Jello dessert. Children were then refused a sample; only adults would be delivered a sample (Intel, 2014).

While the Intel technology aggregates data, some digital signage is being used to actually connect facial recognition to official identification. For example, the Phoenix Airport Check-in Kiosk from Materna ips uses facial recognition to identify passengers and compare their image with passport photos. The kiosk then delivers boarding passes. (New Touchscreen, 2014).

Connection with mobile devices is another feature of digital signage that continues to grow and develop. Clear Channel Outdoor launched its own global interactive mobile advertising platform called Connect in March 2014. According to Clear Channel Outdoor, Connect will "seamlessly" connect DOOH and mobile using QR Codes or NFC Connect tags on digital signage or physical street walls.

An example would be a bus shelter. Viewers would then just tap or scan a sign with their phones to get a more immersive experience. The platform has been rolled out in the UK, Singapore, and parts of Europe. The U.S. platform deployment is planned to be completed by summer 2014 (Digital Signage Federation, 2014).

Mobile payment through NFC and digital signage continues to develop. Vending machines now use NFC technology to allow consumers to pay for items with a mobile device. Diji-Touch interactive vending machines use Kinect gesture control, touchscreen LED monitors, and NFC. The vending machines also have HD motion graphics, nutrition information, and 3D packaging images. Dwell time for these vending machines averages 6 minutes, and purchasing is increased by a third over a traditional vending machine (Digital Signage Powers, 2014).

One very new trend in digital signage is augmented reality. "Augmented reality (AR) is a live, direct or indirect, view of a physical, real-world environment whose elements are augmented by computer-generated sensory input such as sound, video, graphics or GPS data (Augmented Reality, n.d.)".

In one innovative application Pepsi used AR technology to create all kinds of unusual surprises on a street in London. People in a bus shelter could look through a seemingly transparent window at the street, and then monsters, wild animals, and aliens would appear, interacting with people and objects on the street. The transparent sign was actually a live video feed of the street (Pepsi Max Excites, 2014). See Figure 10.2. A video of the deployment can be seen at http://youtu.be/Go9rf9GmYpM.

Another application of augmented reality can be found in clothing stores where customers can try on clothing virtually. The digital sign shows the customer and then allows the customer to choose and virtually try on different clothes. One fun application of AR used animated children's characters in an Australian shopping mall during school holidays. Children could use AR to interact with pirates, fish, and fairies (Augmented Reality Digital, 2014).

Social networking continues to be a popular addition to digital signage. It's fairly common to find digital signage with Twitter feeds, especially with Point of Wait digital signage. News organizations, in particular, use Twitter to keep their digital signage updated with their news feed. Facebook posts can also be part of digital signage. For example, corporate digital signage could feature company Facebook posts.

Digital signage can also call for viewers to interact with social networks and even then display, those posts and tweets. Clear Channel Outdoor' Dunk Tank in New York City's Times Square asked viewers to vote via Twitter on which person should be subject to the dunk tank. Viewers then used gesture control to throw the ball at a target to initiate the dunk (Times Square, 2013).

Figure 10.2
Pepsi Max Augmented Reality

Source: Grand Visual

Text messaging, likewise, is used to connect and engage digital sign viewers. Voting is a common use of SMS—users can poll customers on a range of issues. For example, your university food service could poll students on favorite pizza toppings. Clear Channel Outdoors' Connect program has SMS capabilities in Latin America in addition to QR codes and NFC.

One major part of a digital sign that hasn't been discussed yet is the actual screen and media players. Advances in 4K and Ultra HD screens (discussed in Chapter 6) and displays have made digital signs clearer even at very large sizes.

For example, LG has a 105" Ultra HD display with 5K resolution. Samsung's D series Ultra HD display comes as large as 85". While the resolution of the displays is excellent, the cost is still very high for not only the screen but also the production of 4K content (Lum & Khatri, 2014). In addition, the viewing distance to see the difference in 4K is such that the sign must be very large or the viewer needs to be quite close (Navigating the, 2014).

With Ultra HD screens comes the need for 4K or Ultra HD content. Almost every major digital signage provider now offers a 4K media player. For example, BrightSign's player delivers native 4K H.265 10 bit video files at 60 fps. The player uses HDMI 2.0, has a solid state drive and costs about $1,000 (World's First, 2014).

Curved and transparent screens are also being further developed. For example NanoLumens makes NanoCurve displays, which use LED technology (NanoCurve, n.d). LG has a 55" curved OLED screen, the EA9800. Because OLED screens don't require a backlight, they can very thin (LG Showcases, 2013).

As mentioned at the beginning of this chapter, digital signage is everywhere—fast food restaurants, schools, hospitals, and on the side of the road. These days, though, digital signage is being used in some new ways. For example, digital menu boards are common in fast food restaurants but digital signage is less common in other restaurants. Chili's Grill and Bar, a "fast casual" restaurant chain has installed digital signage at tables that allow customers to order drinks and dessert, and pay their bills without speaking to a server. When tested at 180 restaurants, tips and dessert orders went up, while tables turned over faster. The signs also offer games for children at $.99 a pop. Chili's plans to install the signs throughout its U.S. stores by the end of 2014 (Wong, 2013).

Pizza Hut has gone a little further by making the table itself an interactive digital sign. Customers use the table interface to design their pizza and order it. Then they can play games until the pizza arrives. The table can also use NFC to recognize a customer by his or her cell phone (Pizza Hut Puts, 2014).

Current Status

The digital signage industry is growing steadily. Total industry revenues worldwide in 2013 were $13.9 billion, 5.6% more than in 2012. Industry revenues are expected to grow to $17.1 billion by 2017 (Global Digital Signage, 2013). Figure 10.3 shows the growth in revenue from 2011 to 2016.

Figure 10.3
Global Digital Signage Revenue Forecast (Billions of US Dollars)

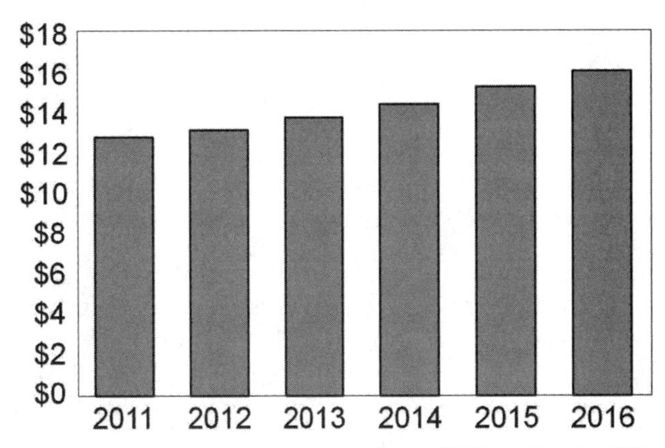

Source: IHS Inc., October 2013

Lyle Bunn's Digital Signage Industry Status 2014 Outlook Report for 2014 noted that 20 million flat panel displays were in use as of the end of 2013, and 2 million of these displays are added each year. In addition, the report noted that ad revenues for digital signage were $1 billion in 2013 with half of the total coming from outdoor digital signage (Bunn, 2014).

As signage is growing, more people are seeing signs and reacting. Research from Arbitron found that 68% of people see a digital sign every day. 44% of people have seen a digital billboard in the past month, while 32% have seen one that week. 37% of bus, taxi, subway and commuter rail users had seen a digital billboard in the past month with 20% having seen one in the past week (Williams, 2013). The study also found that 20% of people who had seen a digital sign had made an unplanned purchase because of the advertising on the sign.

Studies of the effectiveness of digital signage have found that it beats out radio, magazines, traditional billboards, and television for advertising recall (SignAd Network, n.d.).

The list of industries that use digital signage is broad, including corporate communications, quick service restaurants, retail, banking and finance, government, education, automotive, entertainment, healthcare, transportation, hotel and casino, and DOOH/Outdoor (Scala, n.d.).

Factors to Watch

The digital signage industry continues to mature and develop. Trends worth noting include increased interactivity with touch; multi-touch and gesture; more curved, transparent and Ultra HD displays; increased use of mobile technologies; and increased consolidation within the industry.

Interactive features continue to develop at a fast pace, with the challenge being to design signage that effectively utilizes these features.

Whether or not Ultra HD displays take off, they are a very hot topic in the industry as of mid 2014. What remains to be seen is whether people will notice or care about perceived resolution changes, and whether the technology will mature enough to reduce the price of these displays.

The use of mobile technologies like NFC will increase, but. NFC has a challenger. Bluetooth Low Energy (BLE) is another short range wireless communication technology incorporated into some mobile phones and devices. Apple supports BLE and incorporated it into iOS7 with iBeacon. Android devices using version 4.3 or higher also support BLE.

BLE devices listen for BLE beacons that are attached to signs, walls, etc. The range of BLE is much greater than NFC. NFC's range is a few centimeters as opposed to BLE's range of several meters. Some are calling for the demise of NFC although this seems unlikely considering Clear Channel Outdoors' commitment to NFC with Connect (Gurley, 2014).

Finally, the digital signage industry began to see consolidation in 2014. Barco bought X2O Media, and Broadcast International and Wireless Ronin Technologies merged. Some Industry experts see these early signs as a prelude to massive industry consolidation. "Keith Kelsen, author and a longtime digital signage analyst/consultant/executive, indicated that these tremors were early signs that the industry is about to shift dramatically" (Hall, 2014). There are also a number of rumors that Google is going to enter the market.

Whatever happens, digital signage will continue to grow and develop into a mature industry. Innovative methods of reaching audiences will be incorporated into signage, reaching people out of home and through their mobile devices.

Digital Signage Visionary: John Underkoffler

John Underkollfer is best known as the developer of the G-Speak Gestural Technology System, a form of human machine communication best known as the one featured in the film *Minority Report*. Now the fantasy of being able to move data with gestures is a reality.

A Ph.D. from MIT, Underkollfer is now CEO and Chief Scientist for Oblong Industries. The company is known for g-speak and next generation computing interfaces. G-speak is more than simple gesture systems "...they do much more than simply allow users to point and gesture to interact with computers. (And, in fact, gestures themselves are optional.) These are spatial, networked, multi-user, multi-screen, multi-device computing environments" (Our Story, 2014).

G-speak is now used in a multitude of industries including motion pictures, aerospace, video games, shipping, and medical imaging (Inventor of the week, n.d.).

Besides his work at Oblong Industries, Underkoffler is an adjunct professor in the USC School of Cinematic Arts. He has worked as a special advisor on films *Iron Man*, *The Hulk* and *Aeon Flux* in addition to *Minority Report*.

Bibliography

A Brief History of the Sign Industry. (1976). American Sign Museum. Retrieved from http://www.signmuseum.org/a-brief-history-of-the-sign-industry/.

Augmented Reality. (n.d.). Mashable. Retrieved from http://mashable.com/category/augmented-reality/.

Augmented Reality Digital Signage Takes on Pirates in Oz. (2014). Digital Signage Today. Retrieved from http://www.digitalsignagetoday.com/article/229449/Augmented-reality-digital-signage-takes-on-pirates-in-Oz.

Bellis, M. (n.d.). Liquid Crystal Display—LCD. Retrieved from http://inventors.about.com/od/lstartinventions/a/LCD.htm.

Brannon, M. (2009). The Astrodome Scoreboard. Houston Examiner. Retrieved from http://www.examiner.com/article/the-astrodome-scoreboard.

Bunn, L. (2014). Digital Signage Industry Status & 2014 Outlook. Retrieved from http://www.lylebunn.com/Documents/Office365PDFViewer.aspx?file=http://www.lylebunn.com/Documents/2014%20Outlook%20for%20Digital%20Signage%20-%20BUNN.pdf.

Coca-Cola Company. (2012). Sing For Me Vending Machine. Retrieved from http://www.coca-colacompany.com/videos/coca-cola-sing-for-me-vending-machine-ytcjeys8zuxgc.

Dell, L. (2013). Why Dayparting Must Be Part of Your Mobile Strategy. Mashable. Retrieved http://mashable.com/2013/08/14/dayparting-mobile/.

Digital Screenmedia Association. (n.d.). Frequently Asked Questions. Retrieved from http://www.digitalscreenmedia.org/faqs#faq15.

Digital Signage Federation. (n.d.). Digital Signage Glossary of Terms. Retrieved from http://www.digitalsignagefederation.org/glossary#D.

Digital Signage Federation. (2011). Digital Signage Privacy Standards. Retrieved from http://www.digitalsignagefederation.org/Resources/Documents/Articles%20and%20Whitepapers/DSF%20Digital%20Signage%20Privacy%20Standards%2002-2011%20(3).pdf.

Digital Signage Federation. (2014). Clear Channel Launces Interative Mobile Advertising Platform. Retrieved from http://www.digitalsignageconnection.com/clear-channel-launches-global-mobile-interactive-advertising-platform-990.

Digital Signage Powers Next-Gen Vending Machines. (2014). Digital Signage Today. Retrieved from http://www.digitalsignagetoday.com/article/230145/Digital-signage-powers-next-gen-vending-machines.

Eggert, D. (2014, February 1). Mich law may limit growth of digital billboards. *Washington Post.* Retrieved from http://www.washingtontimes.com/news/2014/feb/1/mich-law-may-limit-growth-of-digital-billboards/?page=all.

Federal Highway Administration (n.d.) Outdoor Advertising Control. Retrieved from http://www.fhwa.dot.gov/real_estate/practitioners/oac/oacprog.cfm.

Foot Locker Digital Signage Lets Shoppers Design Their Own Footware. (2013). Digital Signage Today. Retrieved from http://www.digitalsignagetoday.com/article/220097/Foot-Locker-digital-signage-lets-shoppers-design-own-New-Balance.

Global Digital Signage Market Nears $14 Billion Spurred by Real Time Analytics in Retail. (2013). IHS. Retrieved from http://press.ihs.com/press-release/design-supply-chain-media/global-digital-signage-market-nears-14-billion-spurred-real-.

Gurley, S. (2014). NFC vs. iBeacon: How Do They Really Stack Up? Retrieved from http://www.digitalsignagetoday.com/blog/12311/NFC-vs-iBeacon-How-do-they-really-stack-up.

Hall, C. (2013a). Pepsi Soda Machine Trades Soda for Facebook "likes". *Digital Signage Today.* Retrieved from http://www.digitalsignagetoday.com/article/214361/Pepsi-Like-Machine-trades-soda-for-Facebook-likes-Video.

Hall, C. (2013b). Digital Signage Casts a Spell on Commuters in NYC. Digital Signage Today. Retrieved from http://www.digitalsignagetoday.com/article/220589/Digital-signage-casts-a-spell-on-subway-commuters-in-NYC.

Hall, C. (2014). Is the Digital Signage Industry Finally Starting to Consolidate? Digital Signage Today. Retrieved from http://www.digitalsignagetoday.com/article/229777/Is-the-digital-signage-industry-finally-starting-to-consolidate.

History of OOH. (n.d.). Outdoor Advertising Association of America. Retrieved from http://www.oaaa.org/outofhomeadvertising/historyofooh.aspx.

Intel. (2014). Intel Audience Impression Metric Suite. Retrieved from http://www.intel.com/content/www/us/en/retail/retail-aim-suite.html.

Inventor of the Week. (n.d.). Massachusetts Institute of Technology. Retrieved from http://web.mit.edu/invent/iow/underkoffler.html.

Ion, F. (2013). From touch displays to the Surface: A brief history of touchscreen technology. Ars Technica. Retrieved from http://arstechnica.com/gadgets/2013/04/from-touch-displays-to-the-surface-a-brief-history-of-touchscreen-technology/2/.

Kelsen, K. (2010). *Unleashing the Power of Digital Signage.* Burlington, MA: Focal Press.

NanoCurve. (n.d.). NanoLumens. Retrieved from http://www.nanolumens.com/display-solutions/nanocurve/.

National Park Service. (n.d.). Preservation Briefs 24-34. Technical Preservation Services. U.S. Department of the Interior.

Navigating the Seas of Emerging Digital Signage Tech. (2014). Digital Signage Today. Retrieved from http://www.digitalsignagetoday.com/article/230191/Navigating-the-seas-of-emerging-digital-signage-tech.

New Touchscreen Airport Kiosk Features Facial Recognition. (2014). Digital Signage Today. Retrieved from http://www.digitalsignagetoday.com/article/230115/New-touchscreen-aiport-kiosk-features-facial-recognition.

LG Showcases First OLED TV With Curved Screen. (2014). Digital Signage Connection. Retrieved from http://www.digitalsignageconnection.com/lg-showcases-first-oled-tv-curved-screen-913.
Lum, K., & Khatri, S. (2014). 2014 Digital Signage Expo Overview and New Product Launches. IHS Technology. Retrieved from http://technology.ihs.com/492224/2014-digital-signage-expo-overview-and-new-product-launches.
Our Story. (2014). Obling Industries. Retrieved from http://www.oblong.com/our-story/.
Pepsi Max Excites Londoners With Augmented Reality DOOH First. (2014). Grand Visual. Retrieved from http://www.grandvisual.com/pepsi-max-excites-londoners-augmented-reality-digital-home-first/.
Pizza Hut Puts Digital Signage on the Table. (2014). Digital Signage Today. Retrieved from http://www.digitalsignagetoday.com/article/228965/Pizza-Hut-puts-digital-signage-on-the-table?rc_id=166.
Rehabilitation. (2014). GestureTek Health. Retrieved from http://www.gesturetekhealth.com/products-rehab.php.
Richtel, M. (2010, March 2). Driven to Distraction. New York Times. Retrieved from http://www.nytimes.com/2010/03/02/technology/02billboard.html?pagewanted=all&_r=0.
Scala. (n.d.). Discover. Retrieved from http://scala.com/discover/.
SignAd Network. (n.d.). Why. Retrieved from http://www.signadnetwork.com/why-digital-signage/.
Times Square Digital Billboard Transforms Into Interactive Advertainment. (2013). Digital Signage Connection. Retrieved from http://www.digitalsignageconnection.com/times-square-digital-billboard-transforms-interactive-advertainment-651.
Williams, D. (2013). Arbitron Out of Home Advertising Study. Retrieved from http://www.lamar.com/howtoadvertise/~/media/7845006D467B432592A4A1867BBFBA7D.ashx.
Wong, V. (2013). That Table on the Table Will Make You Spend More. Business Week. Retrieved from http://www.businessweek .com/articles/2013-09-17/that-digital-tablet-on-the-restaurant-table-will-make-you-spend-more.
World's First True 4K Player. (2014). Brightsign. Retrieved from http://www.brightsign.biz/digital-signage-products/4k-product-line/4k1240/.

11

Cinema Technologies

Michael R. Ogden, Ph.D. and
Maria Sanders, M.F.A.*

Why Study Cinema Technologies?

- A typical feature-length motion picture costs millions of dollars to make and requires the skills of hundreds of people.
- New digital cinema production as well as projection technologies are changing the workflow and distribution of movies.
- Cinema is the world's most lucrative cultural industry, and watching movies is one of the most popular cultural practices.
- In the U.S. alone, movie box office receipts totaled nearly $11 billion in 2013 and are expected to top $2 trillion globally by 2016.

Introduction

Storytelling is a universally human endeavor. In his book, Tell Me A Story: Narrative and Intelligence (1995), computer scientist and cognitive psychologist Roger Schank conjectures that not only do humans think in terms of stories; our very understanding of the world is in terms of stories we already understand. "We tell stories to describe ourselves not only so others can understand who we are but also so we can understand ourselves... We interpret reality through our stories and open our realities up to others when we tell our stories" (Schank, 1995, p. 44).

With the advent of mass media—and, in particular, modern cinema—the role of storyteller in popular culture was subsumed by the "Culture Industry" (c.f. Horkheimer & Adorno, 1969; Adorno, 1975; Andrae 1979), an apparatus for the production of meanings and pleasures involving aesthetic strategies and psychological processes (Neale 1985) and bound by its own set of economic and political determinants and made possible by contemporary technical capabilities.

In other words, cinema functions as a social institution providing a form of social contact desired by citizens immersed in a world of "mass culture." It also exists as a psychological institution whose purpose is to encourage the movie-going habit by providing the kind of entertainment desired by popular culture (Belton, 2013).

Simultaneously, cinema evolved as a technological institution, its existence premised upon the development of specific technologies, most of which originated during the course of industrialization in Europe and America in the late 19th and early 20th Centuries (*e.g.*, film, the camera, the projector and sound recording). The whole concept and practice of filmmaking evolved into an art form dependent upon the mechanical reproduction and mass distribution of "the story"—refracted through the aesthetics of framing, light and shade, color, texture, sounds, movement, the shot/counter-shot, and the *mise en scène* of cinema.

"[T]here is something remarkable going on in the way our culture now creates and consumes entertainment and media," observes Steven Poster, National President of the International Cinematographers Guild (2012, p. 6).

The name most associated with film, Kodak, after 131 years in business as one of the largest producers of film stock in the world, filed for Chapter 11 bankruptcy in 2012 (De La Merced, 2012). The company was hit

* Ogden is Professor & Director, Film & Video Studies Program. Sanders is Assistant Professor of Film & Video Studies, Central Washington University (Ellensburg, Washington).

hard by the recession and the advent of the RED cameras (One, Epic & Scarlet), the Arri Alexa, and other high-end digital cinema cameras. In the digital world of bits and bytes, Kodak became just one more 20th Century giant to falter in the face of advancing technology.

Perhaps it was inevitable. The digitization of cinema began in the 1980s in the realm of special visual effects. By the early 1990s, digital sound was widely propagated in most theaters, and digital nonlinear editing began to supplant linear editing systems for post-production. "By the end of the 1990s, filmmakers such as George Lucas had begun using digital cameras for original photography and, with the release of *Star Wars Episode I: The Phantom Menace* in 1999, Lucas spearheaded the advent of digital projection in motion picture theatres" (Belton, 2013, p. 417).

As if lamenting the digital shift in cinematic storytelling, Poster (2012) states that, "…this technological overlay I'm talking about has made a century-old process (light meters, film dailies, lab color timing) that was once [invisible] to everyone but the camera team, visible and accessible to virtually anyone… It is a loud wake-up call for the entire industry" (p. 6).

As filmmakers grapple with the transition from film to digital image acquisition, another shift is in the works. Video capable digital single lens reflex (DSLR) cameras equipped with large image sensors; digital cinema cameras with increasing color depth and dynamic range; emerging standards in 2D, 3D, 2K, 4K and high frame rate (HFR) Digital Cinema Packages (DCP) for theater distribution and projection; all promise to provide moviegoers a more immersive cinema experience. These developments hopefully, will re-energize moviegoers, lure them back into the movie theaters, and increase profits.

As Mark Zoradi, President of Walt Disney Studios Motion Pictures Group stated, "The key to a good film has always been story, story, story; but in today's environment, it's story, story, story and blow me away" (cited in Kolesnikov-Jessop 2009).

Background

However, until recently "no matter how often the face of the cinema…changed, the underlying structure of the cinematic experience…remained more or less the same" (Belton, 2013, p. 6). Yet, even that which is most closely associated with cinema's identity—sitting with others in a darkened theater watching "larger-than-life" images projected on a big screen—was not always the norm.

From Novelty to Narrative

The origins of cinema as an independent medium lie in the development of mass communication technologies evolved for other purposes (Cook, 2004). Specifically, photography (1826-1839), roll film (1880), the Kodak camera (1888), George Eastman's motion picture film (1889), the motion picture camera (1891-1893), and the motion picture projector (1895-1896) each had to be invented in succession for cinema to be born.

Early experiments in series photography for capturing motion were an important precursor to cinema's emergence. In 1878, Eadweard Muybridge set up a battery of cameras triggered by a horse moving through a set of trip wires. Adapting a Zoëtrope (a parlor novelty of the era) for projecting the photographs, Muybridge arranged his photograph plates around the perimeter of a disc that was manually rotated. Light from a "Magic Lantern" projector was shown through each slide as it stopped momentarily in front of a lens. The image produced was then viewed on a large screen (Neale, 1985). If rotated rapidly enough, a phenomenon known as *persistence of vision* (an image appearing in front of the eye lingers a split second in the retina after removal of the image), allowed the viewer to experience smooth, realistic motion.

Muybridge called his apparatus the Zoopraxiscope, which was used to project photographic images in motion for the first time to the San Francisco Art Association in 1880 (Neale, 1985).

In 1882, French physiologist and specialist in animal locomotion, Étienne-Jules Marey, invented the Chronophotographic Gun in order to take series photographs of birds in flight (Cook, 2004). Shaped like a rifle, Marey's camera took 12 instantaneous photographs of movement per second, imprinting them on a rotating glass plate coated with a light-sensitive emulsion. A year later, Marey switched from glass plates to paper roll film. But like Muybridge, "Marey was not interested in cinematography… In his view, he had invented a machine for dissection of motion similar to Muybridge's apparatus but more flexible, and he never intended to project his results" (Cook, 2004, p. 4).

In 1887, Hannibal Goodwin, an Episcopalian minister from New Jersey, first used celluloid roll film as a base for light-sensitive emulsions. George Eastman later appropriated Goodwin's idea and in 1889, began to mass-produce and market celluloid roll film on what would eventually become a global scale (Cook, 2004). Neither Goodwin nor Eastman were initially interested in motion pictures. However, it was the introduction of this durable and flexible celluloid film, coupled with the technical breakthroughs of Muybridge and Marey, that inspired Thomas Edison to attempt to produce recorded moving images to accompany the recorded sounds of his newly-invented phonograph (Neale, 1985). It is interesting to note that, according to Edison's own account (cited in Neale, 1985), the idea of making motion pictures was never divorced from the idea of recording sound. "The movies were intended to talk from their inception so that, in some sense, the silent cinema represents a thirty-year aberration from the medium's natural tendency toward a total representation of reality" (Cook, 2004, p. 5).

Capitalizing on these innovations, W.K.L. Dickson, a research assistant at the Edison Laboratories, invented the first authentic motion picture camera, the Kinetograph—first constructed in 1890 with a patent granted in 1894. The basic technology of modern film cameras is still nearly identical to this early device. All film cameras, therefore, have the same five basic functions: a "light tight" body that holds the mechanism which advances the film and exposes it to light; a motor; a magazine containing the film; a lens that collects and focuses light on to the film; and a viewfinder that allows the cinematographer to properly see and frame what he or she is photographing (Freeman, 1998).

Thus, using Eastman's new roll film, the Kinetograph advanced each frame at a precise rate through the camera, thanks to sprocket holes that allowed metal teeth to grab the film, advance it, and hold the frame motionless in front of the camera's aperture at split-second intervals. A shutter opened, exposing the frame to light, then closed until the next frame was in place. The Kinetograph repeated this process 40 times per second. Throughout the silent era, other cameras operated at 16 frames per second; it wasn't until sound was introduced that 24 frames per second became standard, in order to improve the quality of voices and music (Freeman, 1998). When the processed film is projected at the same frame rate, realistic movement is presented to the viewer.

However, for reasons of profitability alone, Edison was initially opposed to projecting films to groups of people. He reasoned (correctly, as it turned out) that if he made and sold projectors, exhibitors would purchase only one machine from him—a projector—instead of several Kinetoscopes (Belton, 2013) that allowed individual viewers to look at the films through a magnifying eyepiece. By 1894, Kinetographs were producing commercially viable films. Initially the first motion pictures (which cost between $10 and $15 each to make) were viewed individually through Edison's Kinetoscope "peep-shows" for a nickel apiece in arcades (called Nickelodeons) modeled on the phonographic parlors that had proven so successful for Edison (Belton, 2013).

It was after viewing the Kinetoscope in Paris that the Lumière brothers, Auguste and Louis, began thinking about the possibilities of projecting films on to a screen for an audience of paying customers. In 1894, they began working on their own apparatus, the Cinématograph. This machine differed from Edison's machines by combining both photography and projection into one device at the much lower (and thus, more economical) film rate of 16 frames per second. It was also much lighter and more portable (Neale, 1985).

In 1895, the Lumière brothers demonstrated their Cinématograph to the *Société d'Encouragement pour l'Industries Nationale* (Society for the Encouragement of National Industries) in Paris. The first film screened was a short actuality film of workers leaving the Lumière factory in Lyons (Cook, 2004). The actual engineering contributions of the Lumière brothers were quite modest when compared to that of W.K.L. Dickson—they merely synchronized the shutter movement of the camera with the movement of the photographic film strip. Their real contribution is in the establishment of cinema as an industry (Neale, 1985).

As pointed out by Jaques Deslandes in his 1966 book, *Histoire Comparée du Cinéma* (Comparative History of Cinema), "This is what explains the birth of the cinema show in France, in England, in Germany, in the United States... Moving pictures were no longer just a laboratory experiment, a scientific curiosity, from now on they could be considered a commercially viable public spectacle" (cited in Neale, 1985, p. 48).

The early years of cinema were ones of invention and exploration. The tools of visual storytelling, though crude by today's standards, were in hand,

and the early films of Edison and the Lumiére brothers were fascinating audiences with actuality scenes—either live or staged—of everyday life. However, an important pioneer in developing film narrative was Alice Guy Blaché. Remarkable for her time, Guy Blaché was arguably the first director of either sex to bring a story-film to the screen with the 1896 release of her one-minute film, *La Fée aux Choux* (The Cabbage Fairy) that preceded the story-films of Georges Méliès by several months.

Figure 11.1

First Publicly Projected Film: Sortie des Usines Lumière à Lyon, 46 seconds, 1895

Source: Screen capture courtesy Lumière Institute

If the cameras and projectors were the hardware, Guy Blaché was film's first software designer. From 1896 to 1920 she wrote and directed hundreds of short films (including over 100 synchronized sound films and 22 feature films) and produced hundreds more (McMahan, 2003). In the first half of her career, as head of film production for the Gaumont Company (where she was first employed as a secretary), Guy Blaché would almost single-handedly develop the art of cinematic narrative (McMahan, 2009) with an emphasis on storytelling to create meaning.

To pursue her career, Alice Guy Blaché had to overcome the confines of a rigid social structure that barely tolerated women in leadership roles, but she persevered to become the first—and so far only—woman to own and run her own film studio; The Solax Studio in Fort Lee, NJ from 1910 to 1914 (McMahan, 2003). In 1922, she divorced her philandering husband—who she had trained in directing—Herbert Blaché, the director of Buster Keaton's first feature film, The *Saphead* (1920). After the divorce, she auctioned off her film studio, returned to France, and never made another film (McMahan, 2009).

By the turn of the century, film producers were beginning to assume greater editorial control over the narrative, making multi-shot films and allowing for greater specificity in the story line (Cook, 2004). Such developments are most clearly apparent in the work of Georges Méliès. A professional magician who owned and operated his own theater in Paris, Méliès was an important early filmmaker, developing cinematic narrative demonstrating a created, cause-and-effect reality. Méliès created and employed a number of important narrative devices, such as the fade-in and fade-out, "lap" (overlapping) dissolve as well as impressive visual effects such as stop-motion photography (Parce qu'on est des geeks! 2013). Though he didn't employ much editing within individual scenes, the scenes were connected in a way that supported a linear, narrative reality.

Figure 11.2

Méliès, Le Voyage Dans La Lune, 1902, 13 minutes

Source: Screen capture, M.R. Ogden.

By 1902, with the premiere of his one-reel film *Le Voyage Dans La Lune* (A Trip to the Moon), Méliès was fully committed to narrative filmmaking. Unfortunately, Méliès became embroiled in two lawsuits with Edison concerning issues of compensation over piracy and plagiarism of his 1902 film. Although he remained committed to his desire of "capturing dreams through cinema" (Parce qu'on est des geeks! 2013) until the end of his filmmaking career—and produced several other ground-breaking films (*Les Hallucinations Du Baron de Münchhausen*, 1911, and *A La Conquête Des Pôles*, 1912)—his legal battles left him embittered and by 1913, Méliès abandoned filmmaking and returned to performing magic.

Middle-class American audiences, who grew up with complicated plots and fascinating characters from such authors as Charles Dickens and Charlotte Brontë, began to demand more sophisticated film narratives. Directors like Edwin S. Porter and D.W. Griffith began crafting innovative films in order to provide their more discerning audiences with the kinds of stories to which theatre and literature had made them accustomed (Belton, 2013).

Influenced by Méliès, American filmmaker Edwin S. Porter is credited with developing the "invisible\"technique" of continuity editing. By cutting to different angles of a simultaneous event in successive shots, the illusion of continuous action was maintained. Porter's *Life of an American Fireman* and *The Great Train Robbery*, both released in 1903, are the foremost examples of this new style of storytelling through crosscutting (or, intercutting) multiple shots depicting parallel action (Cook, 2004).

Figure 11.3

Porter, The Great Train Robbery, 1903, 11 minutes

Source: Screen capture, M.R. Ogden.

Taking this a step further, D.W. Griffith, who was an actor in some of Porter's films, went on to become one of the most important filmmakers of all time, and truly the "father" of modern narrative form. Technologically and aesthetically, Griffith advanced the art form in ways heretofore unimagined. He altered camera angles, employed close-ups, and actively narrated events, thus shaping audience perceptions of them. Additionally, he employed "parallel editing"—cutting back and forth from two or more simultaneous events taking place in separate locations—to create suspense (Belton, 2013).

Even though Edison's Kinetograph camera had produced more than 5,000 films (Freeman, 1998), by 1910, other camera manufacturers such as Bell and Howell, and Pathé (which acquired the Lumière patents in 1902) had invented simpler, lighter, more compact cameras that soon eclipsed the Kinetograph. In fact, "it has been estimated that, before 1918, 60% of all films were shot with a Pathé camera" (Cook, 2004, p. 42).

Nearly all of the cameras of the silent era were hand-cranked. Yet, camera operators were amazingly accurate in maintaining proper film speed (16 fps) and could easily change speeds to suit the story. Cinematographers could crank a little faster (over-crank) to produce slow, lyrical motion, or they could crank a little slower (under-crank) and when projected back at normal speed, they displayed the frenetic, sped-up motion apparent in the silent slapstick comedies of the Keystone Film Company (Cook, 2004).

By the mid-1920s, the Mitchell Camera Corporation began manufacturing large, precision cameras that produced steadier images than previously possible. These cameras became the industry standard for almost 30 years until overtaken by Panavision cameras in the 1950s (Freeman, 1998).

Figure 11.4

Mitchell Standard Model A 35mm camera, circa 1920s

Source: mitchellcamera.com.

In the United States, the early years of commercial cinema were tumultuous as Edison sued individuals and enterprises over patent disputes in an attempt to protect his monopoly and his profits (Neale, 1985). However, by 1908, the film industry was becoming more stabilized as the major film producers "banded together to form the Motion Picture Patents Company (MPPC) which sought to control all aspects of motion picture production, distribution and exhibition" (Belton, 2013, p. 12) through its control of basic motion picture patents.

In an attempt to become more respectable, and to court middle-class customers, the MPPC began a campaign to improve the content of motion pictures by engaging in self-censorship to control potentially offensive content (Belton, 2013). The group also provided half-price matinees for women and children and improved the physical conditions of theaters. Distribution licenses were granted to 116 exchanges that could distribute films only to licensed exhibitors who paid a projection license of two dollars per week.

Unlicensed producers and exchanges continued to be a problem, so in 1910 the MPPC created the General Film Company to distribute their films. This development proved to be highly profitable and "was…the first stage in the organized film industry where production, distribution, and exhibition were all integrated, and in the hands of a few large companies" (Jowett, 1976, p. 34) presaging the emergence of the studio system 10 years later.

The Studio System

For the first two decades of cinema, nearly all films were photographed outdoors. Many production facilities were like that of George Méliès, who constructed a glass-enclosed studio on the grounds of his home in a Paris suburb (Cook, 2004). However, Edison's Black Maria facility in New Jersey was probably the most famous site. Eventually, the industry outgrew these small, improvised facilities and moved to California, where the weather was conducive to outdoor productions. Large soundstages were also built in order to provide controlled staging and more control over lighting.

By the second decade of the 20th Century, dozens of movie studios were operating in the U.S. and across the world. A highly specialized industry grew in southern California, honing sophisticated techniques of cinematography, lighting, and editing. The Hollywood studios divided these activities into preproduction, production, and post-production. During preproduction, a film was written and planned. The production phase was technology intensive, involving the choreography of actors, cameras and lighting equipment. Post-production consisted of editing the films into coherent narratives and adding titles—in fact, film editing is the only art that is unique to cinema.

Figure 11.4

Edison's Black Maria, World's First Film Studio, circa 1890s

Source: Wikimedia Commons.

The heart of American cinema was now beating in Hollywood, and the institutional machinery of filmmaking evolved into a three-phase business structure of production, distribution, and exhibition to get their films from studios to theater audiences. Although the MPPC was formally dissolved in 1918 as a result of an antitrust suit initiated in 1912 (Cook, 2004), powerful new film companies, flush with capital, were emerging. With them came the advent of vertical integration.

Through a series of mergers and acquisitions, formerly independent production, distribution, and exhibition companies congealed into major studios; Paramount, Metro-Goldwin-Mayer (MGM), Warner Bros, and Fox. "All of the major studios owned theater chains; the minors—Universal, Columbia, and United Artists—did not" (Belton, 2013, p. 68), but distributed their pictures by special arrangement to the theaters owned by the majors. The resulting economic system was quite efficient. "The major studios produced from 40 to 60 pictures a year… [but in 1945 only] owned 3,000 of the 18,000 theaters around the country… [yet] these theaters generated over 70% of all box-office receipts" (Belton, 2013, p. 69).

Figure 11.5
Cinema History Highlights

Late 1800s
- Early experiments between 1870s & 1980s in motion photography by Muybridge & Marey inspire Thomas Edison to invent the **Kinetograph** camera in 1894. **Kinetoscope** 'peep-shows' become popular entertainment.
- Eastman markets first commercially available transparent roll film in 1889.
- In 1895, the Lumière brothers demonstrated their **Cinématograph**.

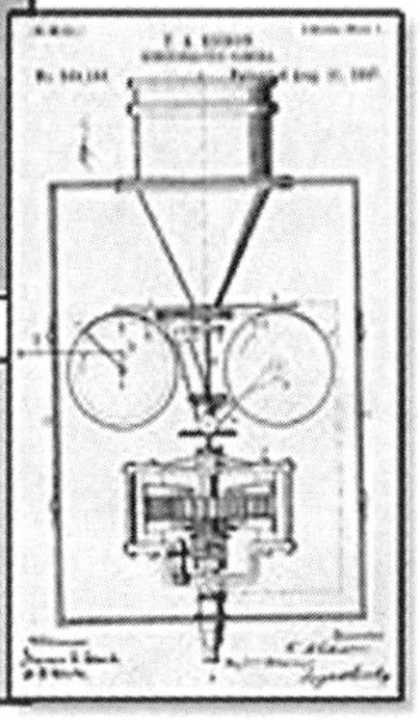

1900–1920s
- Georges Méliès, Edwin S. Porter & D.W. Griffith develop and perfect narrative film techniques enhancing storytelling.
- Pathé camera (based on Lumière patents) becomes the dominant camera.
- In 1927, Warner Brothers released *The Jazz Singer*, with synchronized dialog & music using their **Vitaphone** process .

1930s–1940s
- Invented in the 1920s, the **Moviola** becomes the dominant film editing device in the 1930 through the 1970s (with successive improvements).
- Disney uses the **Technicolor** three-color process in his *Silly Symphonies* cartoon series.
- "Deep focus" cinematography demonstrated by Orson Welles & Gregg Toland in the 1941 film *Citizen Kane*.

1950s–1960s
- **Cinerama** (1952) launches a widescreen revolution.
- **Eastman Color** used on *The Robe* (1953), by 1955, most widescreen films photographed using Eastman Color film.
- Network TV introduced in the United States in 1949, film box office receipts plummet.

1970s–1980s
- 1976 *Futureworld* features 1st use of 3D Computer Generated Imagery (CGI).
- 1970s, Dolby Labs introduced Dolby noise reduction & Dolby Stereo, by 1984 **Dolby 5.1** surround sound technology
- 1980s saw studio consolidations & rise of VHS rental market.

1990s–2000s
- 1st all-CGI animated film, *Toy Story* (1995), released by Pixar.
- *Star Wars Episode II: Attack of the Clones* (2002) is 1st completely digital film (acquired, distributed & projected).
- 2009 James Cameron releases the **digital 3D** blockbuster *Avatar* which grosses $3 billion worldwide.

2010s–Beyond
- Rapid declines in home DVD revenue in 2011 force Hollywood to adopt a new business model—switch to on-demand services & accelerate delivery of movies over the Internet.
- Digital film projection overtakes 35mm film in early 2012, anticipated by 2015, 35mm film will only be 17% of market.
- Filmmakers push for 4K & higher frame rates in movie acquisition/exhibition.

Source: M.R. Ogden

As films and their stars increased in popularity, and movies became more expensive to produce, studios began to consolidate their power, seeking to control each phase of a film's life. However, since the earliest days of the Nickelodeons, moralists and reformers had agitated against the corrupting nature of the movies and their effects on American youth (Cook, 2004). A series of scandals involving popular movie stars in the late 1910s and early 1920s resulted in ministers, priests, women's clubs, and reform groups across the nation encouraging their membership to boycott the movies.

In 1922, frightened Hollywood producers formed a self-regulatory trade organization—the Motion Picture Producers and Distributors of America (MPPDA). By 1930, the MPPDA adopted the rather draconian Hayes Production Code. This "voluntary" code, intended to suppress immorality in film, proved mandatory if the film was to be screened in America (Mondello, 2008). Although the code aimed to establish high standards of performance for motion-picture producers, it "merely provided whitewash for overly enthusiastic manifestations of the 'new morality' and helped producers subvert the careers of stars whose personal lives might make them too controversial" (Cook, 2004, p. 186).

Sound, Color and Spectacle

Since the advent of cinema, filmmakers hoped for the chance to bring both pictures and sound to the screen. Although the period until the mid-1920s is considered the silent era, few films in major theaters actually were screened completely silent. Pianists or organists—sometimes full orchestras—performed musical accompaniment to the projected images. At times, actors would speak the lines of the characters and machines and performers created sound effects. "What these performances lacked was fully synchronized sound contained within the soundtrack on the film itself" (Freeman, 1998, p. 408).

By the late 1920s, experiments had demonstrated the viability of synchronizing sound with projected film. When Warner Bros Studios released The *Jazz Singer* in 1927, featuring synchronized dialog and music using their Vitaphone process, the first "talkie" was born. Vitaphone was a sound-on-disc process that issued the audio on a separate 16-inch phonographic disc. While the film was projected, the disc played on a turntable indirectly coupled to the projector motor (Bradley, 2005). Other systems were also under development during this time, and Warner Bros Vitaphone process had competition from Movietone, DeForest Phonofilm, and RCA's Photophone.

Figure 11.6

Vitaphone Projection Setup, 1926 Demonstration

Source: Wikimedia Commons.

Though audiences were excited by this new novelty, from an aesthetic standpoint, the advent of sound actually took the visual production value of films backward. Film cameras were loud and had to be housed in refrigerator-sized vaults to minimize the noise; as a result, the mobility of the camera was suddenly limited. Microphones had to be placed very near the actors, resulting in restricted blocking and the odd phenomenon of actors leaning close to a bouquet of flowers as they recited their lines; the flowers, of course, hid the microphone. No question about it, though, sound was here to stay.

Once sound made its appearance, the established major film companies acted cautiously, signing an agreement to only act together. After sound had proved a commercial success, the signatories adopted Movietone as the standard system—a sound-on-film method that recorded sound as a variable-density optical track on the same strip of film that recorded the pictures (Neale, 1985). The advent of the "talkies" launched another round of mergers and expansions in the studio system. By the end of the 1920s, more than 40% of theaters were equipped for sound (Kindersley, 2006), and by 1931, "...virtually all films produced in the United States contained synchronized soundtracks" (Freeman, 1998, p. 408).

Movies were gradually moving closer to depicting "real life." But life isn't black and white, and experiments with color filmmaking had been conducted since the dawn of the art form. Finally, in 1932, Technicolor introduced a practical color process that would slowly revolutionize moviemaking. Though aesthetically beautiful color processes were now available, they were extremely expensive and required strict oversight by the Technicolor Company, which insisted on a policy of strict secrecy during every phase of production. As a result, most movies were still produced in black and white well into the 1950s.

The popularity of television and its competition with the movie industry helped drive more changes. In the early years of film, the 4:3 aspect ratio set by Edison (4 units wide by 3 units high, also represented as 1.33:1) was assumed to be more aesthetically pleasing than a square box); it was the most common aspect ratio for most films until the 1960s (Freeman, 1998).

The impetus for widescreen technology, and its adoption throughout the industry, was that films were losing money at the box office because of television. However, the studios' response was characteristically cautious, initially choosing to release fewer but more expensive films (still in the standard Academy aspect ratio) hoping to lure audiences back to theaters with quality product (Belton, 2013). However, "[it] was not so much the Hollywood establishment…as the independent producers who engineered a technological revolution that would draw audiences back" (Belton, 2013, p. 327).

Because of the efforts of independent filmmakers, the 1950s and early 1960s saw the most pervasive technological innovations in Hollywood since the late 1920s. "A series of processes changed the size of the screen, the shape of the image, the dimensions of the films, and the recording and reproduction of sound" (Bordwell, Staiger & Thompson, 1985, p. 358).

Cinerama (1952) launched a widescreen revolution that would permanently alter the shape of the motion picture screen. Cinerama was a widescreen process that required filming with a three-lens camera and projecting with synchronized projectors onto a deeply curved screen extending the full width of most movie theaters. This viewing (yielding a 146° by 55° angle of view) was meant to approximate that of human vision (160° by 60°) and fill a viewer's entire peripheral vision. Mostly used in travelogue-adventures, such as *This is Cinerama* (1952) and *Seven Wonders of the World* (1956), the first two Cinerama fiction films—*The Wonderful World of the Brothers Grimm* and *How the West Was Won*—were released in 1962, to much fanfare and critical acclaim. However, three-camera productions and three-projector system theaters like Cinerama and CineMiracle (1957) were extremely expensive technologies and quickly fell into disuse.

Single-camera and single projector processes like MGM's Arnoldscope (1953) and Paramount Picture's VistaVision (1954) consisted of shooting—and, initially screening—the film on its horizontal axis (90° to the frame's normal orientation on the film strip) to give a wider and less grainy image (Bordwell, Staiger & Thompson, 1985) were unconventional and fared no better.

Anamorphic processes used special lenses to shoot or print squeezed images onto the film as a wide field of view. In projection, the images were unsqueezed using the same lenses, to produce an aspect ratio of 2.55:1—almost twice as wide as the Academy Standard aspect ratio (Freeman, 1998). When Twentieth Century-Fox released *The Robe* in 1953 using the CinemaScope anamorphic system, it was a spectacular success and just the boost Hollywood needed. Soon, other companies began pro-ducing widescreen films using similar anamorphic processes such as Panascope and Superscope. Nearly all these widescreen systems—including Cinema-Scope—incorporated stereophonic sound reproduction.

Figure 11.7
Film Aspect Ratios

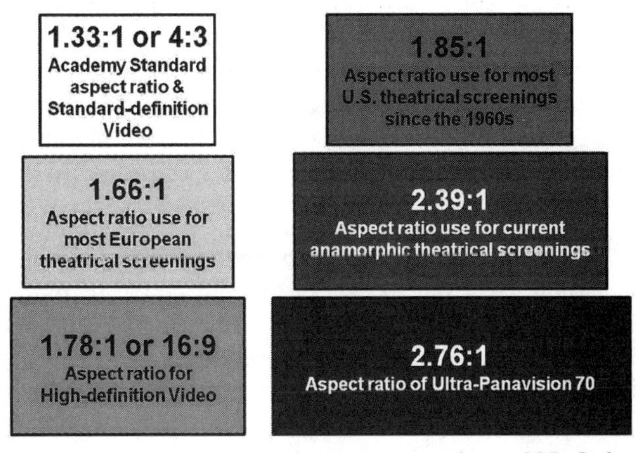

Source: M.R. Ogden

If widescreen films were meant to engulf audiences, pulling them into the action, "3D assaulted audiences—hurling spears, shooting arrows, firing guns, and throwing knives at spectators sitting peacefully in their theatre seats" (Belton, 2013, p. 328). Stereoscopy and three-dimensional (3D) cinematography, until recently, had been relatively unsuccessful at creating depth in an otherwise flat image (Freeman, 1998).

The technology of 3D is rooted in the basic principles of binocular vision. Early attempts at reproducing monochromatic 3D used an anaglyphic system: two strips of film, one tinted red, the other cyan, were projected simultaneously for an audience wearing glasses with one red and one cyan filtered lens (Cook, 2004). When presented with slightly different angles for each eye, the brain processed the two images as a single 3D image. The earliest 3D film using the anaglyphic process was *The Power of Love* in 1922.

In the late 1930s, MGM released a series of anaglyphic shorts, but the development of polarized filters and lenses around the same time permitted the production of full-color 3D images. Experiments in anaglyphic films ceased in favor of the new method. In 1953, Milton Gunzberg released *Bwana Devil*, a "dreadful" film shot using a polarized 3D process called Natural Vision. It drew in audiences and surprisingly broke box office records, grossing over $5 million by the end of its run (Jowett, 1976). Natural Vision employed two interlocked cameras whose lenses were positioned to approximate the distance between the human eyes and record the scene on two separate negatives. In the theater, when projected simultaneously onto the screen, spectators wearing disposable glasses with polarized lenses perceived a single three-dimensional image (Cook, 2004). Warner Bros released the second Natural Vision feature, *House of Wax* (1953), which featured six-track stereophonic sound and was a critical and popular success, returning $5.5 million on an investment of $680,000 (Cook, 2004). "Within a matter of months after the initial release of *Bwana Devil*, more than 4,900 theaters were converted to 3D" (Belton, 2013, p. 329).

Although Hollywood produced 69 features in 3D between 1953 and 1954, most were cheaply made exploitation films. By late 1953, the stereoscopic 3D craze had peaked. Two large budget features shot in Natural Vision, MGM's *Kiss Me Kate* (1953) and Alfred Hitchcock's *Dial M for Murder* (1954) were released "flat" because the popularity of 3D had fallen dramatically. Although 3D movies were still made decades later for special short films at Disney theme parks, 3D was no longer part of the feature-film production process (Freeman, 1998). One reason for 3D's demise was that producers found it difficult to make serious narrative films in such a gimmicky process (Cook, 2004). Another problem was the fact that audiences disliked wearing the polarized glasses; many also complained of eyestrain, headaches and nausea.

But, perhaps, the biggest single factor in 3D's rapid fall from grace was cinematographers' and directors' alternative use of deep-focus widescreen photography—especially anamorphic processes that exploited depth through peripheral vision—and compositional techniques that contributed to the feeling of depth without relying on costly, artificial means. Attempts to revive 3D, until most recently, met with varying degrees of success, seeing short runs of popularity in the 1980s with films like *Friday the 13th, Part III* and *Jaws 3D* (both 1983).

In 1995, with the release of the IMAX 3D film, *Wings of Courage*—and later, *Space Station 3D* in 2002—the use of active display LCD glasses synchronized with the shutters of dual-filmstrip projectors using infrared signals presaged the eventual rise of digital 3D films 10 years later.

Hollywood Becomes Independent

"The dismantling of the studio system began just before World War II when the U.S. Department of Justice's Antitrust Division filed suit against the eight major studios, accusing them of monopolistic practices in their use of block booking, blind bidding, and runs, zones, and clearances" (Belton, 2013, p. 82).

In 1948, the major studios were forced to divorce their operations from one another, separate production and distribution from exhibition, and divest themselves of their theater chains (Belton, 2013). Other factors also contributed to the demise of the studio system, most notably changes in leisure-time entertainment, competition with television, and the rise of independent production (Cook, 2004). Combined with the extreme form of censorship Hollywood imposed upon itself through the Hayes Production Codes—and "after World War II, with competition from TV on the family front, and from foreign films with nudity on the racy front" (Mondello, 2008)—movie studios were unable (or unwilling) to rein in independent filmmakers who chafed under the antiquated Code.

The U.S. Supreme Court ruled in 1952 (the "Miracle decision") that films constitute "a significant medium of communication of ideas" and were therefore protected by both the First and Fourteenth Amendments (Cook, 2004, p. 428). By the early 1960s, supported by subsequent court rulings, films were "guaranteed full freedom of expression" (Cook, 2004, p. 428). The influence of the Hayes Production Code had all but disappeared by the end of the 1960s, replaced by the MPAA ratings system (MPAA, 2011) and instituted in 1968, revised in 1972 and now, in its latest incarnation since 1984.

The major studios delayed the process of divesture as long as possible, but the studio system's previously rigid control over the assembly line of moviemaking was already disintegrating in the 1960s, while up-and-coming independent production companies and the more "free-wheeling" counterculture were poised to influence popular cinema.

Though the 1960s still featured big-budget, lavish movie spectacles, a parallel movement reflected the younger, more rebellious aesthetic of the "baby boomers." Actors and directors went into business for themselves, forming their own production companies, and taking as payment lump-sum percentages of the film's profits (Belton, 2013). The rise and success of independent filmmakers like Arthur Penn (*Bonnie and Clyde*, 1967), Stanley Kubrick (*2001: A Space Odyssey*, 1968), Sam Peckinpah (*The Wild Bunch*, 1969), Dennis Hopper (*Easy Rider*, 1969), and John Schlesinger (*Midnight Cowboy*, 1969), demonstrated that the studio system need not dominate popular filmmaking. Outside of that system, filmmakers were freer to experiment with style and content.

In the early 1960s, an architectural innovation changed the way most people see movies—the move from single-screen theaters (and drive-ins) to multiscreen cineplexes. Although the first multi-screen house with two theaters was built in the 1930s, it was not until the late 1960s that film venues were built with four to six theaters. What these theaters lacked was the atmosphere of the early "movie palaces." While some owners put effort into the appearance of the lobby and concessions area, in most cases the "actual theater was merely functional" (Haines, 2003, p. 91). The number of screens in one location continued to grow; 1984 marked the opening of the first 18-plex theater in Toronto (Haines, 2003).

The next step in this evolution was the addition of stadium seating—offering moviegoers a better experience by affording more comfortable seating with unobstructed views of the screen (EPS Geofoam, n.d.). Although the number of screens in a location seems to have reached the point of diminishing returns, many theaters are now working on improving the atmosphere they create for their patrons. From bars and restaurants to luxury theaters with a reserved $29 movie ticket, many theater owners are once again working to make the movie-going experience something different from what you can get at home (Gelt & Verrier, 2009).

Recent Developments

In August 1895, Kinetoscope audiences were shocked to see the head of the queen chopped off in Edison Studio's historical dramatization, *The Execution of Mary, Queen of Scots* (1895). This simple, one-minute long depiction employed not only the first film edit, but "...ostensibly the first 'visual effect' in cinema [and] would, more than a century later, lead to an industry dominated by films with visual effects" (Rogers, 2014, p. 60).

Since the subsequent films of Georges Méliès thrilled audiences with even more inventive cinematic "tricks," much has changed in the past century of moviemaking. Indeed, as Herbert C. McKay (1927), stated, "[t]here is hardly a single fantastic idea, which cannot be given existence upon the screen... Trickery in one form or another is possibly the greatest single factor in the success of the modern film" (p. 211). One would assume such a statement was in reference to the stunning images presented in the films of the most recent Academy Award nominees, but McKay made his declaration in 1927!

According to Box Office Mojo, of the top ten highest grossing films of all time, all of them featured heavy use of visual effects (Box Office Mojo, 2014a). When adjusted for inflation (Box Office Mojo, 2014b), this list is topped by *Gone With the Wind* (1939), which featured innovative matte shots as well as other "trickery." The second film on the list is *Star Wars* (1977), arguably the most iconic visual effects movie ever made (Bredow, 2014).

Such "trickery," special effects, or more commonly referred to now as "visual effects" (VFX), are divided into mechanical, optical, and computer-generated imagery (CGI). "Mechanical effects include those devices used to make rain, wind, cobwebs, fog, snow, and explosions. Optical effects allow images to be combined... through creation of traveling mattes run through an optical printer" (Freeman, 1998, p. 409).

In the early sound era, miniatures and rear projection became popular along with traveling mattes (Martin, 2014), like those employed in the landmark VFX film of the 1930s, *King Kong* (1933).

This film inspired a young Ray Harryhausen to experiment with stop-motion animation and split-screen action in *The 7th Voyage of Sinbad* (1958) and the invention of "Dynamation" (a technique of rear and front projecting footage one frame at a time, still used by stop-motion animators today) most famously employed in the skeleton sword fight scene in *Jason and the Argonauts* (1963), as well as in two additional *Sinbad* films (1973 & 1977), and the 1981 film, *Clash of the Titans*.

Four years in the making, the 1968 Stanley Kubrick film, *2001: A Space Odyssey*, created a new standard for VFX credibility (Martin, 2014). Kubrick used sophisticated traveling mattes combined with "hero" miniatures of spacecraft (ranging from four to 60 feet in length) and live-action to stunning effect (Cook, 2004). The film's "trip" sequence dazzled audiences with controlled streak photography, macro-photography of liquids, and deliberate misuse of color-records, and throughout the "Dawn of Man" sequence, audiences witnessed the first major application of front projection (Martin, 2014).

Arguably the first movie ever to use computers to create a visual effect—a two-dimensional rotating structure on one level of the underground lab—was *The Andromeda Strain* in 1971. This work was considered extremely advanced for its time.

In 1976, American International Pictures released *Futureworld*, which featured the first use of 3D CGI—a brief view of a computer-generated face and hand. In 1994, this groundbreaking effect was awarded a Scientific and Engineering Academy Award. Since then, CGI technology has continued to progress rapidly.

"In the history of VFX, there is a before-and-after point demarcated by the year 1977 when *Star Wars* revolutionized the industry" (Martin, 2014, p. 71-72). VFX supervisor John Dykstra invented an electronic motion-controlled camera capable of repeating its movements (later called the "Dykstraflex") and developed methodologies for zero-gravity explosions. Likewise, George Lucas' visual effects company, Industrial Light & Magic (ILM), took a big step forward for CGI with the rendering of a 3D wire-frame view of the Death Star trench depicted as a training aid for rebel pilots in Star Wars IV: A New Hope (1977).

Star Trek: The Wrath of Kahn (1982) incorporated a one-minute sequence created by Pixar (a LucasFilm spin-off), that simulated the "Genesis Effect" (the birth and greening of a planet) and is cinema's first totally computer-generated VFX shot. It also introduced a fractal-generated landscape and a particle-rendering system to achieve a fiery effect (Dirks, 2009).

Tron (1982) was the first live-action movie to use CGI for a noteworthy length of time (approximately 20 minutes in the famed Lightcycle sequence) in the most innovative sequence of its 3D graphics world inside a video game.

In *Young Sherlock Holmes* (1985), LucasFilm/ Pixar created perhaps the first fully photorealistic CGI character in a full-length feature film with the sword-wielding medieval "stained-glass" knight who came to life when jumping out of a window frame.

Visual impresario James Cameron has always relied on VFX in his storytelling dating back to the impressive low-budget miniature work on *The Terminator* (1984), later expanded on for *Aliens* (1986). Cameron's blockbuster action film, *Terminator 2: Judgment Day* (1991) received a Best Visual Effects Oscar thanks to its depiction of Hollywood's first CGI main character, the villainous liquid metal T-1000 cyborg (Martin, 2014).

Toy Story (1995), was the first successful animated feature film from Pixar, and was also the first all-CGI animated feature film (Vreeswijk, 2012).

In the *Lord of the Rings* trilogy (2001, 2002 and 2003), a combination of motion-capture performance and key-frame techniques brought to life the main digital character Gollum (Dirks, 2009) by using a motion capture suit (with sensors) and recording the movements of actor Andy Serkis. In the 2004 animated film, *The Polar Express*, the same motion capture technique is used for all its actors (Vreeswijk, 2012)

CGI use has grown exponentially and hand-in-hand with the increasing size of the film's budget it occupies. *Sky Captain and the World of Tomorrow* (2004) was the first big-budget feature to use only "virtual" CGI back lot sets. Actors Jude Law, Gwyneth Paltrow, and Angelina Jolie were filmed in front of blue screens; everything else was added in post-production (Dirks, 2009).

More an "event" than a movie, James Cameron's *Avatar* (2009) ushered in a new era of CGI. Many believe that *Avatar*, a largely computer-generated, 3D film—and the top-grossing movie in history, earning nearly $3 billion worldwide—has changed the moviegoing experience (Muñoz, 2010). New technologies used in the film included CGI Performance Capture techniques for facial expressions, the Fusion Camera System for 3D shooting, and the Simul-Cam for blending real-time shoots with CGI characters and environments (Jones, 2012).

One of the greatest obstacles to CGI has been effectively capturing facial expressions. In order to overcome this hurdle, Cameron built a technology he dreamed up in 1995, a tiny camera on the front of a helmet that was able to "track every facial movement, from darting eyes and twitching noses to furrowing eyebrows and the tricky interaction of jaw, lips, teeth and tongue" (Thompson, 2010).

Multichannel Sound

There's no mistaking the chest-rumbling crescendo associated with THX sound in theaters. It's nearly as recognizable as its patron's famous *Star Wars* theme. Developed by Lucasfilms and named after *THX1138* (George Lucas' 1971 debut feature film), THX is not a cinema sound format, but rather a standardization system that strives to "reproduce the acoustics and ambience of the movie studio, allowing audiences to enjoy a movie's sound effects, score, dialogue, and visual presentation with the clarity and detail of the final mastering session" (THX, 2014).

At the time of THX's initial development in the early 1980s, most of the cinemas in the U.S. had not been updated since World War II. Projected images looked shoddy, and the sound was crackly and flat. "All the work and money that Hollywood poured into making movies look and sound amazing was being lost in these dilapidated theaters" (Denison, 2013).

Even if movies were not being screened in THX-certified theaters, the technical standards set by THX illustrated just how good the movie-going experience could be and drove up the quality of projected images and sound in all movie theaters. THX was introduced in 1983 with *Star Wars Episode VI: Return of the Jedi* and quickly spread across the industry. To be a THX Certified Cinema, movie theaters must meet the standards of best practices for architectural design, acoustics, sound isolation and audio-visual equipment performance (THX, 2014). The most recent self-reported company data states that there are presently about 2,000 THX certified theaters worldwide (THX, 2014).

More recently, THX has delved into home theater equipment certification and "...created a specification for performance and developed software for the post-processing side that helped translate the cinema experience to the home experience through a modification of the actual movie track sound" (Denison, 2013). In early 2013, THX released *THX Tune-Up*, a free iOS app that lets consumers connect their iPhone, iPad or iPod Touch to their TV or projector and perform basic TV calibration tests using custom video test patterns, pictures, and tutorials as well as basic audio tests to check external speakers are working in phase and connected properly for 2-channel stereo or 5.1 sound systems.

While THX set the standards, Dolby Digital 5.1 Surround Sound is one of the leading audio delivery technologies in the cinema industry. In the 1970s, Dolby Laboratories introduced Dolby noise reduction (removing hiss from magnetic and optical tracks) and Dolby Stereo—a highly practical 35mm stereo optical release print format that fit the new multichannel soundtrack into the same space on the print occupied by the traditional mono track (Hull, 1999).

Dolby's unique quadraphonic matrixed audio technique allows for the encoding of four channels of information (left, center, right and surround) on just two physical tracks on movie prints (Karagosian & Lochen, 2003). The Dolby stereo optical format proved so practical that today there are tens of thousands of cinemas worldwide equipped with Dolby processors, and more than 25,000 movies have been released using the Dolby Stereo format (Hull, 1999).

By the late 1980s, Dolby 5.1 was introduced as the cinematic audio configuration documented by various film industry groups as best satisfying the requirements for theatrical film presentation (Hull, 1999). Dolby 5.1 uses "five discrete full-range channels—left, center, right, left surround, and right surround—plus a...low-frequency [effects] channel" (Dolby, 2010a). Because this low-frequency effects (LFE) channel is felt more than heard, and because it needs only one-tenth the bandwidth of the other five channels, it is refered to as a ".1" channel (Hull, 1999).

Dolby also offers Dolby Digital Surround EX, a technology developed in partnership with Lucasfilm's THX that places a speaker behind the audience to allow for a "fuller, more realistic sound for increased dramatic effect in the theatre" (Dolby, 2010b).

Dolby Surround 7.1 is the newest cinema audio format developed to provide more depth and realism to the cinema experience. By resurrecting the full range left extra, right extra speakers of the earlier Todd-AO 70mm magnetic format, but now calling them left center and right center, Dolby Surround 7.1 improves the spatial dimension of soundtracks and enhances audio definition thereby providing full-featured audio that better matches the visual impact on the screen. Movies released using Dolby Surround 7.1 include *Toy Story 3* (2010), *Tron: Legacy* (2010), *Transformers: Dark of the Moon* (2011), *War Horse* (2011), the 3D film *Hugo* (2011), another 55 films released in 2012 including Sound Mixing Academy Award winner *Les Misérables* (Dolby, 2014), and an additional 43 films in 2013 including Best Animated Feature *Frozen*, and *Gravity*, winner of Best Sound Editing and Sound Mixing (Oscars, 2014).

Film's Slow Fade to Digital

With the rise of CGI-intensive storylines and a desire to cut costs, celluloid film is quickly becoming an endangered medium for making movies as more filmmakers use digital cinema cameras capable of creating high-quality images without the time, expense, and chemicals required to shoot and process on film. "While the debate has raged over whether or not film is dead, ARRI, Panavision, and Aaton quietly ceased production of film cameras in 2011 to focus exclusively on design and manufacture of digital cameras" (Kaufman, 2011).

But, film is not dead—at least, not yet. There is compelling evidence that it is still alive and well among A-list filmmakers. Of the 12 Oscar nominees for either Best Picture or Best Cinematography in 2014, six were shot at least partially on film, four were shot exclusively in a digital format, and only two exclusively on film—*Nebraska* and *12 Years a Slave* (both 2013).

"Of course, facts are funny things... Digital cinema has a very short history—*Star Wars Episode II: Attack of the Clones* (2002) was the first full-on 24p [high definition digital] release…and now [10 years later], more than one-third of the films [up for] the industry's highest honors were shot digitally" (Frazer, 2012). This trend seems to be holding steady—for now.

Digital production also presents a significant savings for low-budget and independent filmmakers. Production costs using digital cameras and non-linear editing are a fraction of the costs of film production; sometimes as little as a few thousand dollars, and lower negative costs mean a faster track to profitability.

Indeed, the 2009 horror film *Paranormal Activity* was shot using a Sony FX-1 camcorder and edited on a Dell PC using *Sony Vegas*. The film's initial production budget was around $15,000 (Lally, 2009) and it generated close to $200 million worldwide, making it the number one most profitable movie based on return on investment (The Numbers, 2014a).

High-end digital cinema cameras like the RED and the RED Epic used on nearly 300 feature films since 2008 (RED Digital Cinema, 2014); or the Arri Alexa and Alexa XT, used on over 70 notable features (Arri Group, 2014) since its U.S. debut in the 2011 Disney film *Prom* (Kadner, 2011); along with other cameras like the Sony CineAlta F65 or the Canon EOS C500; are all making serious inroads as the camera of choice.

Smaller high definition digital still cameras (HDSLRs) with large sensors, interchangeable prime lenses, and video recording capabilities are also making inroads in the specialty and independent film markets. Whereas the 2013 Academy Award nominated films for best picture may have been dominated by the Arri Alexa (digital) and the Arriflex (film) cameras, the *2014 Sundance Film Festival* winners employed quite an array of cinema cameras including, DSLRs, ARRI, Canon, Panasonic, and the RED Epic and RED Scarlet (Indiewire, 2014). With the transition to digital, the Academy of Motion Picture Arts and Sciences sounded a clarion call in 2007 over the issue of digital motion picture data longevity in the major Hollywood studios.

In their report, titled *The Digital Dilemma*, the Academy concluded that, although digital technologies provide tremendous benefits, they do not guarantee long-term access to digital data compared to traditional filmmaking using motion picture film stock. Digital technologies make it easier to create motion pictures, but the resulting digital data is much harder to preserve (Science & Technology Council, 2007).

The Academy's update to this initial examination of digital media archiving was published in 2012 as *The Digital Dilemma 2* and focused on the new challenge of maintaining long-term archives of digitally originated features created by the burgeoning numbers of independent and documentary filmmakers. "Independent ('indie') filmmakers operating outside of the major Hollywood studios supply 75% of feature film titles screened in U.S. cinemas, despite facing substantial obstacles in doing so. As digital moviemaking technologies have lowered the barrier to entry for making films, competition among indie filmmakers seeking theatrical distribution has increased" (Science & Technology Council, 2012, p. 3-4).

As new digital distribution platforms have emerged, making it easier for independent filmmakers to connect their films with target audiences (through video-on-demand and pay-per-view) and possible revenue streams, these platforms have not yet proven themselves when it comes to archiving and preservation (Science & Technology Council, 2012). "Unless an independent film is picked up by a major studio's distribution arm, its path to an audiovisual archive is uncertain. If a filmmaker's digital work doesn't make it to such a preservation environment, its lifespan will be limited—as will its revenue-generating potential and its ability to enjoy the full measure of U.S. copyright protection" (Science & Technology Council, 2012, p. 6). For now at least, the "digital dilemma" seems far from over.

As physical film acquisition slowly yields to digital recording, so too has film editing made the digital shift. In 1989, *Lightworks* was introduced as the first and most advanced professional editing system on the market, and it had competition right out of the gate when Avid Technology began taking orders the same year for their prototype *Avid/1* digital nonlinear, computer-based editor.

By 1992, the National Association of Broadcasters (NAB) convention was rich with digital nonlinear editing systems—including a wide array of products from Avid Technology. Avid would continue to dominate the digital non-linear editing market for both video and film. As CGI and other VFX became an increasingly important part of cinematic storytelling, Avid was integrated into the "digital intermediate" workflow.

In 1999, Apple released *Final Cut Pro* 1.0 along with "a new generation of desktop high-quality editing, specifically designed to take advantage both of the new DV formats and FireWire-fitted Macs" (Rubin, 2000, p. 72).

Now that digital cinema cameras like the RED and Arri Alexa are coming into more common use, and the new non-linear editing systems have the capability of working with the digital footage at full resolution; it is now possible to shoot, edit, and project a movie without ever having to leave the digital environment.

Digital 3D

The first digital 3D film released was Disney's *Chicken Little*, shown on Disney Digital 3D (PR Newswire, 2005). Dolby Laboratories outfitted about 100 theaters in the 25 top markets with Dolby Digital Cinema systems in order to screen the film. The idea of actually shooting live-action movies in digital 3D did not become a reality until the creation of the Fusion Camera, a collaborative invention by director James Cameron and Vince Pace (*Hollywood Reporter*, 2005). The camera fuses two Sony HDC-F950 HD cameras "2½ inches apart to mimic the stereoscopic separation of human eyes" (Thompson, 2010). The camera was used to film 2008's *Journey to the Center of the Earth* and 2010's *Tron Legacy*.

Cameron used a modified version of the Fusion Camera to shoot 2009's blockbuster *Avatar*. The altered Fusion allows the "director to view actors within a computer-generated virtual environment, even as they are working on a 'performance-capture' set that may have little apparent relationship to what appears on the screen" (Cieply, 2010).

Another breakthrough technology born from *Avatar* is the swing camera. For a largely animated world such as the one portrayed in the film, the actors must perform through a process called motion capture which records 360 degrees of a performance, but with the added disadvantage that the actors do not know where the camera will be (Thompson, 2010). Likewise, in the past, the director had to choose the shots desired once the filming was completed. Cameron tasked virtual-production supervisor Glenn Derry with creating the swing camera, which "has no lens at all, only an LCD screen and markers that record its position and orientation within the volume relative to the actors" (Thompson, 2010). An effects switcher feeds back low-resolution CG images of the virtual world to the swing camera allowing the director to move around shooting the actors photographically or even capturing other camera angles on the empty stage as the footage plays back (Thompson, 2010).

Figure 11.8

Vince Pace and James Cameron with Fusion 3D Digital Cinema Camera

Source: CAMERON | PACE Group
Photo credit: Marissa Roth

When Hollywood studio executives saw the $2.8 billion worldwide gross receipts for *Avatar* (2009), they took notice of the film's ground-breaking use of 3D technology; and when Tim Burton's 2010 *Alice in Wonderland*—originally shot in 2D—had box office receipts reach $1 billion thanks to 3D conversion and audiences willing to pay an extra $3.50 "premium" up-charge for a 3D experience, the film industry decided that 3D was going to be their savior (Smith, 2013). "Fox, Paramount, Disney, and Universal collectively shelled out $700 million to help equip theaters with new projectors, and the number of 3D releases jumped from 20 in 2009 to 45 in 2011" (Smith, 2013). However, there was a slight drop to 41 in 2012, that fell further to only 35 in 2013 (Movie Insider, 2014a & 2014b respectively), and the release calendar for 2014 mentions 22 films in 3D with a meager 18 films projected for 2015 (Film Releases, 2014), and only 11 in development for 2016 (Box Office Mojo, 2014c).

What accounts for this meteoric rise and precipitous fall of 3D? In part, it could be that the novelty has worn off with film audiences. Likewise, moviegoers may have tired of paying the hefty surcharge for 3D films that failed to meet expectations. The latter could be the result of studios rushing post-production conversions of 2D films into 3D resulting in inferior products (Smith, 2013).

Perhaps because James Cameron helped kick off the 3D craze with *Avatar* (2009), he's also become the format's harshest critic—especially with regard to 3D conversions. In particular, Cameron has slammed the habit of hastily converting films from 2D in order to pad box office grosses (Lang, 2012). "The conversion process is a creative process that uses technology. It is not a technological process, [Jon] Landau said. When you try to convert film in six weeks, it can only be a technological process" (Lang, 2012).

So, when Cameron decided to convert *Titanic* (1997) to 3D, he spent 60 weeks going over more than 260,000 frames of the movie to make sure the conversion looked "right" (Lang, 2012). When *Titanic 3D* (2012) was released, to critical acclaim, it made more than $25 million over its first five days of release, more than justifying the $18 million it cost to retrofit the film for 3D screens (Lang, 2012).

Digital Theater Conversions

In 1999, *Star Wars I: The Phantom Menace* was the first feature film projected digitally in a commercial cinema—although there were few theaters capable of doing so at the time.

In January 2012, the film industry marked the crossover point when digital theater projection surpassed 35mm. Indeed, demand for 35mm cinema film has declined from a peak level of 13 billion feet a year in 2008 to as little as 4 billion in 2012 (Hancock, 2011). Meanwhile, the cost of producing celluloid film is soaring due to rising prices for a key raw material, silver. It is projected that by 2015, 35mm film will be used in only 17% of global movie theaters, relegating it to a niche projection format (Hancock, 2011).

Nearly 90,000 movie theaters were converted to digital by the end of 2012, demonstrating a growth rate of 40 percent over the 63,825 digital movie theaters in 2011 (Hancock, 2013). Today, the vast majority of movie theaters in the U.S. (nearly 93 percent) have already switched from film to digital projection (Hancock, 2014).

Roger Frazee, Regal Theater's Vice President of Technical Services, having observed the changes, stated, "Films are now 200-gigabyte hard drives, and projectors are those big electronic machines in the corridor capable of working at multiple frame rates, transmitting closed-captioned subtitles and being monitored remotely. The benefit of digital is, you don't have damaged film. You don't have scratched prints, and it looks as good in week six as it does on day one" (cited in Leopold, 2013).

When Digital Cinema Implementation Partners (DCIP) was formed in 2007 in cooperation with the exhibition companies AMC Entertainment Inc., Cinemark USA, Inc., and Regal Entertainment Group, it was charged with bring the major studios on-board to help provide incentives for exhibitors to take on the challenge—and massive cost—of converting their theaters from film to digital projection systems (Digital Cinema Implementation Partners, n.d.).

Early adopters of the digital systems numbered about 5,000 between 2005 and 2006 (Mead, 2008). But the largest exhibitors—instrumental in the formulation of the DCIP—were in the second wave. Although growth in the digital theater market is slowing in North America, it is approaching the endgame 15 years after the first digital film screened commercially and nearly nine years after the DCI Specification was first released (Hancock, 2014).

Current Status

Box office revenues remained relatively flat between 2000 and 2011, averaging $8.6 billion over the period, dropping slightly between 2004 and 2005, with a slow recovery beginning in 2006 that peaked at $10.6 billion in 2009, before dropping off again. By 2011, the average price of a movie ticket had risen to $7.93 even as ticket sales dropped by just over 200 million from the year before and revenue fell by nearly $357 million (The Numbers, 2014b). One year later, ticket sales were back up (1.36 billion) with total box office gross receipts topping out at $10.8 billion (up nearly half a billion) while the average price of a ticket rose by only 3 cents (The Numbers, 2014c).

By 2013, the average price of a movie ticket had risen to $8.16 with more than 1.34 billion sold. The highest grossing film was *Iron Man 3* ($409 million), Warner Bros was the top distributor with just over 17 percent share and $1.9 billion in gross revenues. PG-13-rated films accounted for 52 percent of all rated film releases (second place went to R-rated films at 28 percent). Adventure and Action films were the top genres (26 percent and 20 percent respectively) and total box office gross receipts for all films totaled almost $11 billion (The Numbers, 2014d).

The 10 percent increase in PG-13 movies since 2000 is responsive to the industry trend in producing for the adolescent market (17 and younger) that typically contributes about one quarter of the industry's revenues. On average, PG-13 movies make more than three times what R-rated films do at the domestic box office (Cunningham, 2013). Also, of the 149 PG-13 films released in 2013, 15 of the top 20 in box office revenues were Adventure and Action films (The Numbers, 2014d) and none of 2013's top grossing films were rated R.

As the number of primary and secondary school-age children rises, industry pressure to produce more PG-13 films will likely also increase (IBISWorld, 2014). Finally, except for brief aberrations in the mid-2000s, Hollywood revenues have recovered from the economic downturn and are back on a modest upward trend.

The steady rise in the price of movie tickets—often greater than inflation—corresponds with big changes in the movie business. For Hollywood studios, the switch to mostly CGI-intensive Adventure and Action genre blockbuster movies has raised production costs and pushed up marketing expenses, which often reach as high as $70 million worldwide to publicize and distribute the biggest movies (Block, 2012).

For exhibitors, the cost of building new multiplex theaters and switching to digital projection and sound has also been a factor in the rising costs. Purchasing and operating capital equipment such as digital projection systems, screens, and speakers can be expensive and the costs of establishing auditoriums on par with the quality offered by the industry's major players is quite high (IBISWorld, 2014). Couple this with increasing pressure from cable and satellite TV, home theaters, online streaming, and various second-screen platforms and other entertainment products appearing in households and lowering demand for screening films in movie theaters and it is just as perilous not to upgrade to digital. Therefore, film exhibitors have invested heavily in upgrading their projection systems so that, by the end of 2013, there were nearly 20,000 digital screens added to movie theaters worldwide. This brought the total to 109,176 screens and boosted the penetration rate of digital screens to 90 percent globally.

Today, Denmark, Hong Kong, Luxembourg, Norway, the Netherlands, Canada, and South Korea have achieved full conversion. Meanwhile, Belgium, Finland, France, Indonesia, Switzerland, Taiwan, the U.K., and the U.S. are all well above 90 percent penetration for digital screens, while China accounts for over half of the 32,642 digital theater screens in the Asia-Pacific (Hancock, 2014).

As part of the second phase of the digital cinema technology rollout, cinema operators are now turning their attention to the electronic distribution of movies. Since 2012, a plethora of services have cropped up around the world to support electronic distribution, mainly by satellite, but also using fiber optic cable (Hancock, 2013).

The steady distribution of 3D movies titles, though showing recent signs of contraction, is still a major source of box office revenue for Hollywood due to higher ticket prices. In 2009, up-charged 3D tickets for *Avatar* accounted for 71 percent of the opening weekend gross receipts, even *The Final Destination* (2009) made 70 percent of its gross revenues from 3D ticket sales. In 2010, *Alice in Wonderland* brought in 70 percent through 3D ticket sales and *TRON: Legacy* (2010) got a major boost from 3D tickets at 82 percent of gross revenue (Smith, 2013). However, even though there were more 3D releases than ever in 2011, the average receipts for a 3D movie were down from the previous year. 3D box office receipts took a serious tumble as audiences increasingly opted for cheaper 2D tickets (Smith, 2013).

By 2013, revenues were down substantially for 3D ticket sales. Family movies including *Monsters University* (31 percent), *Despicable Me 2* (27 percent) and *Turbo* (25 percent), each in turn led 3D ticket sales on a downward trend. Even summer releases that tend to do better at the box office, plunged; *World War Z* (34 percent) and *The Great Gatsby* (33 percent) only slightly out performing their animated counterparts whereas *Pacific Rim* earned 2013's highest share of summer 3D ticket sales at 50 percent total box office gross revenue. According to Ray Subers, editor of *Box Office Mojo*, "2013 is the first year in which movies are consistently seeing 3D shares below 40 percent" (cited in Smith, 2013) and 3D ticket sales have dropped by a third over since 2012.

It is obvious, even as 3D cinema infrastructure is expanding; revenues from 3D movies are trending downward. Still, the film industry has invested too much capital into 3D to allow it to fail. Likewise, theater owners have already invested in digital equipment and love the extra income they make from screening "premium" 3D films. As Patrick Corcoran of the National Association of Theatre Owners observes, "Essentially, it's an add-on… [e]xhibitors are making more money per ticket on those 3D showings. There isn't a big extra expense for [them] in it" (cited in Smith, 2013). Likewise, at $10 million to $20 million per 3D conversion, Hollywood is not likely to give up on 3D, and if a film suffers poor 3D returns, a studio can earn back its investment fairly easily.

Whereas, the U.S. domestic audience may be eschewing 3D, "the booming international box office can't be ignored. Developing markets like China, Russia, and Brazil have seen major growth since *Avatar* in 2009" (Smith, 2013) and 3D is still a massive draw in the overseas markets, or at least accepted as normal for viewing Hollywood exports. Therefore, "[i]n today's marketplace, where a big-budget film's financial fate is often decided by overseas dollars, it's almost fiscal self-injury not to make the call [for 3D]" (Mendelson, 2013).

It is not all doom and gloom for domestic 3D cinema, however. Writing for *Wired*, Jennifer Wood observed that, after wringing their hands and lamenting whether 3D was even viable any longer in the domestic movie marketplace, a film comes along to silence the naysayers (Wood, 2013). "In 2009 it was James Cameron's *Avatar*. With 2011 came Martin Scorsese's *Hugo*… [and in 2013], Alfonso Cuarón's *Gravity* became the most recent reminder of the genuine storytelling power that stereoscopic filmmaking holds" (Wood, 2013).

One of the things that made *Gravity* standout is that Cuarón embraced 3D as a full fledged storytelling tool, and audiences validated that decision. In its opening weekend, 80 percent of all *Gravity* ticketbuyers chose to see the film in 3D. The following weekend, that number rose to 82 percent and remained steady through the third weekend (Wood, 2013). This is good news for the industry, but the take away should be that 3D must assume the role of supporting element and enabler of an otherwise outstanding story. If this is so, audiences will come and awards will follow—witness the 7 out of 10 Academy Awards given to *Gravity*—including Best Cinematography—at the 2014 Oscars as validation. As stated by Keith Simanton, managing editor of IMDb, "The perception of depth does not lead the perception of good taste astray" (cited in Wood, 2013).

Factors to Watch

The National Association of Broadcaster's (NAB) convention is the annual trade show for broadcast, cinema, and video production professionals. In 2012, the buzz was all about 4K-plus digital cinema cameras (Reeve, 2012) with faster frame rates and greater dynamic range (ratio of the brightest to the darkest portions in a frame measured in *f*-stops). That's great, but what is 4K?

At the moment, there is no real "standard," but Digital Cinema Initiatives, LLC (DCI)—whose primary charge comes from the big studios that support it—seeks to establish and document voluntary specifications for a new generation of high-resolution digital cinema (DCI, 2011). As such, DCI defines 4K as having pixel dimensions of 4096x2160 at the top-end—the actual height being determined by the aspect ratio.

For example, RED Digital Cinema's 4K cameras are either 4096x2304 or 4096x2048, depending on aspect ratio. Other camera manufacturers are adopting the Quad-HD format—3840x2160—exactly twice the height and width of 1080p HD which preserves the 16:9 aspect ratio.

The first high-profile 4K cinema release was *Blade Runner: The Final Cut* (2007), a new cut and print of the 1982 Ridley Scott film. Unfortunately, at that time, very few theaters were able to show it in its full resolution (Pendlebury, 2012). Still, camera manufacturers including Arri and RED already delivered their version of a 4K camera, while Canon, JVC and Sony introduced prototype 4K cameras at NAB 2012

Although 3D was noticeably absent from the television market at NAB 2013, if the hype over 4K technology was generating all the buzz at NAB 2012, it was obvious the industry was ready to embrace 4K in 2013—or UltraHD, the general marketing term for 4K and 8K video. UltraHD is four to eight times the data of present HD standards.

Also making their debut were larger and faster SSD storage solutions designed for 4K, 5K, and 3D workflows; new technologies in cloud computing; platforms for editorial collaboration that promise to make computationally intensive graphics work (like 3D CGI) across internal and external networks faster and easier; and Thunderbolt connectivity for transporting the "fat" 4K-plus files. It was easy, to see that 4K is here to stay (Frazer, 2013).

With the advent of cloud computing in the second decade of the 21st Century, the entertainment industry has taken advantage of solutions provided by working "in the cloud." Whereas the music industry has used cloud storage to record, produce and distribute millions of tracks, the film industry has been a little late capitalizing on the benefits of cloud technology—but recent innovations have paved the way for present implementation of distributed post-production.

According to Steve Andujar, the CIO at Sony Pictures, cloud computing represents "[a] great opportunity to drive down costs and improve implementation of services" (Doperalski, 2012). Professor Norman Hollyn, Editing Track Head, and Michael Kahn, Endowed Chair at the University of Southern California's School of Cinematic Arts, stated that the last three films he cut were done remotely; "I can work with people all over the world who I never would have had the opportunity to work with before" thanks to the technology of cloud computing and distributed post-production (Kaufman, 2014a). While the cloud offers anytime/anywhere access to studio content, others have caution that the potential of the cloud should be undertaken with caution.

Bringing consistency and stability to the entire digital workflow—from production to post-production—is one of the goals of the Academy Color Encoding System (ACES). Achieving this goalwill positively impact cinematographers, VFX supervisors and content distributors, among others, (Kaufman, 2014a) and plug, in seamlessly with a distributed post-production model. Such consistency and potential stability in the digital workflow is of paramount importance to filmmakers as 4K and HFR productions become the norm.

However, with Hollywood's understandable concerns over digital piracy and the protection of the incredible investment in intellectual property each film represents, the film industry wants some assurances from the cloud computing community. For years, pirated movies, television shows, and music have been on the Internet, and, for just as long, Hollywood and the entertainment business have been trying—and failing—to stop it (*The New York Times*, 2012). If the results of a Google search for a film pops up a piracy site first (like The Pirate Bay or Megaupload), Hollywood is concerned.

If 4K seems inevitable, then apparently, so are higher frame rates (HFR). Ever since the late 1920s and the introduction of synchronous sound in movies the standard frame rate has been 24 frames per second (fps). In the shift from the hand-cranked 16fps of the silent days to the new technical demands of the "talkies," a constant playback speed was necessary to keep the audio synchronized with the visuals.

But, as Peter Jackson, director of the *Lord of the Rings* trilogy (2001, 2002 & 2003), pointed out while speaking at CinemaCon in 2012 (via a videotaped message), "with digital, there's 'no reason' to stay with 24[fps]... higher frame rates can result in smoother, more lifelike pictures while producing fewer motion artifacts" (Giardina, 2012, April 16). Speaking to the same group one year earlier, James Cameron emphasized that, "if watching 3D in cinemas is like looking through a window—making the jump to 60fps was removing that window" (Billington, 2011).

As a leading proponent of HFR, Jackson has justified the release of his *Hobbit* film trilogy, "science tells us that the human eye stops seeing individual pictures at about 55 fps. Therefore, shooting at 48 fps gives you much more of an illusion of real life. The reduced motion blur on each frame increases sharpness and gives the movie the look of having been shot in 65mm or IMAX. One of the biggest advantages is the fact that your eye is seeing twice the number of images each second, giving the movie a wonderful immersive quality. It makes the 3D experience much more gentle and hugely reduces eyestrain. Much of what makes 3D viewing uncomfortable for some people is the fact that each eye is processing a lot of strobing, blur and flicker. This all but disappears in HFR 3D" (Jackson, 2014).

Filmmakers like Cameron and Jackson, who are advocating for higher frame rates in pursuit of sharper, more realistic images on the screen, contend that films shot and delivered in 24fps have persistent visual problems. At 24fps, fast panning and sweeping camera movements are severely limited by visual artifacts and motion-blur that result from such movement.

When a film is shot and shown in 3D, the flaws of 24fps are even more obvious because of the technical challenges and the sheer volume of visual data being projected. According to Jackson, "shooting and projecting at 48fps does a lot to get rid of these issues. It looks much more lifelike, and it is much easier to watch, especially in 3D" (Singer, 2011).

However, HFRs has its detractors who opine that 24fps films deliver a depth, grain and tone that is unique to the aesthetic experience and not possible to recreate with digital video—this lack of "graininess" is often jarring and uncomfortable to first-time viewers of HFR 3D. "Such adjectives as "blurry," "plastic-y" and "weirdly sped-up" were thrown around a lot," in response to the screenings of The Hobbit: An Unexpected Journey (2012) projected in 48fps 3D. Responding to critics of HFR, James Cameron observed earlier that the "filmic" style critics so love comes from the angle of the shutter and lighting in the scene, not necessarily from the frame rate (Billington, 2011).

Some think the complaints about HFR might be due to a matter of taste and that the 48fps and other HFR formats being discussed are still in the early stages of development (Schaefer, 2014). Most digital cinemas are already using existing "Generation 2" projectors—those manufactured in 2010 or later—and all they would need is a software upgrade to be able to screen movies (2D or 3D) at 48fps, 60fps or 96fps. These movie theaters are just waiting for the technology to be sorted out and standardized.

Another potential game-changer is the introduction of light-field technology—also referred to as plenoptic cameras. Whereas, ordinary cameras are capable of receiving 3D light and focusing this on an image sensor to create a 2D image, a plenoptic camera samples the 4D light field on its sensor by inserting a microlens array between the sensor and main lens (Ng, Levoy, et al., 2005). Not only does this effectively yield 3D images with a single lens, but the creative opportunities of light-field technology will enable such typical production tasks as refocusing, virtual view rendering, shifting focal plains, and dolly zoom effects could all be accomplished during post-production.

There is little doubt that emerging technologies will continue to have a profound impact on cinema's future. This will also force the industry to look at an entirely new kind of numbers game that has nothing to do with weekend grosses, 3D, or 48fps.

As Steven Poster, ASC and President of the International Cinematographers Guild stated, "Frankly I'm getting a little tired of saying we're in transition.

I think we've done the transition and we're arriving at the place we're going to want to be for a while. We're finding out that software, hardware and computing power have gotten to the point where it's no longer necessary to do the things we've always traditionally done… [a]nd as the tools get better, faster and less expensive… [what] it allows for is the image intent of the director and director of photography to be preserved in a way that we've never been able to control before" (Kaufman, 2014b).

Indeed, moviegoers of the future might look back on today's finest films as quaint, just as silent movies produced a century ago seem laughably imperfect to moviegoers today (Hart, 2012). Cinematographer David Stump, ASC, noting the positive changes brought about by the transition from analog to digital, states that "[t]he really good thing that I didn't expect to see… is that the industry has learned how to learn again… We had the same workflow, the same conditions and the same parameters for making images for 100 years. Then we started getting all these digital cameras and workflows and… [now] we have accepted that learning new cameras and new ways of working are going to be a daily occurrence" (Kaufman, 2014a).

Cinema Technologies Visionary: James Cameron

No single individual has driven contemporary cinematic technology as far or as fast as filmmaker **James Cameron**. From his man-sized robotic puppet in 1984's *The Terminator*, to the computer-generated 3D world in 2009's *Avatar*, Cameron has relentlessly pursued the means to make the fantastic visions of his imagination a reality on the screen. Manohla Dargis of the *New York Times* called him "a filmmaker whose ambitions transcend a single movie or mere stories to embrace cinema as an art, as a social experience and a shamanistic ritual, one still capable of producing the big WOW" (Dargis, 2009).

In 1982, Cameron wrote a sci-fi-action script that would eventually become *The Terminator*. Looking for funding, Cameron and his producer approached Hemdale Pictures. Always a visual innovator and aggressive salesman, Cameron impressed the executives by bringing actor Lance Hendriksen, in full Terminator costume and makeup, to the pitch meeting (Keegan, 2010). *The Terminator* was an unexpected success; produced for $6.4 million, the film grossed over $38 million in 1984 (Box Office Mojo, 2014d). This innovation launched Cameron's career as not just a writer-director, but a developer of cutting-edge technology designed to serve his stories and characters.

After the success of *The Terminator,* Cameron was recruited to direct *Aliens* (1986), the sequel to Ridley Scott's successful *Alien*. An enormous insect-like alien queen was needed for the story, so Cameron dreamed up another complex puppet, this time fourteen feet high and controlled by two puppeteers suspended inside. In *The Futurist: The Life and Films of James Cameron*, author Rebecca Keegan notes: "Rare among directors, he has the artistic ability to sketch, paint, and build things himself—and he prefers to… Cameron knows better than anyone else what he envisions in his own mind." (Keegan, 2010; p. 79). *Aliens* grossed over $85 million in 1986 (Box Office Mojo, 2014e).

In 1994 Cameron released another expensive and ambitious production, the romantic comedy/action thriller *True Lies* (1997), but soon departed from sci-fi and action to direct what Manohla Dargis of the *New York Times* called a "megamelodrama" entitled *Titanic* (1997). The film production was a huge undertaking, requiring a massive set and about $200 million from script to screen. At the time, it was the most expensive film ever made.

Titanic was released in December 1997, six months late and way over budget. It became the highest-grossing film of the year in only 13 days, earning over $600 million (Box Office Mojo, 2014f), and won the 2014 Academy Award for Best Picture, cementing James Cameron's status as a giant of the industry.

In the decade after *Titanic*, Cameron was waiting for 3D/CGI technology to catch up with a project he was envisioning, set almost entirely in a computer-generated world and featuring the most advanced CG creatures ever seen. This became *Avatar* (2009), the story an ex-marine who inhabits the body of an alien on the planet Pandora. The Na'vi—the creatures of Pandora—are tall and blue, with humanoid features.

Cameron's goal was to not only create Pandora with as many detailed 3D images as possible, but also to use motion capture technology to depict authentic emotions on CG faces. Cameron told Jackson in 2009, "The experience of creating a soulful performance is through the eyes: knowing how to rig eyes, how to light for eyes, get the reflections and refractions in the eyes… We couldn't accomplish the characters we're doing in Avatar through any kind of makeup means" (Setoodeh, 2009).

At the 2010 Academy Awards, *Avatar* won the categories of Best Art Direction, Best Cinematography, and Best Visual Effects. It eventually earned over $2.7 billion worldwide, making it the number one grossing film of all time (Box Office Mojo, 2014g). Development for a second and third *Avatar* is well underway.

Often asked about his role as a technological innovator, Cameron told Peter Jackson in 2009: "I think the simple answer is that filmmaking is not going to ever fundamentally change. It's about story-telling. It's about humans playing humans. It's about close-ups of actors. It's about those actors somehow saying the words and playing the moment in a way that gets in contact with the audience's hearts. I don't think that changes" (Setoodeh, 2009).

Bibliography

Adorno, T. (1975). Culture industry reconsidered. *New German Critique*, (6), Fall. Retrieved from http://libcom.org/library/culture-industry-reconsidered-theodor-adorno.

Andrae, T. (1979). Adorno on film and mass culture: The culture industry reconsidered. *Jump Cut: A Review of Contemporary Media*, (20), 34-37. Retrieved from http://www.ejumpcut.org/archive/onlinessays/JC20folder/AdornoMassCult.html.

Arri Group. (2014). *Digital Camera Credits*. Retrieved from http://www.arri.com/camera/digital_cameras/credits/.

Belton, J. (2013). *American cinema/American culture* (4th Edition). New York: McGraw Hill.

Bernstein, P. (2014, January 22). Here's what Sundance cinematographers think of shooting film vs. digital. *Indiewire*. Retrieved from http://www.indiewire.com/article/heres-what-sundance-cinematographers-think-of-shooting-film-vs-digital.

Billington, A. (2011, April 4). CinemaCon: James Cameron demos the future of cinema at 60fps. FirstShowing.net. Retrieved from http://www.firstshowing.net/2011/cinemacon-james-cameron-demos-the-future-of-cinema-at-60-fps/.

Block, A. (2012, February 9). Movie ticket prices hit all time high in 2011. *The Hollywood Reporter*. Retrieved from http://www.hollywoodreporter.com/news/movie-ticket-prices-increase-2011-288569.

Bordwell, D., Staiger, J. & Thompson, K. (1985). *The classical Hollywood cinema: Film style & mode of production to 1960*. New York: Columbia University Press.

Box Office Mojo. (2014a). *All Time Box Office Worldwide Grosses*. Retrieved from http://boxofficemojo.com/alltime/world/.

Box Office Mojo. (2014b). *All Time Box Office Worldwide Grosses: Adjusted for Ticket Price Inflation*. Retrieved from http://boxofficemojo.com/alltime/adjusted.htm.

Box Office Mojo. (2014c). *Genres: 3D, 1980-Present*. Retrieved from http://boxofficemojo.com/genres/chart/?id=3d.htm.

Box Office Mojo. (2014d). *1984 Domestic Grosses*. Retrieved from http://www.boxofficemojo.com/yearly/chart/?view=widedate&view2=domestic&yr=1984&p=.htm.

Box Office Mojo. (2014e). *1986 Domestic Grosses.*. Retrieved from http://www.boxofficemojo.com/yearly/chart/?yr=1986&view=widedate&view2=domestic&sort=gross&order=DESC&&p=.htm.

Box Office Mojo. (2014f). *1997 Domestic Grosses*. Retrieved from http://boxofficemojo.com/yearly/chart/?yr=1997&p=.htm.

Box Office Mojo. (2014g). *Avatar*. Retrieved from http://www.boxofficemojo.com/movies/?id=avatar.htm.

Bradley, E.M. (2005). *The first Hollywood sound shorts, 1926-1931*. Jefferson, NC: McFarland & Company.

Bredow, R. (2014, March). Refraction: Slight of hand. *International Cinematographers Guild Magazine*, 85(03), p. 24 & 26.

Cameron, J. (2010, February). *James Cameron: Before Avatar … a Curious Boy*. TED Conference. Retrieved from http://www.ted.com/talks/james_cameron_before_avatar_a_curious_boy.

Cieply, M. (2010, January 13). For all its success, will "Avatar" change the industry? *The New York Times* , C1.

Cook, D.A. (2004). *A history of narrative film* (4th ed.). New York, NY: W.W. Norton & Company.

Cunningham, T. (2013, December 4). If PG-13 Is the Moneymaker, Why Is Hollywood Cranking Out So Many R-Rated Movies? *The Wrap*. Retrieved from http://www.thewrap.com/pg-13-movies-dominated-2013-box-office-r-rated-comedies-clicked/.

Dargis, M. (2009, December 17). *A New Eden, Both Cosmic and Cinematic*. Review of *Avatar* by James Cameron. *New York Times*. Retrieved from http://www.nytimes.com/2009/12/18/movies/18avatar.html?pagewanted=all&_r=0.

DCI. (2011). *Digital Cinema Initiatives: About DCI*. Retrieved from http://www.dcimovies.com/.

De La Merced, M. (2012, January 19). Eastman Kodak files for bankruptcy. *The New York Times*. Retrieved from http://dealbook.nytimes.com/2012/01/19/eastman-kodak-files-for-bankruptcy/.

Denison, C. (2013, March 1). THX wants to help tune your home theater, not just slap stickers on it. *Digital Trends*. Retrieved from http://www.digitaltrends.com/home-theater/thx-wants-to-help-tune-your-home-theater-not-just-slap-stickers-on-it/.

Dirks, T. (2009, May 29). Movie history—CGI's evolution From *Westworld* to *The Matrix* to *Sky Captain and the World of Tomorrow. AMC Film Critic.* Retrieved from http://www.filmcritic.com/features/2009/05/cgi-movie-milestones/.
Digital Cinema Implementation Partners. (n.d.). *About Us.* Retrieved from http://www.dcipllc.com/aboutus.xml.
Dolby. (2014). *Big Sound for the Big Screen: Dolby Surround 7.1 for Movies.* Dolby Laboratories. Retrieved from http://www.dolby.com/us/en/consumer/content/movie/release/dolby-surround-7-1-movies.html.
Dolby. (2010a). *Dolby Digital Details.* Retrieved from http://www.dolby.com/consumer/understand/playback/dolby-digital-details.html.
Dolby. (2010b). 5.1 Surround sound for home theaters, TV broadcasts, and cinemas. Retrieved from http://www.dolby.com/consumer/understand/playback/dolby-digital-details.html.
Doperalski, D. (2012, March 2). Studios maneuver into cloud technology. *Variety.* Retrieved from http://www.variety.com/article/VR1118051006?refcatid=1009.
EPS geofoam raises stockton theater experience to new heights. (n.d.). Retrieved from http://www.falcongeofoam.com/Documents/Case_Study_Nontransportation.pdf.
Epstein, E.J. (2005). How to finance a Hollywood blockbuster. *Slate.* Retrieved from http://www.slate.com/articles/arts/the_hollywood_economist/2005/04/how_to_finance_a_hollywood_blockbuster.html.
Feltman, R. (2012, December 13). *Seeing double: How do 3-D movies really work? Scienceline.org.* Retrieved from http://scienceline.org/2012/12/seeing-double-how-do-3-D-movies-really-work/.
Film Releases. (2014). *Movies in 3-D for 2014 and beyond.* Retrieved from http://www.film-releases.com/film-releases-schedule-3-D.php#.
Frazer, B. (2013, May 1). Eight Technology Trends from NAB 2013. *Studio Daily.* Retrieved from http://www.studiodaily.com/2013/05/eight-trends-spotted-at-nab-2013/.
Frazer, B. (2012, February 24). Oscars favor film acquisition, but digital looms large. *Studio Daily.* Retrieved from http://www.studiodaily.com/2012/02/oscars-favor-film-acquisition-but-digital-looms-large/.
Freeman, J.P. (1998). Motion picture technology. In M.A. Blanchard (Ed.), *History of the mass media in the United States: An encyclopedia,* (pp. 405-410), Chicago, IL: Fitzroy Dearborn.
Gelt, J. and Verrier, R. (2009, December 28) "Luxurious views: Theater chain provides upscale movie-going experience." *The Missoulian.* Retrieved from http://www.missoulian.com/busi ness/article_934c08a8-f3c3-11de-9629-001cc4c03286.html.
Giardina, C. (2012, April 16). NAB 2012: James Cameron and Vince Pace aiming for 3-D profitability. *The Hollywood Reporter.* Retrieved from http://www.hollywoodreporter.com/news/james-cameron-nab-vince-pace-3-D-312312.
Haines, R. W. (2003). *The Moviegoing Experience, 1968-2001.* North Carolina: McFarland & Company, Inc.
Hancock, D. (2014, March 13). Digital Cinema approaches end game 15 years after launch. *IHS Technology.* Retrieved from https://technology.ihs.com/494707/digital-cinema-approaches-end-game-15-years-after-launch.
Hancock, D. (2013, July 17). Press Release: Digital Projection Domination Grows, with 90 Percent Penetration of Cinema Screens This Year. *IHS Technology.* Retrieved from https://technology.ihs.com/438699/digital-projection-domination-grows-with-90-percent-penetration-of-cinema-screens-this-year.
Hancock, D. (2011, November 15). The end of an era arrives as digital technology displaces 35mm film in cinema projection. *IHS Screen Digest Cinema Intelligence Service.* Retrieved from http://www.isuppli.com/Media-Research/News/Pages/The-End-of-an-Era-Arrives-as-Digital-Technology-Displaces-35-mm-Film-in-Cinema-Projection.aspx.
Hart, H. (2012, April 25). Fast-frame Hobbit dangles prospect of superior cinema, but sill theaters bite? *Wired.* Retrieved from http://www.wired.com/underwire/2012/04/fast-frame-rate-movies/all/1.
Hollywood Reporter. (2005, September 15). *Future of Entertainment.* Retrieved from http://www.hollywoodreporter.com/hr/search/article_display.jsp?vnu_content_id=1001096307.
Horkheimer, M. & Adorno, T. (1969). *Dialectic of enlightenment.* New York: Herder & Herder.
Hull, J. (1999). *Surround sound: Past, present, and future.* Dolby Laboratories Inc. Retrieved from http://www.dolby.com/uploadedFiles/zz-_Shared_Assets/English_PDFs/Professional/2_Surround_Past.Present.pdf.
IBISWorld. (2014, Feburary). Movie Theaters in the US. *IBIS World Industry Risk Rating Report 51213.* Retrieved from https://www.ibisworld.com/industry/default.aspx?indid=1244.
Indiewire. (2014, January 27). *How'd They Shoot That? Here's the Cameras Used By the 2014 Sundance Filmmakers.* Retrieved from http://www.indiewire.com/article/how-they-shot-that-heres-what-this-years-sundance-filmmakers-shot-on.
Jackson, P. (2014). *See It in HFR 3D: Peter Jackson HFR Q&A.* Retrieved from http://www.thehobbit.com/hfr3d/qa.html.
Jones, B. (2012, May 30). New technology in AVATAR—Performance capture, fusion camera system, and simul-cam. AVATAR. Retrieved from http://avatarblog.typepad.com/avatar-blog/2010/05/new-technology-in-avatar-performance-capture-fusion-camera-system-and-simulcam.html.
Jowett, G. (1976). *Film: The Democratic Art.* United States: Little, Brown & Company.

Kadner, N. (2011, May). First Dance: Arri's Alexa makes its U.S. feature debut on Prom, shot by Byron Shah. *American Cinematographer*. Retrieved from http://www.theasc.com/ac_magazine/May2011/Prom/page1.php.

Kaplan, D.A. & Duignan-Cabrera, A. (1991, July 22). *Lights! Action! Disk Drives!* Review of *Terminator 2: Judgment Day* by James Cameron. *Newsweek*. p. 54.

Karagosian, M. & Lochen, E. (2003). *Multichannel film sound*. MKPE Consulting, LLC. Retrieved from http://mkpe.com/publications/d-cinema/misc/multichannel.php.

Kaufman, D. (2014a). Technology 2014 | Production, Post & Beyond: Part ONE. *Creative Cow Magazine*. Retrieved from http://library.creativecow.net/kaufman_debra/4K_future-of-cinematography/1.

Kaufman, D. (2014b). Technology 2014 | Production, Post & Beyond: Part TWO. *Creative Cow Magazine*. Retrieved from http://library.creativecow.net/kaufman_debra/4K_future-of-cinematography-2/1.

Kaufman, D. (2011). Film fading to black. *Creative Cow Magazine*. Retrieved from http://magazine.creativecow.net/article/film-fading-to-black.

Keegan, R. (2010). *The Futurist: The Life and Films of James Cameron*. New York: Three Rivers Press.

Kindersley, D. (2006). *Cinema Year by Year 1894-2006*. DK Publishing.

Kolesnikov-Jessop, S. (2009, January 9). Another dimension. *The Daily Beast*. Retrieved from http://www.thedailybeast.com/newsweek/2009/01/09/another-dimension.html.

Koo, R. (2012, April 16). BlackMagic Design's Cinema Camera is a 2.5K RAW shooter with built-in monitor and recorder for $3K. *No Film School*. Retrieved from http://nofilmschool.com/2012/04/blackmagic-designs-cinema-camera-2-5k/.

Lally, K. (2009, September 24). Paranormal Activity—Film Review. *The Hollywood Reporter*, Retrieved from http://www.hollywoodreporter.com/review/paranormal-activity-film-review-93579.

Lang, B. (2012, April 9). Titanic 3-D: How James Cameron became a convert to 3-D conversion. *The Wrap*. Retrieved from http://www.thewrap.com/movies/column-post/titanic-3-D-team-dishes-re-release-everything-makes-conversion-difficult-was-there-36855.

Leitner, D. (2913, April 5). Back to the Future with 4K: Large-Sensor Motion Picture Cameras in 2013. *Filmmaker Magazine*. Retrieved from http://filmmakermagazine.com/68031-back-to-the-future-with-4k-large-sensor-motion-picture-cameras-in-2013/#.Uy4ITFxQm0t.

Leopold, T. (2013, June 3). Film to digital: Seeing movies in a new light. *CNN Tech*. Retrieved from http://www.cnn.com/2013/05/31/tech/innovation/digital-film-projection/.

Martin, K. (2014, March). Kong to Lift-Off: A history of VFX cinematography before the digital era. *International Cinematographers Guild Magazine*, 85(03), p. 68-73.

McKay, H.C. (1927). *The Handbook of Motion Picture Photography*. New York: Falk Publishing, Co. Retrieved from https://archive.org/details/handbookofmotion00herb.

McMahan, A. (2009). *Alice Guy Blaché: Inventing the Movies*. Retrieved from http://www.aliceguyblache.com/inventing-the-movies/home.

McMahan, A. (2003). *Alice Guy Blaché: Lost Visionary of the Cinema*. New York: Continuum 2003.

Mead, B. (2008). The rollout rolls on: U.S. digital conversions nearing critical mass. *Film Journal International*, 40.

Mendelson, S. (2013, August 6). *Pacific Rim, The Lone Ranger*, And The Necessary Evil Of 3D. *Forbes*. Retrieved from http://www.forbes.com/sites/scottmendelson/2013/08/06/pacific-rim-the-lone-ranger-and-the-necessary-evil-of-3d/.

Mondello, B. (2008, August 12). Remembering Hollywood's Hays Code, 40 years on. *NPR*. Retrieved from http://www.npr.org/templates/story/story.php?storyId=93301189.

Motion Picture Association of America (MPAA). (2011). *What each rating means*. Retrieved from http://www.mpaa.org/ratings/what-each-rating-means.

Movie Insider. (2014a). *3-D Movies 2012*. Retrieved from http://www.movieinsider.com/movies/3-D/2012/.

Movie Insider. (2014b). *3-D Movies 2013*. Retrieved from http://www.movieinsider.com/movies/3-D/2013/.

Muñoz, L. (2010, August). James Cameron on the future of cinema. *Smithsonian Magazine*. Retrieved from http://www.smithsonianmag.com/specialsections/40th-anniversary/James-Cameron-on-the-Future-of-Cinema.html.

Neal, S. (1985). *Cinema and technology: Image, sound, colour*. Bloomington, IN: Indiana University Press.

Ng, R., Levoy, M., Brüdif, M., Duval, G. , Horowitz, M. & Hanrahan, P. (2005, April). *Light Field Photography with a Hand-Held Plenoptic Camera*. Retrieved from http://graphics.stanford.edu/papers/lfcamera/.

Oscars. (2014). *The Winners – Recognizing the year's best films*. Retrieved from http://oscar.go.com/nominees.

Parce qu'on est des geeks! [Because we are geeks!]. (July 2, 2013). *Pleins feux sur – Georges Méliès, le cinémagicien visionnaire* [Spotlight—Georges Méliès, the visionary cinema magician]. Retrieved March 16, 2014 from http://parce-qu-on-est-des-geeks.com/pleins-feux-sur-georges-melies-le-cinemagicien-visionnaire/.

Pendlebury, T. (2012, January 24). What is 4K? Next-generation resolution explained. *c | net*. Retrieved from http://reviews.cnet.com/8301-33199_7-57364224-221/what-is-4k-next-generation-resolution-explained/.

Poster, S. (2012, March). President's letter. *ICG: International Cinematographers Guild Magazine*, 83(03), p. 6.

PR Newswire. (2005, June 27). The Walt Disney Studios and Dolby Bring Disney Digital 3-D(TM) to Selected Theaters Nationwide With CHICKEN LITTLE. Retrieved from http://www.prnewswire.co.uk/cgi/news/release?id=149089.

RED Digital Cinema. (2014). Shot on RED. Retrieved from http://www.red.com/shot-on-red?genre=All&sort=release_date_us:desc.

Reeve, D. (2012, April 14). 4K and the future. *Edit Geek*. Retrieved from http://dylanreeve.com/videotv/2012/4k-and-the-future.html.

Rogers, P. (2014, March). Worlds Asunder. *International Cinematographers Guild Magazine*, 85(03), p. 58-67.

Rubin, M. (2000). *Nonlinear* (4th Edition). Gainesville, FL: Triad Publishing Company.

Schubin, M. (2910, April 30). The elephant in the room: 3-D at NAB 2010. *Schubin Cafe*. Retrieved from http://www.schubincafe.com/2010/04/30/the-elephant-in-the-room-3-D-at-nab-2010/.

Science & Technology Council. (2007). *The Digital Dilemma*. Hollywood, CA: Academy of Motion Picture Arts & Sciences. Retrieved from http://www.oscars.org/science-technology/council/projects/digitaldilemma/index.html.

Science & Technology Council. (2012). *The Digital Dilemma 2*. Hollywood, CA: Academy of Motion Picture Arts & Sciences. Retrieved from http://www.oscars.org/science-technology/council/projects/digitaldilemma2/.

Schaefer, S. (2014, February 16). *The Hobbit: An Unexpected Journey*: Peter Jackson Addresses 48 FPS Controversy. *Screen Rant*. Retrieved from http://screenrant.com/peter-jackson-hobbit-48-fps-controversy/.

Schank, R. (1995). *Tell me a story: Narrative and intelligence*. Evanston, IL: Northwest University Press..

Setoodeh, R. (2009, December 28). *James Cameron - Peter Jackson*. *Newsweek*. p. 91.

Singer, M. (2011, April 12). Projecting the future of movies at 48 frames per second. *IFC*. Retrieved from http://www.ifc.com/fix/2011/04/will-the-future-of-movies-run.

Smith, G. (2013, August 9). Is the 3-D fad over? *Entertainment Weekly*. Retrieved from http://insidemovies.ew.com/2013/08/09/3-D-movies-box-office/.

Stevens, M. (2013). 2013 NAB Show recap. *Thunder::tech*. Retrieved from http://chatter.thundertech.com/post/2013_NAB_Show_recap.aspx.

The New York Times. (2012, February 8). Copyrights and Internet Piracy (SOPA and PIPA Legislation). *Times Topics*. Retrieved from http://topics.nytimes.com/top/reference/timestopics/subjects/c/copyrights/index.html.

The Numbers. (2014a). *Movie Budget and Financial Performance Records*. Retrieved from http://www.the-numbers.com/movie/budgets/.

The Numbers. (2014b). *Domestic Theatrical Market Summary for 2011*. Retrieved from http://www.the-numbers.com/market/2011/summary.

The Numbers. (2014c). *Domestic Theatrical Market Summary for 2012*. Retrieved from http://www.the-numbers.com/market/2012/summary.

The Numbers. (2014d). *Domestic Theatrical Market Summary for 2013*. Retrieved from http://www.the-numbers.com/market/2013/summary.

The Numbers. (2012a). *Jaws*. Retrieved http://www.the-numbers.com/movies/1975/0JWS.php.

The Numbers. (2012b). *Star Wars Ep. IV: A New Hope*. Retrieved from http://www.the-numbers.com/movies/1977/0STRW.php.

The Numbers. (2012c). *US Movie Market Summary for 2000*. Retrieved from http://www.the-numbers.com/market/2000.php.

Thompson, A. (2010, January). How James Cameron's innovative new 3-D tech created Avatar. *Popular Mechanics* Retrieved from http://www.popularmechanics.com/technology/digital/visual-effects/4339455.

THX. (2014). *THX Certified Cinemas*. Retrieved from http://www.thx.com/professional/cinema-certification/thx-certified-cinemas/.

Vreeswijk, S. (2012). A history of CGI in movies. *Stikkymedia.com*. Retrieved from http://www.stikkymedia.com/articles/a-history-of-cgi-in-movies .

Wood, J. (2013, October 22). What *Gravity's* Box Office Triumph Means for the Future of 3-D Film. *Wired*. Retrieved from http://www.wired.com/underwire/2013/10/gravity-future-3d-movies/.

Wootten, A. (2003, April 13). *James Cameron – part two*. *The Guardian*. Retrieved from http://www.theguardian.com/film/2003/apr/13/guardianinterviewsatbfisouthbank1.

III

Computers & Consumer Electronics

12

Personal Computers

Chris Roberts, Ph.D., and
Michael Andrews, M.A.*

Why Study Personal Computers?

- As business and technology continue to evolve, computers remain fundamental to the development and economics of every communications technology highlighted in this book, often disrupting business and technology.
- The dramatic improvements in size, power, and price of computers make their development techniques worthy of application to other fields.
- As computer usage continues rising worldwide, there are profound implications for education, communication, economics, and politics.

Introduction

Ruggeri's restaurant, which closed years ago in St. Louis, is less famous for its food than for what former head waiter and baseball legend Yogi Berra said about it in the 1950s: "Nobody goes there anymore. It's too crowded." Yogi's twisted logic also makes sense in the contemporary life of the personal computer industry, which is seeing declines in sales of its core products even as it experiences a renaissance in growth with newer and smaller products. Even as the computing industry continues to sell hundreds of millions of conventional personal computers each year, the rise of smartphones and tablets has fundamentally changed what it means to make, sell, own, and use computers. The biggest development for the personal computing industry since the last edition of this book is simple: Sales are falling for bigger, heavier desktop, laptop, and notebook computers, while sales are soaring for smaller, lighter mobile devices.

The U.S. Census Bureau (2013) estimates that nearly 45% of Americans were born since 1980, meaning that within the next few years more than half of Americas will be "digital natives" (Palfrey & Gasser, 2008) who have never known a world without personal computers. During their lifetimes, digital natives have seen PCs evolve from bulky, difficult-to-operate machines to increasingly smaller, faster, less expensive, and easier-to-use devices. For a growing number of people worldwide, especially in developing nations, their first computer will be a smartphone or tablet, not a PC (King, 2014).

Industry data show dramatic increases in the adoption and use of computers led to the boom for tablets and for sub-notebooks, an emerging category of machines with functionality and size between a laptop and a tablet. Consider that:

- The computing industry shipped 496 million desktops, laptops, sub-notebooks, and tablets in 2013—nearly one new device for every 14 people on Earth. (Include the 1.8 billion mobile phones shipped, and it's nearly one device sold per three people on the planet (Gartner, 2014a).
- While sales of non-phone computing devices rose 6% in 2013, the real growth is in tablets and other smaller, portable machines. Tablet sales rose 50% in 2013, and by 2015 the industry will sell more tablets than desktops or laptops. But desktop and laptop sales fell 12% in 2013, and the slide is expected to continue as tablet sales keep rising (Gartner, 2014a).
- The number of computer users—and computers in use—continues to grow. Sometime in 2014, the

* Roberts is an Associate Professor of Journalism at the University of Alabama (Tuscaloosa, Ala.). Andrews is a doctorial student there.

world is expected to top 2 billion computers in use—double the number in 2008 (Reuters, 2008a).

- Prices continue to fall, even as the speed and quality of computing products improve. Hardware makers in 2013 sold $622 billion worth of "smart" devices that could be connected to the Internet, but nearly 68% of the sales were for phones, tablets, and PCs selling for less than $350 (IDG, 2013). A decade ago, few computers sold for below that price point, even after accounting for inflation.
- The rise of computers and the Internet continues to aid the U.S. economy. The United States delivered $628 billion in digitally-deliverable services in 2011, and it imported another $222 billion (United States Economics and Statistics Administration, 2014).
- 79% of American households owned a computer in 2012, up from 77% in 2010, 62% in 2003 and 8% in 1984 (U.S. Census Bureau, 2014). Nearly 80% of urban households have computers, nearly 10 percentage points more than rural households.

Figure 12.1

Percentage of U.S. Households with a Computer

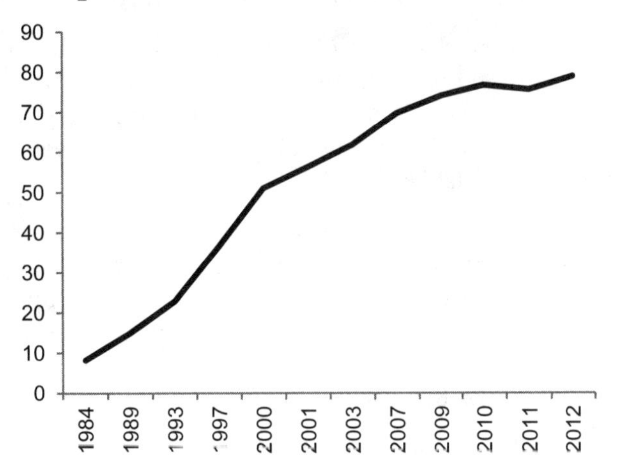

Source: U.S. Census Bureau, Population Survey, 2014

- Nearly 93 of every 100 Americans live in places that offer wired broadband access (U.S. Economics and Statistics Administration, 2013). At least 70% of Americans have broadband Internet access, up from 4% in 2000 (Pew Internet Research Project, 2013).
- Computers are moving more traffic on the Internet. Networking company Cisco estimated that the Internet moved 56 exabytes of traffic per month in 2013—the equivalent of 45.4 billion DVDs of information. That's up from 5 exabytes per month in 2005 (Cisco, 2013).
- More traffic is going mobile. The Internet moved 1.3 exabytes of data via mobile devices each month in 2013. That number is expected to rise to 11.2 exabytes per month by 2017. Put another way, mobile traffic online is expected to move from about 2% of all Internet traffic to more than 10% by 2017 (Cisco, 2013).
- Few consumer products have seen such dramatic increases in quality and declines in price. The IBM 5150, the original "personal computer" that hit the market in 1981, ran at 4.77 megahertz, had a monochrome monitor, no hard drive, and not quite enough memory to hold all the words in this chapter. It cost about $3,000, the equivalent of $7,720 after adjusting for inflation (IBM, n.d.). Today, that much money could buy dozens of computers and tablets that run hundreds of times faster and hold billions of times more data.

Background

A Brief History of Computers

As America boomed during the late 1800s, the U.S. Bureau of the Census needed a new way to meet its constitutional mandate to conduct a decennial headcount. It took seven years to compile 1880 data, too late for useful decision making. The agency hired employee Herman Hollerith, who built a mechanical counting device based on how railroad conductors punched tickets of travelers. The punched-card tabulation device helped the agency compile its 1890 results in about three months (Campbell-Kelly & Aspray, 1996) and was the first practical use of a "computer."

Hollerith, whose company later became a founding part of International Business Machines Corporation, owed a debt to 1800s British inventor Charles Babbage. While his "difference engine" was never built, Babbage's idea for a computer remains a constant regardless of the technology: Data is input into a computer's "memory," processed in a central unit, and the results delivered.

Figure 12.2
Percentage of U.S. Households with Computers by State, 2011

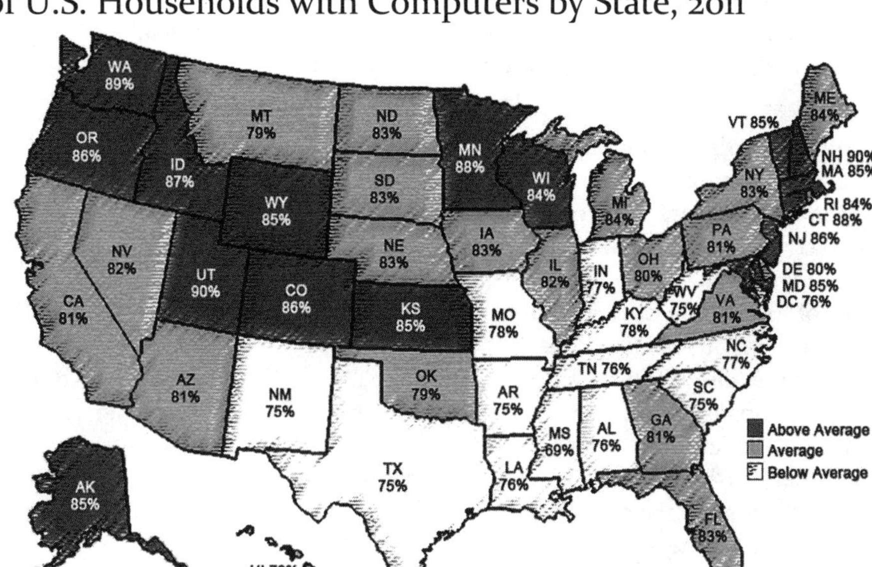

Data Source: U.S. Department of Commerce

Early computers were first called "calculators," because "computers" were people who solved math equations. The early machines were built for a specific task—counting people, calculating artillery firing coordinates, or forecasting weather. The first general-purpose computers emerged at the end of World War II in the form of the 30-ton ENIAC, the "electronic numerical integrator and computer" that could count to 50 in a second (Aguilar, 1996). The next technological leap was Sperry-Rand's UNIVAC, or "UNIVersal Automatic Computer," which reached the market in 1950.

IBM controlled two-thirds of the computing market by 1975, when the introduction of the MITS Altair 8800 became the first practical PC on the market. It used an Intel chip and software designed by a new company called Microsoft (Freiberger & Swaine, 2000). An assembled box started at $600, or $2,600 in 2014 dollars. A year later, Apple demonstrated its first computer—which, unlike the Altair, came with a keyboard. The company's Apple II machine arrived in 1977, and the company made half of the world's PC sales within three years. The early 1980s were marked by a Babel of personal computing formats, but a standard emerged when the IBM PC went on sale in August 1981. The machine was powered by an Intel chip and MS-DOS (Microsoft-Disk Operating System), the programs that manage all other programs in a computer (Campbell-Kelly & Aspray, 1996).

The two powers behind the IBM PC took different paths. IBM controlled less than one-fourth of the computer market share in 1985, but its influence in smaller computers faded as competitors delivered machines with lower prices and higher performance. The company in 2004 refocused on its server-based markets and sold its PC business to Lenovo (Lorh, 2004), which became the world's largest PC maker by 2013 (Ranii, 2013). Soon after introduction of the IBM PC in 1980s, Microsoft controlled the market for text-based operating systems and held off competitors (including IBM) in the transformation to operating systems with graphical user interfaces (GUI), which harness a computer's video capability to make the machine simpler to operate. In 1984 Apple debuted the Macintosh, which like its 1983 predecessor, the Lisa, was built upon a GUI that had an ease-of-use advantage over text-based systems and a premium price. Microsoft began selling a GUI operating system as an overlay to MS-DOS in late 1985, but its Windows software did not reach widespread adoption until its third version shipped in mid-1990. The mass acceptance of Windows gave Microsoft further dominance in the business of selling operating systems, and a Microsoft OS ran 93% of all PCs that accessed the Internet in January 2014 (Netmarketshare.com, 2014a).

Portable computing began with the 1981 introduction of the Osbourne 1, a 24-pound machine with a 5-inch screen and $1,795 price tag (or $4,600 in 2014 dollars), which first brought the notion of a portable computer to the public (Old-computers.net, n.d.). Laptops have overcome heat and weight issues to become at least as powerful as desktops, and laptops began outselling desktops in the mid-to-late 1990s (Reuters, 2008b).

Mobile tablet computers—flat-screen devices with screens larger than a cellphone and touchscreens—were first introduced in the 1990s and supported by a Windows operating system, but never reached wide acceptance. Apple's iPad introduction in 2010 led to widespread adoption of "post-PC" devices, but not without competition. The arrival of Google in 1996, first as an Internet search engine but now as both a software and hardware company, has led to competing operating systems for tablets and phones—and even notebook computers, where the company's Chromebook uses its own proprietary operating system and generally aims at customers seeking lower-priced machines.

How Computers Work

Understanding the technology ecosystem of computers introduced in Chapter 1 requires differentiating between hardware, software, and content. Hardware describes the physical components of a computer, such as the central processing unit, power controllers, memory, storage devices, input devices such as keyboards or touchscreens, and output devices such as printers and video monitors. Software is the term that describes the instructions regarding how hardware manipulates the information (data) (Spencer, 1992). The software, in turn manipulates the content processed by the computer, from documents and spreadsheets to movies and music.

Hardware

The most important piece of hardware is the central processing unit (CPU), or microprocessor, that is the brain of the computer and performs math and logic tasks. To do its work, the CPU's memory holds the data and software instructions. The memory is based upon a series of switches that, like a light switch in a house, are toggled on or off to signify the presence or absence of data.

Original memory devices required vacuum tubes, which were expensive, bulky, fragile, and ran hot. The miniaturization of computers began in earnest after December 1947, when scientists perfected the first "transfer resistor," better known as a "transistor." Nearly a decade later, in September 1958, Texas Instruments engineers built the first "integrated circuit"—a collection of transistors and electrical circuits built on a single "crystal" or "chip." Thanks to miniaturization, CPUs once the size of buildings now can rest on a fingernail. Today, circuit boards hold the CPU and the electronic equipment needed to connect the CPU with other components. The "motherboard" is the main circuit board that holds the CPU, sockets for random access memory, expansion slots, and other devices. "Daughterboards" attach to the motherboard to provide additional components, such as extra memory or graphics cards.

The CPU needs two types of memory: random access memory (RAM) and storage memory. RAM chips hold the data and instruction set to be dealt with by the CPU. Before a CPU can do its job, data quickly load into RAM—and can be quickly wiped away. RAM, until a few years ago routinely measured in "megabytes" (the equivalent of typing a single letter of the alphabet one million times) is now stated in "gigabytes" (roughly one billion letters). Microsoft's Windows 7 operating system claims to function with as little as 1 gigabyte of RAM (Microsoft, 2009). Few PCs ship with less than 1 gigabyte of RAM; standard consumer-aimed computers ship with at least 2 gigabytes. Some higher-performing tablets include 1 gigabyte of RAM.

Think of RAM as "brain" memory, a quick-but-volatile memory that clears when a computer's power goes off or when a computer crashes. Think of storage memory as "book" memory—information that takes longer to access but is stored even after a computer's power is turned off. Storage memory devices use magnetic or optical media to hold information. Major types of storage memory for today's computers include:

- *Hard drives* are rigid platters holding vast amounts of information. The platters spin at speeds of 5,400 to 15,000 revolutions per minute, and "read/write" heads scurry across the platters to move information into RAM or to put new data on the hard drive. The drives, which can move dozens of megabytes of information each second, are permanently sealed in metal cases to protect the sensitive platters. A drive's capacity is measured in gigabytes, and only the most basic desktop computers today ship with less than 500 gigabytes of hard-drive capacity. The drives almost always hold the operating system for a computer, as well as key software programs (PC World, 2007). Although nearly every computer has a built-in hard drive, external drives that plug into computers using universal serial bus (USB) or other ports are increasingly common. Many external drives are powered through the USB port, meaning the drive need not be plugged into a traditional electrical socket.

- *Solid-state drives (SSD)* use "flash" memory with no moving parts. These drives weigh less, load data into RAM at hundreds of megabytes per second, are quieter, and use less energy than traditional hard drives. But SSDs generally have less storage capacity and cost more, limiting their sales to higher-priced machines. Tablets and smartphones almost exclusively use flash memory, typically ranging from 8 gigabytes to 128 gigabytes.
- *Hybrids* are single device with small-capacity SSDs to hold operating systems to help the machine book quickly, and large-capacity hard drives to hold software and data.
- *Keydrives* (also called flashdrives or thumbdrives) are tiny storage devices that earned the name because they are the size of a thumb and can attach to a keychain. They use flash memory and plug into a computer's USB port, which also powers the drive. Some larger-capacity keydrives hold 512 gigabytes of data. Prices have plunged, with 16 GB devices selling for less than $15 as of early 2014. Prices are higher for higher capacity and faster devices that use the USB 3 standard.
- *Other flash memory devices* can be connected to PCs. Most computers ship with devices that access the small memory cards used in digital cameras, music players, and other devices. Storage capacities and access speeds continue to increase.
- *Compact discs (CD),* introduced in the early 1980s, are 12 centimeter wide, one millimeter thick discs that hold nearly 700 megabytes of data, or more than an hour of music. They ship in three formats: CD-ROM (read-only memory) discs that come filled with data and can be read from, but not copied to; CD-R discs that can be written to once; and CD-RW discs that can be written to multiple times.
- *DVDs,* known as "digital versatile" or "digital video" discs, continue to replace CDs as the storage medium of choice. They look like CDs but hold much more information—typically 4.7 gigabytes of computer data, which is more than six times the capacity of a conventional CD. DVD players and burners are standard equipment with new computers because DVD video has reached critical mass acceptance and because DVD players and burners are backward-compatible with CDs. The high-capacity standard is Blu-ray, which delivers double-layer capacity of 50 GB (Fackler, 2008). Many PCs can both read and write to Blu-ray discs. A growing number of PCs save on cost and weight by foregoing on optical drives all together.

Input and output devices are the final key category of computer hardware. Input devices—which deliver information to the computer—include keyboards, mice, microphones, scanners, and touch-sensitive screens. Output devices that deliver information from the computer to users include monitors, printers, and speakers. Other devices, such as network cards or wireless systems, let computers communicate with other devices by sending and receiving high-speed digital data signals across computer networks. Most computers and tablets come with Bluetooth, a protocol that uses short range wireless technology to connect computers to devices that include microphones, headphones, and speakers, and specially designed systems in some vehicles.

Software

Computers need software—the written commands and programs that load in the computer's random access memory and are performed in its central processing unit. The most important software is the operating system, which coordinates with hardware and manages other software applications. The operating system controls a computer's "look-and-feel," stores and finds files, takes input data and formats output data, and interacts with RAM, the CPU, peripherals, and networks. Microsoft's Windows operating systems reigns supreme in sales against competing operating systems such as Apple's OS X, Google's Chrome, UNIX, and various versions of GNU/Linux. Tablets tend to run on other operating systems, such as Apple's iOS or Google's Android.

Operating systems provide the platform for applications—programs designed for a specific purpose for users. Programmers have created hundreds of thousands of applications that let users write, make calculations, browse the Web, create Web pages, send and receive e-mail, play games, edit images, download audio and video, and program other applications. Some of those applications are designed to be operated from a Web browser. Some apps work across platforms; others are native to a single operating system.

Table 12.1

PC Operating Systems

Microsoft operating systems continue to run more than 90% of PCs in the United States

Windows 7	48%
Windows XP	29%
Windows 8/8.1	10%
Windows Vista	3%
Mac OS (multiple versions)	3%
Linux	2%
Other	5%

Source: Netmarketshare.com, January 2014

Recent Developments

The emergence of the sub-notebook and tablet markets began just a few years ago, but their sales boom is fundamentally changing the personal computer industry. The sub-notebook business has grown in both higher-end and lower-end machines. The tablet business—which barely existed in 2009—is expected to make up half of all computer sales in 2014 (Canalys, 2013). The implications are huge for both hardware and software makers, as companies in both businesses enter and exit the market.

Hardware

Sub-notebook computers generally fall into one of two general categories—small, slow, and cheap, or small, faster, and pricey.

At the lower end are products such as Google's Chromebook, which debuted in June 2011 running the Chrome operating system. Chromebooks, which generally cost between $200 and $400, require an Internet connection to access most of their data. While they remain a niche player for most users, they have found a market in schools and in other commercial channels (NPD Group, 2013) and are the lone bright spot in the low-end sub-notebook market, which held 10% of the laptop market share in 2010 before plunging when tablets took over.

At the higher end are machines such as Apple's MacBook Air and "ultrabooks," the Intel brand name for smaller computers with roughly the same power as traditional laptops and desktops. The new computers include chips that deliver the same power with fewer worries about overheating, and they are lighter because they use SSDs and omit disc drives and other heaver accessories. Intel in June 2013 released its third version of Ultrabook chips, called "Shark Bay" and these chips are able to handle both voice command and touchscreen input (Cunningham, 2013). Higher-priced machines continue to sell, but some premium PC makers have struggled. Sony held 8% of the world market share in 2013 but sold off its popular Vaio line to a Japanese consortium in early 2014 to concentrate on phones and tablets (Crothers, 2014).

The tablet has entrenched itself in a sweet spot between the too-tiny screens of smartphones and still-too-awkward laptops. Contributing to their success is near-ubiquitous Internet access provided by Wi-Fi networks and cellular companies, so users can easily move information and entertainment to and from their portable devices. Internet access is particularly vital for tablets, because most are closed systems with few or no physical connections to peripheral devices.

Apple originally dominated the market for general-use tablets, and it remains the single-largest seller of tablets. Its iPad Mini, with a 7.9-inch diagonal screen, debuted in November 2012 and shipped its second edition a year later. The original iPad, with a 9.7-inch screen, moved into its fifth generation with the October 2013 introduction of the iPad Air. But competitors are gaining on Apple by offering lower-priced devices, many of which use Google's competing Android operating system. Microsoft holds a sliver of the market with its own tablet OS. Among the top Android-based competitors is Amazon, whose Kindle series includes the reading-focused "Paperwhite" product released in 2012 and the general-purpose "Fire" series, which debuted in 2011 and likely reached $4 billion in 2013 sales (Del Rey, 2013). Other reading-focused devices, such as Barnes & Noble's Nook, are struggling in the market (Bosman, 2014). (For more on e-readers, see Chapter 14.)

Continued improvements in central processing units and other computer chips have led to the boom in smaller computing devices such as tablets and smartphones. The tablet has required new generations of chips designed exclusively for these low-powered devices, as well as heightened demand for flash storage that replace hard drives and touch-screens to replace external keyboards and other input devices.

Nearly all of these PC makers rely on Intel, which sells three-quarters of the central processing units in PCs and also has a hand in chips used for big machines such as servers and small machines such as tablets and smartphones. Its current consumer-focused chips are called "Haswell," branded as the fourth generation of i3, i5, and i7 lines and designed to support Windows, Macintosh, and UNIX operating systems (Intel, 2013). Like the "Ivy Bridge" chips they supersede, they have 22-nanometer transistors, smaller than the 45-nanometer devices that shipped just as late as 2010. (The smaller transistors use less power and create less heat.)

In the six years since laptops began outselling desktops (Eddy, 2008), desktop makers have worked to hold market share by producing more mini-desktops and more all-in-one units. New products, such as the NUC, and Apple's long-selling Mac mini are not much bigger than a paperback and require an external monitor. They are sought by companies and others that want computers that have small footprints but are not designed to be portable. Meanwhile, all-in-one units, machines that include a monitor, accounted for 11.4% of the desktop market in 2013. Led by the iMac and Lenovo products, all-in-one sales are growing as the overall PC market is falling (Chien, 2013).

Software

The emergence of tablets, the rise of stand-alone "apps," Google's challenges to Apple and Microsoft, and the slow acceptance of Microsoft's newest operating system have led to more upheaval than usual in the always-evolving software industry.

Microsoft continues to control the PC market for operating systems, with more than 90% of desktops and laptops running a version of a Microsoft platform. But the news has not been all good for Microsoft, which has seen soft acceptance of the Windows 8 OS released in October 2012. The company sold 200 million copies of Windows 8 in the first 15 months—an impressive figure, but just two-thirds the sales that Windows 7 generated in the same time period (Protalinski, 2014). As of early 2014, Windows 8 versions held just 10% of the world's OS market share. Microsoft also is facing competition from Google's Chrome OS, from open-source Linux systems, and from Apple Macintosh systems. Despite Microsoft's declines, its competitors hold less than 10% of the desktop OS market share.

Part of Microsoft's worries can be attributed to overall decline in PC sales, but also to tepid reviews of Windows 8. This operating system was designed with touch-sensitive screens in mind, but touchscreens do not ship with a large majority of PCs. Windows 8 also was designed to provide a similar look and feel across PC, tablet, and phone platforms, but the cross-pollination has not caught on. Microsoft lags well behind Apple's iOS and Google's Android mobile operating systems and is pinning its hopes on the "Surface" tablet, which sold poorly in its first generation but has received better reviews for the second-generation of Surface tablets it shipped in late 2013 (Wollman, 2013). As of January 2014, Apple controlled 55% of mobile operating systems, and Android held nearly 35% (Netmarketshare, 2014b).

Both iOS and Android serve as the backbone for applications, or "apps," that serve specific functions on portable devices. Both companies offer hundreds of thousands of third-party apps available at their online stores; Apple retains approval over apps offered through its iTunes store while Google exercises less control of the Android apps available through its Google Play store.

The rise of online stores has meant dramatic drops in physical sales of software that come on DVDs in boxes. The ability to download software may be tied to the thinking required to upload data, too, with online storage and processing systems known as "cloud computing." Cloud storage can provide much greater capacity and efficiency for file storage, but keeping information in the cloud requires a fast Internet connection and extra attention to security.

The move to the cloud continues to change the face of application software, such as the Microsoft Office suite that dominates the market for word processing, spreadsheets, and visual presentation programs. The Office 13 suite, released in January 2013, shipped with 12 editions that range from free, limited use online to enterprise sets for large organizations, and was designed with more cloud-based computing and document sharing in mind. It competes against Google Docs, a 7-year-old "software as a service" suite that runs on Web browsers with free and paid versions, depending upon the number of users in an organization and the amount of storage space sought.

Web browsers continue to be free to nearly all computer users—and continue to become more powerful. In addition to handling moving around the Web, today's browsers are designed to better handle graphics and to be the shell to run other software, including Google Docs. Microsoft's Internet Explorer versions hold nearly three-fifths of the market share, mostly because many users do not update their browsers to newer editions. Competitors such as Firefox and Google Chrome may surpass any single version of Explorer. On mobile devices, Android's browser and Safari on Apple's iOS remain supreme (Wikimedia, 2014).

Table 12.2

Web Browser Market Share

Microsoft's browser remains the leader, but many PCs users have not updated to newer versions.

Browser	Feb. 2010	Feb. 2012	Feb. 2014
Microsoft Internet Explorer	62%	53%	58%
Firefox	24%	21%	19%
Google Chrome	5%	19%	16%
Apple Safari	5%	5%	6%
Opera	2%	2%	1%
Others	2%	1%	1%

Source: NetMarketshare, February 2014

Finally, new ways to input and output continue to gain traction. For inputting information, more versions of tablets and smartphones use voice recognition software to speak commands to computers. Dragon Systems was among the first to bring voice recognition software to the market in 1985, and it has speech-to-text and vocal computer command applications for Windows, Mac, and tablet/phone systems. For output, the big news is the drastic reduction in prices and boosts in acceptance of 3D printers, robot-like devices that use digital models to print three-dimensional objects using materials ranging from plastic to metal to foodstuff. An expired patent and cheaper manufacturing costs led to a 50% jump in 2013 sales (Gartner, 2013). 3D printers remain a technology to watch.

Current Status

Americans spent more than $225 billion on computers and equipment in 2013, according to federal economic statistics. Sales of domestic computers rose to $95.7 billion in 2013, up nearly 40% from 2010 as the industry increased production within the United States. The computing industry is responsible for nearly 2% of all the goods and services produced in America. Americans imported another $133 billion in computing equipment in 2013, up 11% from 2010 (U.S. Bureau of Economic Analysis, 2014).

Regardless of PC manufacturer, Intel remains the leader in computer chip sales, making chips used in 80% of notebook and 70% of desktop units (Trefis Team, 2013). The company reported $9.6 billion in profits on $57.7 billion in sales for 2013 (Intel, 2014). Its chief PC competitor is AMD, which lost $83 million on $5.4 billion in sales during 2013 but has been boosted by sales of graphics equipment and sales for gaming machines such as Microsoft's Xbox One (Hruska, 2014).

Table 12.3

Personal computer market share, 2013

Sales of desktops and laptops worldwide (not including tablets) continue to fall worldwide. Lenovo replaced Hewlett Packard as the world's largest PC seller.

Company	Shipments	Market Share	Change from 2011
Lenovo	51,252,229	16%	-15%
HP	53,272,522	17%	17%
Dell	36,788,285	12%	-14%
Acer	25,689,496	8%	-35%
ASUS	20,030,837	6%	-4%
Others	128,934,147	41%	-10%
Total	315,967,516	100%	-10%

Source: Gartner Inc., January 2014

Among PC makers, Lenovo became the world's top maker of PCs during 2013 with 17% of the market, and since has been the only major PC maker to report a rise in sales (Gartner, 2014b). In second place was Hewlett Packard, which trailed by a single percentage point. Third place was Dell, which had been the top PC maker until 2005. The company has struggled in recent years, and hopes to right itself after a $24.4 billion buyout, the largest in tech history. The deal took the company private again after 25 years as a publicly traded company, and founder Michael Dell regained more control of its operations (Worthen, Sherr, and Ovide, 2013).

Microsoft remains the world's largest software seller, even as sales and profits are slipping. It reported profits of $16.8 billion on sales of $63 billion for the year ending December 2013, down from $23.5 billion on sales of $72 billion in 2011 (Microsoft, 2011, 2013). Those figures include sales of Xbox consoles and advertising revenue from Bing, its search engine.

Google reported profits of $12.9 billion on revenue of $59.8 billion in 2013, up from profits of $9.7 billion on revenue of $38 billion in 2011 (Google, 2013, 2011.) Nearly all of Google's revenue comes from advertising, and its 2013 revenue was hit by its $2.9 billion sale of parts of Motorola to Lenovo—less than two years after buying Motorola for $12.5 billion in hopes of becoming a bigger player in the mobile phone hardware business (Miller & Gelles, 2014).

Apple remains the largest general computing company—and among the largest in the world. It reported $171 billion in sales and profits of $37 billion in the fiscal year ending September 2013, up from $108 billion in sales and profits of $25.9 billion for 2011. Sales of Macintosh computers declined slightly between 2011 and 2013, to a little more than 16 million each year. But thanks to booming sales of its phones and tablets, Mac sales fell to just 12% of Apple's total sales, down from 20% two years ago. The company sold 150 million iPhones, 71 million iPads, and 26.4 million iPods. (Apple, Inc., 2013, 2011). The company's market capitalization fell in the past two years, to $465 billion in February 2014 from $600 billion in April 2012. If Apple were a nation, it would have the world's 31st-largest economy, nearly the value of all the goods and serviced produced each year in Malaysia, according to CIA Factbook comparisons.

Factors to Watch

Keeping up with the computer industry—and the products it makes—is like following water along a fast-moving river. No matter where you step in, the river keeps flowing as technology and the players change. Given that, these questions may help guide thinking about the industry during the next few years:

Will tablets overtake bigger personal computers? Lots of people think so. While tablet sales are predicted to double between 2013 and 2017 to account for nearly two-thirds of sales for all non-smartphone computing devices, sales for standard notebooks are expected to decline to 100 million, down by nearly half (DisplaySearch, 2014).

Not everyone is convinced. The former head of Netfix's mobile business says tablets have inherent limits that make them the wrong tool for many computing tasks. Their connectivity is limited, too, as just 12% of tablets have cellular connections, making them overly reliant on Wi-Fi networks. Moreover, there is duplication in purpose and applications of tablets and smartphones. Some consumers may split the difference with "phablets," larger-screened smartphones that serve as tablets (Bilimoria, 2014).

What happens with the world's top computer and chip makers? As PC sales decline, makers such as Sony are getting out of the business, and the rest are seeking ways to both maximize sales while looking for market niches. Sales for lower-cost PCs may be gutted by tablets, but some remain in the market while others sell high-performance models that bring higher prices. Others, such as Google, are expected to continue selling lower-cost PCs.

As for chip makers such as Intel, finding niche markets also is the plan, as the company announced efforts to make hardware to run inside wearable computing devices.

Will wearable computers catch on? Google Glass, a voice-activated Android-based computer that users wear as glasses, is expected to be sold commercially during 2014. At least half of Americans had heard of the technology as of late 2013 (NPD Group, 2014). It could be the first of many such devices or a fluke; regardless, its existence is raising ethical questions (when is it OK to record what you're watching?) and legal questions (can you drive with them?) (Klopott & Selway, 2014).

Also reaching the market the past two years are other portable devices, such as smart watches that can work in conjunction with cellphones, and wristbands that capture information about the user's health, movements, and even sleep patterns.

How will Microsoft overcome Windows 8? After scoring a hit with Windows 7, the 2012 introduction of Windows 8 was seen as a letdown. Microsoft cut prices and improved non-touch controls in its Windows 8.1 update, which was scheduled to ship in Spring 2014 as this book was going to press (Ausick, 2014).

What's next for input? Touchscreen continues to be the primary input method for tablets, but the lack of PCs with such screens contributed to the stumbling introduction of Windows 8. More PCs are expected to ship with touchscreens in the coming years, and also with software better able to recognize voice commands similar to standard technology built into many tablets and smartphones.

Eye tracking and gestural input technologies are currently showing potential promise. Both eliminate the need for a mouse and keyboard, thereby simplifying the human/computer interaction. Eye tracking was originally designed as a means to allow physically challenged people to interact with a computer. Companies such as Tobii Technology have developed systems that combine eye tracking with minimal use of keyboards (Marks, 2014). This can be built into a laptop, or attached to a desktop, laptop, or tablet. The EyeX Dev Kit is scheduled to be released during 2014.

Gestural input is a step beyond touchscreen. Rather than actually touching a PC or tablet screen, hand movements and gestures within a specified distance from the screen are read and interpreted as commands. Microsoft's Xbox has proven that the consumer-focused technology works. A potential drawback to gestural input is fatigue after prolonged use, but research continues to look for ways to alleviate this problem (Song et al, 2014).

Will the cloud reach mass acceptance outside of the business world? While online storage has become a convenient and popular way to back up and access critical files and documents, flash and hard drives continue to sell. More companies are providing cloud services, such as Apple's iCloud, the eponymous Box and Dropbox, Google Drive, and Microsoft's recently renamed OneDrive, starting with free, limited accounts and subscription services for more robust services (Personal Cloud Computing, 2014). Security remains the primary concern with cloud storage.

Will solid-state drives take command? Solid-state drives have performance and power advantages over conventional hard drives, but their per-gigabyte costs remain much higher than spinning drives per gigabyte. Solid-state drives accounted for just 6% of the half-billion drives sold in 2012, but IRH iSuppli predicts that percentage will rise sevenfold by 2017 (IHS, 2013). That same report predicts sales of CD and DVD drives will fall by half by 2017, again thanks to rising sales of tablets and smaller PCs.

Will your next printer be a 3D machine? Probably not, but Gartner predicts that lowered prices will lead more retailers to offer the machines at price points acceptable to a growing number of consumers (Lomas, 2013).

Personal Computers Visionary: Bill Gates

He's back—sort of. Microsoft co-founder **Bill Gates**, who is to personal computer software what the late Steve Jobs was to Apple hardware and style, is returning to the company as it looks for ways to return to its powerhouse status of the 1990s and early-2000s. He left day-to-day operations in 2006 to work with his philanthropic foundation, but he remained as chairman. In February 2014, he announced plans to become more active with the company, as product and technology adviser to the company's new chief executive officer (Wingfield, 2014).

Any history of personal computers would be irrelevant without mentioning Gates, who led Microsoft from its modest 1975 beginning to 2000, when the company seemed to reach its apex of power.

He was still a teenager in 1974, when Micro Instrumentation and Telemetry Systems (MITS) developed what is now known as the first general purpose personal computer. Gates and future business partner Paul Allen saw the need for software to let the average person use these computers. The program they developed for the Altair was successful, and Microsoft was launched in 1975.

By 1980, as IBM sought to enter the personal computer business, Microsoft delivered MS-DOS for the IBM PC and its clones, and it became the de-facto standard worldwide as Microsoft made billions of dollars. Its Word word-processing program, along with other business applications in its Office suite of programs, also became the biggest seller in the market. To compete against Apple's graphics-driven Macintosh operating system, Microsoft delivered the graphic interface system Windows 1.0 in 1985, and around the 1988 release of the second version Microsoft had become the world's largest PC software company (Windows, 2013). Gates was a billionaire by age 31, and stepped down as CEO in 2000.

The company responded late to the Internet, and while it still sells billions of dollars in products, it has been eclipsed (at least in popular culture and "coolness") by companies such as Apple and Google. Personal computing has evolved rapidly and drastically in the 14 years since Gates led Microsoft. It remains to be seen what direction he will take the company in this new age of mobile, Internet-connected devices.

Bibliography

Aguilar, R. (1996, February 14). ENIAC hits 50. Cnet. Retrieved from http://news.cnet.com/ENIAC-hits-50/2100-1023_3-204736.html.

Apple, Inc. (2011). 10-K Annual Report 2011. Retrieved from http://investor.apple.com/sec.cfm#filings.

Apple, Inc. (2013, October 30.) 10-K Annual Report 2013. Retrieved from http://files.shareholder.com/downloads/AAPL/2986048781x0xS1193125-13-416534/320193/filing.pdf.

Ausick, P. (2014, February 23). Microsoft Windows 8.1 improves non-touch controls, slashes prices.

Bilimoria, Z. (2014, February 6). Our love affair with the tablet is over. Recode.net. Retrieved from http://recode.net/2014/02/06/our-love-affair-with-the-tablet-is-over/.

Bosman, J. (2014, February 26). Barnes & Noble Nook unit continues to sputter. Retrieved from http://www.nytimes.com/2014/02/27/business/media/barnes-noble-nook-unit-continues-to-sputter.html.

Campbell-Kelly, M. & Aspray, W. (1996). Computers: A history of the information machine. New York: Basic Books.

Canalys. (2013, November 26). Tablets to make up 50% of PC market in 2014. Retrieved from http://pulse2.com/2013/12/03/canalys-tablets-report-97974/.

Chien, J. (2013, December 3). Digitimes Research: Global AIO PC shipments to grow 4.9% in 2014. Retrieved from http://www.digitimes.com/news/a20131127PD211.html.

Cisco. (2013, May 29). The zettabyte era—trends and analysis. Retrieved from http://www.cisco.com/c/en/us/solutions/collateral/service-provider/visual-networking-index-vni/VNI_Hyperconnectivity_WP.html.

Crothers, B. (2014, February 9). Death of the PC, Sony style: Sony's exit from the PC business may be a lesson for rivals: don't bet the farm on consumer. Retrieved from http://news.cnet.com/8301-1001_3-57618596-92/death-of-the-pc-sony-style/.

Cunningham, A. (2013, June 4). The U is for Ultrabook: Intel's low-power, dual-core Haswell CPUs unveiled. Retrieved from http://arstechnica.com/gadgets/2013/06/the-u-is-for-ultrabook-intels-low-power-dual-core-haswell-cpus-unveiled/.

Del Rey, J (2013, August 12). How big is Amazon's Kindle business? Morgan Stanley takes a crack at it. Retrieved from http://allthingsd.com/20130812/amazon-to-sell-4-5-billion-worth-of-kindles-this-year-morgan-stanley-says/.

DisplaySearch. (2014, February 6). Global tablet PC shipments to reach 455 million by 2017, according to NPD DisplaySearch. Retrieved from http://www.displaysearch.com/cps/rde/xchg/displaysearch/hs.xsl/140206_global_tablet_pc_shipments_to_reach_455_million_by_2017.asp.

Duan, H., Miller, T.R., Gregory, J., & Kirchain, R. (2013, December). Quantitative characterization of domestic and transboundary flows of used electronics. Massachusetts Institute of Technology and the National Center for Electronics Recycling. Retrieved from http://www.step-initiative.org/tl_files/step/_documents/MIT-NCER%20US%20Used%20Electronics%20Flows%20Report%20-%20December%202013.pdf.

Eddy, N. (2008, December 24). Notebook sales outpace desktop sales. eWeek. Retrieved from www.eweek.com/c/a/Midmarket/Notebook-Sales-Outpace-Desktop-Sales.

Fackler, M. (2008, February 20.) Toshiba concedes defeat in the DVD battle. New York Times, C2.

Freiberger, P. & Swaine, M. (2000). Fire in the valley: The making of the personal computer. New York: McGraw-Hill.

Gartner Inc. (2013, October 2). Gartner says worldwide shipments of 3D printers to grow 49 percent in 2013. Retrieved from http://www.gartner.com/newsroom/id/2600115.

Gartner Inc. (2014a, January 7). Gartner says worldwide traditional PC, tablet, ultramobile and mobile phone shipments on pace to grow 7.6 percent in 2014. Retrieved from http://www.gartner.com/newsroom/id/2645115.

Gartner Inc. (2014b, January 9). Gartner says worldwide PC shipments declined 6.9 percent in fourth quarter of 2013. Retrieved from http://www.gartner.com/newsroom/id/2647517.

Green Electronics Council. (2014, n.d.) Retrieved from greenelectronicscouncil.org .

Hruska, J. (2014, January 21.) AMD beats earnings estimates thanks to console sales, but APU outlook is bleak. Retrieved from http://www.extremetech.com/gaming/175190-amd-beats-earnings-estimates-thanks-to-console-sales-but-apu-outlook-is-bleak.

IDG. (2013, September 11). Tablet shipments forecast to top total PC shipments in the fourth quarter of 2013 and annually by 2015, according to IDC. Retrieved from https://www.idc.com/getdoc.jsp?containerId=prUS24314413.

IHS Inc. (2013, May 10). SSDs to be one-third of WW PC storage shipments by 2017 – IHS iSuppli. Retrieved from http://www.storagenewsletter.com/rubriques/market-reportsresearch/ssds-hdds-2017-ihs-isuppli/.

Intel. (2013) 4th generation Intel processors information. Retrieved from http://www.intel.com/content/www/us/en/processors/core/4th-gen-core-processor-family.html.

Intel. (2014, January 16). Intel reports full-year revenue of $52.7 billion, net income of $9.6 billion. Retrieved from http://newsroom.intel.com/community/intel_newsroom/blog/2014/01/16/intel-reports-full-year-revenue-of-527-billion-net-income-of-96-billion.

International Business Machines, Inc. (n.d.). The IBM PC's debut. Retrieved from http://www-03.ibm.com/ibm/history/exhibits/pc25/pc25_intro.html.

King, R. (2014, January 9). Gartner analysts suspect PC slump has "bottomed out" despite Q4 decline. Retrieved from http://www.zdnet.com/gartner-analysts-suspect-pc-slump-has-bottomed-out-despite-q4-decline-7000024993/.

Klopott, F., & Selway, W. (2014, February 25). Bloomberg News. Retrieved from http://www.bloomberg.com/news/2014-02-26/google-glass-faces-driving-bans-as-states-move-to-bar-use.html.

Lohr, S. (2004, December 8). Sale of IBM PC unit is a bridge between companies and cultures. *New York Times*, A1.

Lomas, N. (2013, October 13). The much-hyped 3D printer market is entering a new growth phase, says Gartner. Techcrunch. Retrieved from http://techcrunch.com/2013/10/02/gartner-3d-printer-market-forecast/.

Marks, P. (2014, February 18.) Fitbit for the mind: Eye-tracker watches your reading. Retrieved from http://www.newscientist.com/article/mg22129563.700-fitbit-for-the-mind-eyetracker-watches-your-reading.html.

Microsoft. (2009, n.d.). Windows 7 system requirements. Retrieved from www.microsoft.com/windows/windows-7/get/system-requirements.aspx.

Microsoft, Inc. (2011, July 20.) 10-K Annual Report 2011. Retrieved from https://onedrive.live.com/view.aspx?cid=59E8B947004DDBEC&resid=59E8B947004DDBEC!107.

Microsoft, Inc. (2013, July 18.) 10-K Annual Report 2013. Retrieved from http://www.microsoft.com/global/Investor/RenderingAssets/Downloads/FY13/Q4/MSFT_FY13Q4_10K.docx.

Miller, C., & Gelles, D. (2014, January 29). After big bet, Google is to sell Motorola unit. *The New York Times*. Retrieved from http://dealbook.nytimes.com/2014/01/29/google-seen-selling-it-mobility-unit-to-lenovo-for-about-3-billion/.

Netmarketshare.com (2014a, February). Desktop operating system market share. Retrieved from www.netmarketshare.com.

Netmarketshare.com (2014b, January). Mobile/tablet operating system market share. Retrieved from www.netmarketshare.com.

NPD Group. (2013, December 23). U.S. commercial channel computing device sales set to end 2013 with double-digit growth, according to NPD. Retrieved from https://www.npd.com/wps/portal/npd/us/news/press-releases/u-s-commercial-channel-computing-device-sales-set-to-end-2013-with-double-digit-growth-according-to-npd/.

NPD Group. (2014, January 7). Wearable tech device awareness surpasses 50 percent among US consumers, according to NPD. Retrieved from https://www.npd.com/wps/portal/npd/us/news/press-releases/wearable-tech-device-awareness-surpasses-50-percent-among-us-consumers-according-to-npd/.

Oldcomputers.net. (n.d.) Osbourne 1. Retrieved from http://oldcomputers.net/osborne.html.

Palfrey, J. and Gasser, U. (2008). Born digital: Understanding the first generation of digital natives. New York, NY: Basic Books.

PC World. (2007, July 23). How to buy a hard drive. *PC* World. Retrieved from www.pcworld.com/article/id,125778-page,3/article.html.

Personal Cloud Computing. (2014, n.d.). Personal cloud computing companies. Retrieved from http://www.personalcloudstorage.org/personal-cloud-computing-companies/.

Pew Research Internet Project. (2013, August 26). Home broadband 2013. Retrieved from http://www.pewinternet.org/2013/08/26/home-broadband-2013/.

Protalinski, E. (2014, February 13). After 15 months, Windows 8 has sold 100 million fewer copies than Windows 7 did. The Next Web. Retrieved from http://thenextweb.com/microsoft/2014/02/13/15-months-windows-8-sold-100-million-fewer-copies-windows-7/.

Ranii, D. (2013, July 10). Lenovo ranks first in worldwide PC market. *Raleigh (NC) News & Observer*. Retrieved from http://www.newsobserver.com/2013/07/10/3022367/lenovo-vaults-to-top-of-pc-market.html.

Reuters. (2008a, June 23). Computers in use pass 1 billion mark: Gartner. Retrieved from http://www.reuters.com/article/2008/06/23/us-computers-statistics-idUSL2324525420080623.

Reuters. (2008b, December 24). Laptops outsell desktops for first time. Retrieved from http://www.pcpro.co.uk/news/244450/laptops-outsell-desktops-for-first-time.

Song, J., Choa, S., Baekb, S.B., Lee, K. and Banga, H. (2014). GaFinC: Gaze and Finger Control interface for 3D model manipulation in CAD application. *Computer-Aided Design*, 46, 239-245.

Spencer, D. (1992, n.d). Webster's new world dictionary of computer terms, 4th ed. New York: Prentice Hall.

Trefis Team. (2013, January 4). A review of why Intel is worth $33 billion. Forbes.com. Retrieved from http://www.forbes.com/sites/greatspeculations/2013/01/04/a-review-of-why-intel-is-worth-33/.

U.S. Bureau of Economic Analysis. (2014, January 31). Final sales of domestic computers. Retrieved from www.bea.gov.

U.S. Census Bureau (2014, n.d.) Computer use. Retrieved from http://www.census.gov/hhes/computer/.

U.S. Census Bureau. (2013, n.d.) Age and Sex Composition in the United States: 2012. http://www.census.gov/population/age/data/2012comp.html.

U.S. Economics & Statistics Administration (2014, January). Digital economy and cross-border trade: The value of digitally-deliverable services. Retrieved from http://www.esa.doc.gov/sites/default/files/reports/documents/digitaleconomyandtrade2014-1-27final.pdf.

U.S. Economics and Statistics Administration (2013, June). Exploring the digital nation: America's emerging online experience. Retrieved from http://www.esa.doc.gov/sites/default/files/reports/documents/digitalnation-americasemergingonlineexperience.pdf.

Wikimedia. (2014, January 31). Wikimedia traffic analysis report – Browsers e.a. Retrieved from http://stats.wikimedia.org/archive/squid_reports/2014-01/SquidReportClients.htm.

Windows.Microsoft.com. (2013, n.d.). A history of Windows. Retrieved from http://windows.microsoft.com/en-US/windows/history#T1=era1.

Wingfield, N. (2014, February 5). A different Gates is returning to Microsoft. *The New York Times*. Retrieved from www.nytimes.com/2014/02/06/technology/a-different-gates-is-returning-to-microsoft.html.

Wollman, D. (2013, October 20). Microsoft Surface 2 review: a second chance for Windows RT? Engadget. Retrieved from http://www.engadget.com/2013/10/21/microsoft-surface-2-review/.

Worthen B., Sherr, I, and Ovide, S. (2013, February 5). Dell to sell itself for $24.4 billion. The Wall Street Journal. Retrieved from http://online.wsj.com/news/articles/SB10001424127887324900204578285582125381660.

13

Automotive Telematics

Denise Belafonte-Young, M.F.A.*

Why Study Automotive Telematics?

- With local building and road changes, locations are often hard to find. GPS is an integral part of wayfinding for drivers.
- Tracking devices and quick response times can alleviate stressful situations, enhance safety, and limit stolen property threats.
- Smart phone apps will provide light and horn alerts to help find a lost car, unlock the car, direct the driver to the nearest coffee shop, and much more.
- Hands-free communication and enhanced Bluetooth technology will increase "infotainment" in the car.

Introduction

Automotive telematics can be defined as "the blending of computers and wireless telecommunications technologies" (Rouse, 2007). Telematics enables drivers to get information about the location, movement, and state of their vehicles. It also enables vehicles to communicate wirelessly, which opens up a wide range of services.

Telematics is essentially a range of different features, options, and devices that are brought together by a single principle—data and communication. (Coe, Prime, & Jest, 2014b). To provide the above services, telematics products may include GPS (Global Position System), inter-vehicle Wi-Fi connections, digital audio and video solutions, wireless telecommunication modules, and car navigation systems (Cho, Bae, Chu, & Suh, 2006).

* Assistant Professor, Lynn University (Boca Raton, Florida)

Background

Ford Motor began a manufacturing revolution with mass production assembly lines in the early 20th century, and today it is one of the world's largest automakers (Ford Motor Company, 2014). The history of telematics can revert back to Henry Ford's idea in 1903 to create easy transportation for everyday people. Automobiles evolved from strictly a means of transportation to luxury items as time went on. The development of in-vehicle telecommunications, entertainment, and "infotainment" are the landmarks of today's automotive environment.

Table 13.1
Evolution of Automotive Telematics

In a GSMA study, the evolution of automotive telematics was outlined:

Telematics 1.0	Hands-free calling and screen-based navigation
Telematics 2.0	Portable navigation and satellite radio
Telematics 3.0	Introduction of comprehensive connectivity to the vehicle
Telematics 4.0	Seamless integration of mobility and the Web

Source: SBD & GSMA (2012)

The Birth of the Car Radio

The first technological breakthrough in electronic devices was the car radio. According to Gray (n.d.). "The first radios appeared in cars in the 1920s, but it wasn't until the 1930s that most cars contained AM radios." William Lear, who created the Learjet, also created the first mass market car radio. The first FM car radio was created in 1952 by Blaupunkt, a pioneer in audio equipment and systems. By 1953, the blended AM/FM car radio was developed. The first eight-track tape players appeared in cars in 1965, and from 1970 through 1977, cassette tape players made their way into automobiles. By 1984, the first automobile CD player was introduced, representing a major breakthrough. The next big turn was the introduction of DVD entertainment systems, which allowed audio entertainment options to branch into visual displays by 2002. (Gray, n.d.).

Eventually, as the digital age emerged, terrestrial radio took a back seat to a new way of listening. MP3 players, docks for smartphones and portable media players, satellite radio, Bluetooth, HDMI inputs, Wi-Fi, and voice activation technology are now common features in new automobiles.

The Influence of Telecommunications

The history of car phones began in 1946, when a driver extracted an earpiece from under his console and completed a call. The first wireless service was created by AT&T providing service within 100 cities and highway strips. Professional and industrial users such as utility workers, fleet operations, and news personnel were the first customers (AT&T, 2014). Other uses of technology in cars followed, including:

- Power steering and brake systems
- The "Idiot" light (dashboard warning lights)
- Digital and computerized fuel injection and odometer systems
- Cruise control
- Backup cameras
- Innovations in vehicle diagnostics: Onstar, FordSYNC, etc.

These innovations in telecommunications and "social" connectivity listed in Table 13.2 have expanded the definition of what "driving" is today.

Recent Developments

Phillips, Mcgee, Kristinsson, and Yu (2013), electrical engineers from the Ford Motor Company, claim, "The automobile of the future is electric. It is connected. It is smart" (par. 1). They go on to say, "Clearly, we see that the automotive industry is in the midst of a major migration toward electrified vehicle technology. This extends into the automobile where technologies such as Cellular, Bluetooth, Wi-Fi, Vehicle-to-Vehicle and Vehicle-to-Infrastructure, (V2V/V2I), and Power Line Communication (PLC) are already being integrated into new vehicles" (p. 4).

Table 13.2

Developments in Automotive Telematics

Year	Development
1998	The first hands-free car gateways were introduced
2000	The first GSM & GPS systems were brought to market
2002	Bluetooth hand free voice gateways with advanced voice integration features
2003	Integrated GSM phone with Bluetooth
2007	Multimedia handset integration is introduced
2009	Fully-integrated mobile navigation using a car GSM system
2010	3G multimedia car entertainment system
2011	Telematics and Infotainment systems introduced based on Linux

Source: Coe, Prime & Jest (2013)

Built in connectivity and GPS technologies are paving the way to hands-free safety, vehicle tracking, and diagnostic features. General Motor's OnStar system was the pioneer for subscription services in a market that now includes the Mercedes mbrace, BMW Assist, Lexus Enform, Toyota Safety Connect, Ford Sync, Hyundai BlueLink, and Infiniti Connection.

Each of these systems is used for similar applications including connectivity, navigation, and diagnostic analysis. The website Edmunds.com has sorted out the common structures on these devices which include:

"Automatic Collision Notification, Concierge Services, Crisis Assistance, Customer Relations/Vehicle Info, Dealer Service Contact, Destination Download, Destination Guidance Emergency Services, Fuel/Price Finder, Hands-Free Calling, Local Search, Location Sharing POI Communication, POI Search, Remote Door Lock, Remote Door Unlock, Remote Horn/

Lights, Roadside Assistance, Sent-To Navigation Function, Sports/News Information Stock Information, Stolen Vehicle Tracking/Assistance, Text Message/Memo Display, Traffic Information, Vehicle Alarm Notification, Vehicle Alerts (speed, location, etc), Vehicle Diagnostics, Vehicle Locator, and Weather Information." (Edmunds.com., n.d., par. 1).

Premiums and Services

Packages and customer services can be obtained through bundled basics, pay-per use, "freemium" services, and premium subscription services. (Chandna, Desai, Ernst & Young LLP, 2013).

General Motors' OnStar telematics unit has grown into a "profit margin superstar" (Guilford, 2013, May 27). As a long time innovator in wireless technology, AT&T is partnering with GM's On-Star on an arrangement to make 4G LTE high speed Internet access available in vehicles to allow a broader set of offerings such as custom apps, streaming entertainment, and enhanced diagnostic links to dealers" (Guilford, 2013, May 27).

This endeavor was thought to be an innovative way to stay ahead of the competition until 2014 when Apple announced its new CarPlay technology in Ferrari, Mercedes and Volvo vehicles. (Gagnier & Ahmed, 2014, March 3). According to Gagnier and Ahmed (2014, March 3), Carplay mimics the user interface of touch screen technology and Apple devices such as the iPhone 5 or later prototypes can be connected to the vehicle allowing the driver to utilize cellphone service, GPS navigation, and entertainment through their center console. The Apple voice command system "Siri" can be accessed to complete commands.

These innovations have created controversy, as Original Equipment Manufacturers (OEM's) are fearful that many liability issues will occur when the laws of "2 hands on the wheel" are compromised. "They're also concerned about their liability for reliability and safety issues, such as driving distractions. Automakers such as GM or Ford have lengthy testing procedures for their technology to ensure that drivers keep their eyes on the road" (Gagnier & Ahmed, 2014, March 3). It is not known as of this writing whether Apple has addressed any of these concerns.

Advanced Driver Assistance Systems and Vehicle Connectivity

According to the Journal of Intelligent Manufacturing, Advanced Driver Assistance Systems (ADAS) are advancing using camera-based systems to enhance road safety and accident avoidance. These in-vehicle display systems provide visible information on the surrounding environment, specifically to display "blind-spots" and back-up perception. Vision-based integrated ADAS systems are inevitable as a standard feature in future automobiles (Akhlaq, Sheltami, Helgeson, & Shakshuki, 2012).

What auto manufacturers consider "connected vehicles" predominately use cellular and satellite networks. This infrastructure is being analyzed by the United States Department of Transportation (USDOT) to confirm vehicle-to-vehicle safety applications. GM and Ford are paving the way to the new connected vehicle environment, which is an important factor for younger consumers who mandate connectivity in their daily lives. App developers are hard at work expanding these communication tools moving quickly to create a new connected vehicle ecology. (Row, 2013).

Mobile Internet

According to Filev, Lu, Lutetium, & Hrovat, (2013) mobile Internet is made possible by network connectivity through cellular services.. Distance from cellphone towers and location factors into the quality of reception and connections. (Filev et al., 2013). Car manufacturers are ensuring this smart technology through "in-vehicle networking and cloud-computing." As advancement in this connectivity increases, vehicles will be able to share location data and engage in vehicle-to-vehicle communication. (Mearian, 2014, January 6).

Old style navigation was passive-only; new navigation systems not only know maps but also know traffic and can route you in real-time. Navigation in phones is replacing stand-alone navigation devices because of networking capabilities. For example, Apple's Maps App uses Siri to give drivers turn-by-turn directions. Why use a stand-alone navigation system when a smartphone can do the same thing?

New federal regulations will mandate car-to-car communication (for safety) (The University of Michigan Transportation Research Institute, 2014, February 5). Services will automatically be able to call for help in an accident and keep track of the vehicle operator's driving habits (Progressive and a few other insurance companies offer discounts if you put a monitor in your car). Also, central car computers, built into most cars, are the equivalent of airplane "black boxes" that can read back speed and a lot of other factors when a car is in a collision. Conversely there are big privacy issues here. Questions may arise on who owns the data and whether drivers automatically give up this data when they take their car in for servicing and system services take place.

Another area to study is technology that enables vehicle to vehicle communication. "The U.S, government is moving on new rules requiring future vehicles come with Wi-Fi technology necessary for cars to communicate with one another. The goal is to prevent accidents, but networked cars could become much more" (Fitchard, 2014, Feb. 3).

Current Status

The Consumer Electronic Show held in Las Vegas January 2014 hosted an entire display of current trends in automotive telematics. Engineering has brought the future of transportation to our doorstep in the form of connected vehicles and self-driving cars. It is time to take connected vehicles seriously. Shirley Rowe (2013) proclaims in a report that connected vehicles are to at the forefront of automotive technology. She goes on to report a panel discussion at the ITE 2013 Annual Meeting and Exhibit, which addresses the status of "smart" vehicles to transportation industry professionals. Current trends were highlighted:

- Ford SYNC and MyFord Touch were installed on 79% of new 2013 Fords.
- By 2015 the number of networked devices will be twice the global population.
- GM plans to install 4G LTE on all new models in 2015.
- Google purchased Waze (50 million subscribers) for more than $1 billion and has incorporated Waze data into Google maps.

(Row, 2013, p. 24). With statistics on automobile safety and traffic accidents, and existing laws on "no texting and driving" rules and regulations doing little to make driving safer, a new solution has emerged...the driverless vehicle.

Driverless Cars

The ultimate use of technology in automobile technology is the driverless car. Maggie Clark (2013, July 29), from *USA Today*, surmises the future of driverless cars: "As of early 2014, the Google car is the most prominent, but almost every major auto manufacturer is building elements of the technology in the form of radar (to prevent collisions), optical sensors (to help keep a car in its lane)." Clark states that California, Nevada, Florida, and the District of Columbia are in the forefront of implementing laws and procedures for the use of the cars, executing more than 50 rules and regulations that will help enhance the expansion of driverless cars.

According to Google's research team, self-driving cars could reduce road fatalities (currently 30,000 per year in the U.S.) and injuries (more than 2 million a year in the U.S.) by up to 90 percent (Clark, 2013, July 29). The U.S. Department of Transportation is also researching the introduction of driverless cars onto American roadways and soliciting input from researchers and manufacturers.

Figure 13.1
Driverless Car

At a test track at the 2014 *Consumer Electronics Show*, cutouts of children moved in front of cars to test automatic braking systems.

Source: A. Grant

One of the targets of this innovation is commercial vehicles, including buses and taxis. Cargo units and storage trucks are also being studied as a target for self-driving technology. Imagine a world where many causes of accidents are reduced, including fatigue and night-vision hazards.

Yet current trends still include the human driver, who is embracing the life of luxury with driver entertainment and infotainment solutions. Smart cars have been in the works for quite some time. One example is the Infiniti LE prototype released at the 2012 New York International Auto Show, which included an elaborate electronics system offering a combination of navigation, gesture recognition, pupil tracking, fingerprint ignition "keys," and entertainment, along with cloud computing, vehicle tracking, auto concierge, and voice activation (Williams, 2012).

Infotainment, Entertainment, and Internet in Cars

In-car technology is thriving with manufacturers' inclusion of devices and tools that enhance entertainment and infotainment. As "connected" cars, these gadgets achieve a variety of roles to link drivers to their social media, music, movies, and more. Filev, Lu, and Hrovat (2013) describe some of the ways that telematics will change the world of infotainment:

- *Cloud Computing*—Integrated wireless technology structure enhances driving experiences.
- *The Social Car*—Tweeting at a red light? Traffic delays? Social media could play a big role in sharing information among drivers, but concerns about distracted driving need to be alleviated first.
- *Direct Streaming*–Services such as Sky On Demand and Netflix will allow passengers to download and stream movies in their vehicles. DVD technology will be a thing of the past.
- *Voice Recognition*—With the development of cellphone and tablet voice recognition systems, automobiles will use the same technology to avoid driver button pushing and touch display distractions. (Filev, et. al., 2013).

Coe, Prime & Jest (2014a) address the economics of these goods and services: "What is uncertain is how car manufacturers and Internet/media companies will charge for these services. They may build the cost into the new cars or charge monthly fees for people to use selected services from the suite of infotainment offerings. What is certain, is that to have a customer's attention in the car is worth a fortune in advertising revenue to major brands and as such there could be free offerings that are subsidized by adverts. Google has signed a deal with Kia to offer connected car features and it is predicted that Google will be at the forefront of the car infotainment industry for a long time to come."

The most significant development in this battle is Apple's announcement of its CarPlay service that is preparing to give Google and the Android system competition for control of these automotive technologies (Gagnier & Ahmed, 2014, March 3).

More manufacturers are offering the ability to deliver Internet radio and other audio services in cars; iHeartRadio is a leader in this market as it competes to supplement traditional terrestrial radio stations by offering their content online. In the meantime, Pandora the leading Internet radio service, announced it will begin rolling out in-car advertising solutions, with a client list including BP, Ford Motor Company, State Farm, and Taco Bell taking advantage of the first-to-market opportunity ("Pandora launches advertising on automotive platforms," 2014, Jan 06).

It is estimated that more than four million Pandora customers have utilized Pandora on a system offered by one of the 23 car manufacturing companies (so far) that have included this service in new cars. The resulting in-car advertising revenue subsidizes the billions of listening hours for the service. As of 2014, Pandora's Chief Marketing Officer, Simon Fleming-Wood, reported that almost half of all radio listening takes place in the car, with Pandora poised to aggressively compete in this listening environment (Pandora, 2014, Jan 06).

Other entertainment options that are being considered for in-vehicle go beyond audio and radio, including in-vehicle gaming consoles such as the Xbox, PlayStation, and other systems, as well as complete Internet access. Consider the opportunity offered for media consumption if self-driving cars become a reality and users can spend commuting time with any technology that could be built into an automobile.

Factors to Watch

The growing population in major cities and dwellings will necessitate automobile "intelligence" to help navigate through such crowded environments (Filev, Lu, and Hrovat, 2013). Drivers navigating their way in such vehicles will also use personalized information channels to find the best services and sources. Such needs will lead to further advancement of automotive controls.

It is expected that approximately 104 million vehicles will contain some form of connectivity with integrated telematics reaching 88% by 2025. Smartphone technology and new laws, structures, and regulations for drivers may ensure safer roads. The United States will continue to lead sales with approximately 16 million new cars by that time. The countries in the EU, Japan, and other nations are expected to follow suit, increasing the potential for innovation. (Ernst & Young Global Limited, n.d.)

The United States government is creating regulations for future vehicles, mandating them to be equipped with Wi-Fi technology beginning in 2016. The National Highway Traffic and Safety Administration (NHTSA) sees this automotive connectivity as a safety technology similar to seatbelt and airbag laws. Cars can tell each other when a driver is "slamming on brakes," making a turn, or signaling for a lane change. Other drivers—or the cars themselves—can react to these cues to help avoid potential accidents. Driverless cars will in turn be equipped to communicate directly with each other, bypassing human driver involvement (Fitchard, 2014, Feb. 3).

Lastly, these "vehicle monitoring systems" will pave the way to a major breakthrough in the auto insurance industry. Driving data can be used to personalize insurance premiums and produce automatic claims when accidents occur. Insurance companies will certainly use this technology to identify and court safe drivers, utilizing competitive pricing, and strengthening close customer-based relationships to help ensure privacy. (Reifel, Hales, Xu, & Lala, 2010).

In summary, automotive telematics will continue to enhance two major trends: mobility and data analytics. Dhawan (2013, December 23) explains the relationships and alliances of car manufacturers and other technological entities, "Automobile companies will start collaborating with data companies; mobile device makers will partner with healthcare institutions. Industry lines will blur and experts in one field will soon be experts in another. Not only will technology change as a result, but also our world will change." (p. 1, 7)

Bibliography

Akhlaq, M., Sheltami, T. R., Helgeson, B., & Shakshuki, E. M. (2012). Designing an integrated driver assistance system using image sensors. *Journal of Intelligent Manufacturing*, 23(6), 2109-2132. doi: http://dx.doi.org/10.1007/s10845-011-0618-1 Retrieved from http://search.proquest.com/docview/1197064297?accountid=36334.

AT&T. (2014). 1946: First mobile telephone call. Retrieved from http://www.corp.att.com/attlabs/reputation/timeline/46mobile.html.

Chandna, A., Desai, B. & Ernst & Young LLP. (2013). The quest for telematics 4.0: Creating sustainable value propositions supporting car-web integration. Retrieved from http://www.ey.com/Publication/vwLUAssets/The_quest_for_Telematics_4.0/ $File/The_quest_for_Telematics_4_0.pdf.

Cho, K. Y., Bae, C. H., Chu, Y., & Suh, M. W. (2006). Overview of telematics: A system architecture approach. Int. J. *Automotive Technology*, 7(4), 509-517.

Coe, J. Prime, R., & Jest, R. (2014a). In car infotainment. Retrieved from http://www.telematics.com/in-car-infotainment/.

Coe, J. Prime, R., & Jest, R. (2014b). Telematics fleet tracking. Retrieved from http://www.telematics.com/fleet-telematics/.

Coe, J. Prime, R., & Jest, R. (2013, May 30). Telematics history and future predictions Retrieved from http://www.telematics.com/guides/telematics-history-future-predictions/.

Clark, M. (2013, July 29). States take the wheel on driverless cars. *USA Today*. http://www.usatoday.com/story/news/nation/2013/07/29/states-driverless-cars/2595613/.

CORDIS. (n.d.). History of the deployment of transport telematics. Retrieved from http://cordis.europa.eu/telematics/tap_transport/intro/benefits/history.htm.

Dhawan, S. (2013, December 23). Three sectors to drive innovation in 2014. PCQuest. Retrieved from http://www.pcquest.com/pcquest/column/204769/three-sectors-drive-innovation- 2014.

Edmunds.com.(n.d.). Telematics chart. Retrieved from http://www.edmunds.com/car-technology/telematics.html.

Ernst & Young Global Limited. (n.d.). Integrated services for automotive manufacturers, suppliers and retailers. Retrieved from http://www.ey.com/GL/en/Industries/Automotive.

Filev, D., Lu, J., & Hrovat, D. (2013). Future mobility: Integrating vehicle control with cloud computing. *Mechanical Engineering*, 135(3), S18-S24. Retrieved from http://search.proquest.com/docview/1316626710?accountid=36334.

Fitchard, K. (2014, Feb. 3) The networked car is no longer just an idea; it will be mandated in future vehicles. Retrieved from http://gigaom.com/2014/02/03/the-networked-car-is-no-longer-just-an-idea-it-will-be-mandated-in-future-vehicles/.

Ford Motor Company. (2014). Heritage. http://corporate.ford.com/our-company/heritage.

Gagnier, S. & Ahmed, S. (2014, March 3). Apple rolls out CarPlay technology in Ferrari, Mercedes and Volvo vehicles. *Automotive News*. Retrieved from https://www.autonews.com/article/20140303/OEM06/140309988/apple-rolls-out-carplay-technology-in-ferrari-mercedes-and-volvo.

Gray, C. (n.d.). Technology in automobiles. Retrieved from http://www.qwiktag.com/index.php/knowledge-base/152-technology-in-automobiles-Content.

Guilford, D. (2013, May 27). Not satisfied with OnStar's steady profits, GM wants to create a global 4G powerhouse. *Automotive News*. Retrieved from http://www.autonews.com/article/20130527/OEM/305279958/not-satisfied-with-onstars—profits-gm-wants-to-create-a-global-.

Mearian, L. (2014, January 6). By 2018, cars will be self-aware. *Computerworld*. Retrieved from http://www.computerworld.com/s/article/9245206/By_2018_cars_will_be_self_aware.

"Pandora launches advertising on automotive platforms." (2014, Jan 06). *Business Wire*. Retrieved from http://search.proquest.com/docview/1474188620?accountid=36334.

Phillips, A. M., McGee, R. A., Kristinsson, J. G., & Hai, Y. (2013). Smart, connected and electric the future of the automobile. *Mechanical Engineering*, 135(3), 4-9. Retrieved from http://search.ebscohost.com/login.aspx?direct=true&db=bth&AN=90499459&site=ehost-live.

Reifel, J., Hales, M., Xu, G. & Lala, S. (2010). Telematics: The game changer. Retrieved from http://www.atkearney.com/documents/10192/19079b53-8042-43ea-b870-ef42b1f033a6.

Rouse, M. (2007). Telematics. Retrieved from http://whatis.techtarget.com/contributor/Margaret-Rouse.

Row, S. (2013). The future of transportation. *ITE Journal*, 83(10), 24-25. Retrieved from http://search.proquest.com/docview/1445002139?accountid=36334.

SBD & GSMA. (2012). 2025 every car connected: Forecasting the growth and opportunity. Retrieved from http://www.gsma.com/connectedliving/wp-content/uploads/2012/03/gsma2025everycarconnected.pdf .

The University of Michigan Transportation Research Institute. (2014, February 5). USDOT gives green light to connected-vehicle technology. Retrieved from http://umtri.umich.edu/what-were-doing/news/usdot-gives-green-light-connected-vehicle-technology.

Williams, S. (2012). Smarter cars offer glimpse of future for autos, marketing. *Advertising Age*, 83(19), 4-105.

14

E-books

Ashley F. Miller, M.F.A.*

Why Study The E-book?

- Amazon sells more e-books than traditional titles
- More titles are released as e-books than in print
- Nearly half of Americans have read an e-book

Introduction

The e-book is a digitized version of a book, meant to be read on a computer, e-reader, or tablet. Many of the best-selling e-book titles are digital versions of popular print books without any significant structural changes from the print form, but some also introduce animation, video, hypertext, and social networking into the reading process. The goal of the e-book is to provide the ability to read books through an onscreen interface rather than on paper.

Using the communications ecosystem perspective to analyze e-books technology, it is clear that text and images are the content. Just as in printed books and periodicals, what is written is the message being communicated. What e-book technology brings us is a new way to distribute and consume the message.

An important co-development with the e-book is the e-reader, a lightweight, handheld device designed specifically for the reading of e-books and other digital text, such as newspapers and magazines. Unlike computers, many of which are stationary or heavy to carry, and some books, which can be bulky and heavy, e-readers can hold libraries of thousands of books and only weigh half a pound. E-readers also offer the ability to download new content at any time of the day and anywhere one can connect to a computer or the Internet.

Though an e-reader is not necessary to read an e-book, and many e-book users read on computer screens, e-readers have been key to the rising commercial success of the e-book (Rainie, Zickuhr, Purcell, Madden, & Brenner, 2012). Unlike a print book, which can be read without special devices, an e-book is entirely dependent upon the user having a computer, tablet, mobile device, or a specialized e-reader to access the content.

Background

The e-book predates the e-reader by about 20 years. Michael Hart, widely considered to be the father of the e-book, began Project Gutenberg, a digital book library, in 1971, in an attempt to catalogue and make accessible all human knowledge ("Project Gutenberg," n.d.). For 20 years, the e-book continued being developed without any specialized reader and was only used by the small number of people who had access to computers.

By the 1980s, an infrastructure of personal computers was in place that could enable a large number of individuals to read e-books using their personal computers, but the e-book did not become popular. The spread of e-books would be limited until a large-scale diffusion of personal computers, the Internet, and online retailing created the right environment.

The first portable e-reader was introduced in the early 1990s. Sony's Data Discman used the compact disc as a method of storing data, which allowed for the reading of books without a bulky hard drive (Hoffelder, 2011). The Data Discman was similar to a portable gaming system—it accessed information from miniature CDs and displayed the information on

* Doctoral Candidate, School of Journalism and Mass Communications, University of South Carolina (Columbia, South Carolina)

a screen. Although primitive compared to current e-readers, it is easy to see the origins of current designs when looking at the first Data Discman. Sony continued to develop the Discman until the end of the decade, but it was never widely adopted (Hoffelder, 2011). From the creation of the first e-reader, it would take another 17 years before one was introduced to the public market that became popular.

One of the biggest technological advances in e-readers was the development of electronic ink (E Ink). A spin-off of MIT's Media Lab, E Ink Corporation combined chemistry, physics, and electronics to create electronic ink which, when used in an e-reader or tablet, is so paper-like it can fool the reader's eye. Electronic ink is really millions of small capsules, each containing black and white particles. A device such as an e-reader sends an electric charge to each capsule making it appear either black or white thus forming words or images on the screen; in color E-ink, the black and white capsules appear under a color filter (Carmody, 2010).

Sony again led the market in 2006, when it offered the Sony Reader, the first e-reader to offer an E Ink display. Unlike a normal backlit computer screen, E Ink screens behave more like normal paper and require external light to be read, making the experience much more like reading a paper book. Even with this advance, it was not until Amazon entered the scene the following year that the e-reader took off (Rainie et al., 2012).

Amazon, the online retailer that began as a book store and remains primarily invested in book sales, released the Kindle Reader in 2007, just in time for the holiday season. The release marked the first time that an e-reader was a popular success (Patel, 2007). The Kindle was not markedly different from the Sony Reader; the primary difference that allowed the Kindle to be successful was the existence of a company that already sold many books through online interactions. Though the public at large did not warm immediately to the e-reader, Amazon's customers had been primed for the e-book by purchasing books online and reading onscreen text for years. Amazon had a large enough reading customer base that, unlike Sony, they had a target audience that was easy to advertise to through their normal, day-to-day sales. Also, in 2008 Oprah Winfrey claimed the Kindle as one of her "favorite things," sparking sales (Murph, 2008).

Figure 15.1
DATA Discman

Source: Peter Harris (Creative Commons Photo)

(Sony's failure and Amazon's success are an excellent example of the importance of the pre-diffusion theory discussed in Chapter 4. While Sony's focus and expertise was producing and distributing hardware, Amazon had expertise in distributing both hardware and software, making it much easier for the Kindle to achieve the diffusion threshold.)

With the Kindle came something new for e-books: specialized formatting for different devices. Before the Kindle, e-books were in universal formats such as PDF or plain text, but with the release of the Kindle, Amazon introduced a proprietary format that can only be read through the correct kind of device or software. Barnes & Noble's Nook followed in 2009 with a format that could not be read on the Kindle. Finally, setting the stage for the current e-reader marketplace, Apple released its iPad tablet in 2010. The iPad, like a computer, allows the reader to access both Kindle and Nook formats, but it also enables readers to download proprietary e-books from Apple's iBookstore that cannot be read on the other machines.

With Amazon.com, Barnes & Noble, and Apple all having significant online retail experience and the improvements in bandwidth making downloading e-books a simple and fast process, consumers have almost immediate access to any content they want. Early adopters of the e-reader had to download an e-book to their computer before they could transfer it to their e-reader, but e-readers now have Wi-Fi and

3G/4G broadband access, which allows consumers to search for, buy, download and read e-books no matter where they may be (Enderle, 2012).

As we look at recent developments in e-readers, we have to look at two categories: dedicated e-readers and multifunctional tablets. After several false starts with Sony's Data Discman in the 1990s and early 2000s, dedicated e-readers finally hit their stride in 2007 when Amazon launched the Kindle (Kozlowski, 2010). Initially e-readers were limited to storing and displaying e-books , but that changed in 2010 with Apple's introduction of the iPad.

The iPad is a tablet computer that has many functions, only one of which is the ability to read e-books . In fact, the iPad is much more similar to a laptop than to an e-reader in every way except that it is much lighter than a laptop (Buchanan, 2010). The iPad lacks the E Ink display, so the reading experience is much more like reading on a computer (Reardon, 2011). Nevertheless, the iPad altered the landscape of the e-reader marketplace by introducing more functionality to a device that has almost all of the benefits of an e-reader.

The iPad was a game changer. Amazon and Barnes & Noble realized that they needed to keep pace by offering consumers a product that had at least some of the advanced capabilities of the iPad. The race for larger screens, email, and Web browsing on e-readers began, and soon both companies offered tablets (Kendrick, 2012). Before the 2011 holiday season, Amazon introduced the Kindle Fire, and Barnes & Noble introduced the Nook Tablet. Both of these devices offer capabilities similar to those of the iPad, despite coming from the lineage of their E ink, book-reading based predecessors (Reardon, 2011).

Recent Developments

E-readers

Though the Kindle has dominance over the e-reader market, the tablet has surpassed the e-reader in overall sales (Rainie, 2014). While the number of dedicated e-readers owned continues to rise, the growth rate is on the decline; tablets, on the other hand, continue to expand their market share (Bensinger, 2013). Furthermore, those who read e-books are nearly as likely to do so on a tablet as on an e-reader (Rainie, 2014) See Figure 14.1.

In April 2012, Barnes and Noble introduced a Nook that had LED lighting, the Nook Simple Touch Glow Light. Amazon followed this in October 2012 with the release of the Kindle Paperwhite, which also had a built-in light. In September of 2013, Amazon released the second generation of the Kindle Paperwhite, with higher contrast e-ink. The following month, Barnes and Noble released the Nook Glowlight, which offers twice as much memory as the current Paperwhite. Amazon bought book rating system Goodreads and announced at the end of 2013 that it would be integrating it with Kindle, though how and what effect it will have on the marketplace is unclear (Cramer, 2014).

Figure 14.1

As Tablet Ownership Grows, More Use Them for E-books

Among all e-book readers age 18 and older, the percent to read e-books on each device.

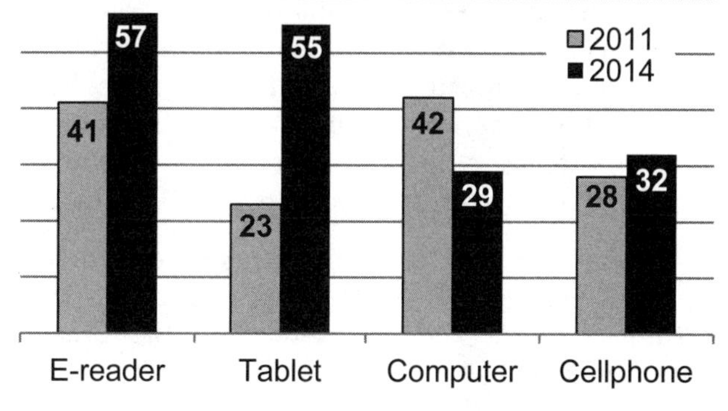

Source: Pew Research Center

Format

When Michael Hart began Project Gutenberg in 1971, his desire was to provide 99% of the public with access to free or very low-cost versions of books and documents that were already in the public domain. His goal was to select texts that would be of interest to the general public and make those texts accessible, readable and searchable on 99% of the hardware that anyone, anywhere might run (Hart, 1992). Selecting which texts to input to reach this 99% goal was a "best guess" process of selection by Hart and the volunteer staff working on the project. Making those texts available to the largest possible number of users was accomplished by selecting a ubiquitous format that ran on every computer and operating system, from DOS and Apple to UNIX and mainframe computers: plain ASCII text.

Table 14.1
Different E-reader Formats

e-reader	Format										
	DOC	EPUB	HTML	Kindle (AZW, TPZ)	MOBI,	Nook	PDB	PDF	PRC	RTF	TXT
Amazon (Kindle)	✦*		✦*	✦	✦			✦	✦		✦
Apple (iPad, iPod, iPhone)	✦	✦	✦	✦**		✦**		✦			
Barnes & Noble (Nook)		✦					✦	✦			
Sony Reader	✦*	✦						✦		✦	✦

*Through conversion **Using Apps

Source: Technology Futures, Inc. *Data:* (BookFinder.com, n.d.)

As e-books became a venture for publishers and booksellers, the formats for accessing the content became less universal. In addition to the various devices which are used for reading e-books , over the past several years, the various formats have fragmented e-book consumers into groups that own a particular device and buy content from the supplier(s) which delivers content readable on that device.

To give an idea of the hodgepodge of formats, the four leaders in the e-book market provide e-readers capable of reading e-books in the formats shown in Table 14.1.

All readers support PDF files, which might identify this as a possible universal format, but PDF does not offer many of the benefits of the e-reader, including resizing and searching the text; it just shows up as an image. Almost all of the platforms support the EPUB format, the closest to a universal format that is offered, but the largest of the suppliers, Amazon, does not support this standard. Although there have been some attempts and pressure to create a universal standard, none is forthcoming (Lee, Guttenberg & McCrury, 2002).

This forcing of formats onto users is a matter of convenience and cost (Schofield, 2010). It is convenient for a reader to stay with one supplier of content because switching to another supplier would mean switching to another e-reader and either converting original files to a format readable by the new device or in some cases having to repurchase entire libraries of books, which dramatically increases the cost beyond the initial reinvestment in new hardware. Even if a reader buys an e-reader or tablet which has the capability of accessing content from the different suppliers through apps, there is the inconvenience of switching between apps to view desired e-book and an inability to easily search across an entire library of e-books .

Availability

Sales of e-books have grown enormously over the last several years (Rainie et al., 2012). This increase is primarily due to e-book retailers overcoming perhaps the biggest obstacles that kept e-books from taking off—content availability. Because of the low numbers of people with e-readers, retailers did not want to commit to large-scale availability of content until there were enough reading devices in circulation to warrant the expense of converting the content into electronic format; consumers did not want to invest in a reading device, which can cost upwards of $300, if there was limited content.

While there are many e-books that are not available through any individual bookseller, millions of titles are available on each. Furthermore, there is content that is available only in e-book format. In fact, there are far more books published in e-book form now than there are books published in print (Bradley et al., 2011).

Another major factor in availability was, and is, cost. In general, adopters of e-readers and e-books are wealthier than the average American (Rainie et al., 2012), but the cost of e-readers has dropped precipitously over time, and, as of 2014, the least expensive

Kindle is $69. It is still a major investment to people who can get books free from the library, borrow them from friends, or sell books after they have been read.

With the capability to carry hundreds of books in a device smaller and lighter than one paperback book, e-textbooks seem like a natural choice for students not only in the U.S., but also around the world. E-textbooks are not only lighter, but are far easier to update and keep current. However, using a textbook is different than reading a novel or newspaper. Students want to be able to highlight, take notes, and flip through pages easily. Current e-reader technology addresses these issues with some basic features. With improvement in these features, the market for e-textbooks should increase. There are also economic factors. Since they can't resell their used books as they can with printed textbooks, students need to be convinced that they will save money over time when considering the combined cost of the e-reader and the e-textbooks.

With schools giving away iPads and tablets to students, full integration of e-books into the classroom looks more possible (Jones, 2014; Kaneshige, 2014). In addition, open source free e-texts have been developed to cut the cost. In California, two laws were passed to create free online and low cost hard copy ($20) texts for 50 core lower division courses at the University of California, California State University, and California Community College system. A state council made up of faculty will determine the courses and review the texts. The California Digital Open Source Library will host the texts (Garber, 2012).

Cost of content for e-readers is an important factor in the continued diffusion of e-reader technology. Realizing the importance of content to stimulate demand for e-readers, Amazon.com lowered the cost of its extensive catalog of titles to $9.99 or less in 2010. This made the content a loss leader, which led to an increase in Kindle purchases. Self-published authors are selling their books at even lower prices and the price of the average e-book bestseller is $6.50 (Cramer, 2014). Also important is the fact that many of the best-selling e-books are not being released in any other format. Self-published e-books are among the most successful books sold by Amazon (Pressman, 2014).

The e-book not only has made books more easily accessible and usable, but also has increased access of the bookselling market to self-published authors, circumventing the publishing mechanism of print. E-books have led to greater availability of content and greater availability of the market to authors. The Kindle has been opened up to allow users to borrow e-books from their local libraries and to lend e-books they have bought electronically to friends who also have an e-reader. For the moment, these exchanges are limited to people who share the same kind of device—a Kindle can loan to a Kindle, but not to a Nook. Public libraries also loan e-books to patrons, who can load them on their e-readers. Like library books, the content is available for a limited amount of time before having to be returned and, unlike physical books, technical difficulties and complexities can make them difficult to access (Bradford, 2013).

In 2010, when unveiling its iBookstore for the new iPad, Apple joined with five publishers to offer e-books on the "agency" model. This model of e-commerce, which Apple and other online retailers have used to sell a large variety of applications and products, allows the publisher to set the price with Apple being paid a percentage of each sale. Under pressure from publishers, Amazon.com agreed in 2010 to work under the agency model as well. But the agency model has been tied to an increase in the cost of books to the end user. After nearly two years under this model, the U.S. Department of Justice (DOJ) filed a suit against Apple and five publishers for colluding to keep the cost of e-books artificially high. The DOJ claims that this price fixing is harmful to consumers (Catan, Trachtenberg, & Bray, 2012). In 2013, Apple lost their initial court case but indicated that they are planning to appeal (Kessenides, 2014).

Current Status

2013 was an interesting year for the e-book - tablet growth outpaced dedicated e-reader growth, Apple lost a major court case for price-fixing (though they are appealing), and the growth of e-books as a percentage of all books stagnated. The place of e-books continues to grow and change, with readers leaving computers as places of reading in favor of other formats, including cell phones. Of those who read e-books , tablet use for e-reading (55%) is just behind dedicated e-readers (57%) (Rainie, 2014).

The holiday season of 2013 led to major growth in both the tablet and e-reader market; as of January 2014, half of Americans own either a tablet or an e-reader, up from 43% in September 2013 (Rainie,

2014). A quarter of Americans own both a tablet and a dedicated e-reader (Rainie, 2014).

E-readership is up among every demographic except 65+, but there are major differences in the reading habits between these age groups. Nearly half of young people have read an e-book in the last year, suggesting that e-books may become more popular over time. (Rainie, 2014). Despite their popularity, however, e-books remain most popular with people who are avid readers and those who still read print as well (Rainie, 2014).

Figure 14.2:

Almost Half of Readers Under 30 Read an E-book in the Past Year

Among those in each age group who read at least one book in the past year, the percent who read an e-book during that time

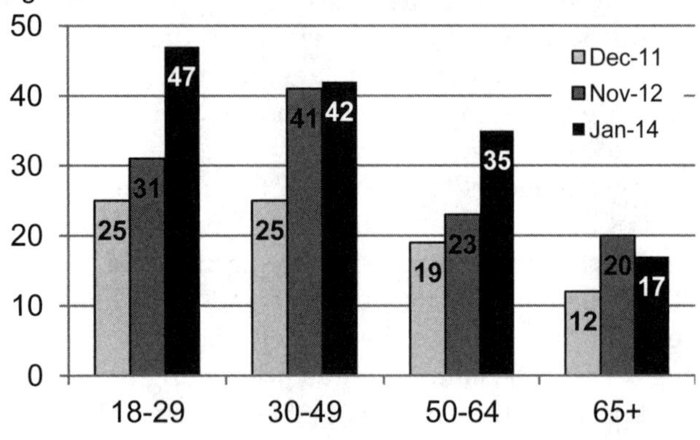

Source: Pew Research Center

E-book sales are stagnant at about 30% of all book sales; they also account for only 15% of the money made from book sales (Greenfield, 2013). In previous years, growth had been explosive, but it is unclear whether e-books are suffering a temporary lull or if, as a percentage of the book market, e-books have found saturation.

Factors to Watch

Clearly further diffusion of e-book and e-reader technology would benefit from development of a standard format for the content, or the acceptance of an existing one. The interoperability achieved could help e-books reach the diffusion level predicted. However, the issue of digital rights management (DRM) still presents a drawback for the technology, introducing restrictions that make e-books less attractive than print books. We've all read a print book and passed it along to a friend. The ability to do so with an e-book is severely limited.

The continued war among e-readers is also important. It looks like devices that have more to offer than just e-books are going to ultimately be the winner here. This probably means e-books that are much more interactive and complicated than just text will become standard. The e-ink reader is likely to become obsolete as people who have grown up reading on cellphone screens and tablets become the primary market. Marketing for e-readers recommends them as an addition to a tablet for reading in the sunlight.

The huge growth in the number of titles published shows no signs of slowing, meaning that the availability of content is going to be overwhelming. Soon we will need better ways of sorting through all of the available content to be able to find what we are looking for. There has been a lot of progress in terms of matching content to consumer through sites such as Netflix and YouTube, which have always had an enormous amount of content, and undoubtedly the online system needs to get better at finding books for people. Though there has been some development in this area, there is no leader in the marketplace or service with a large number of users.

It will also be interesting to watch the sales figures of traditional book titles and their relationship with e-book titles. Though e-book sales figures grew an incredible 72.3% in January 2012 over the previous year, traditional print titles also saw overall growth, though it was much smaller (Souppouris, 2012). The e-book seems to be expanding the number of books purchased overall, without necessarily taking market share away from traditional print books (Rainie et al., 2012), but prolonged expansion of both print books and e-books seems unsustainable. As of 2013, however, the percentage of book sales that e-books make up has stagnated at 30% of units and only 15% of dollars (Greenfield, 2013).

Finally, it will be important to monitor the legal arena as manufacturers, publishers, distributors, and consumers battle over the structure of the market and the revenues that will be produced. The DOJ's anti-trust suit against Apple and other publishers may be just an early skirmish in a battle to control consumer access to content. That Apple has lost the first round could be good news for consumers, but the outcome of Apple's appeal could have a major impact on pricing.

E-books Visionary: Michael Hart

If there is one figure who looms large over the world of e-books , it is **Michael Hart**, the creator of Project Gutenberg. Generally credited as the inventor of the e-book, Hart spent 40 years of his life devoted to making e-books . Initially, gaining access to a very early computer network in 1971, when he was an undergrad, he began typing books by hand and posting them for people to download at will (Jensen, 2011). He started with the Declaration of Independence, and moved on to other books free from copyright, like the works of Shakespeare and Mark Twain (Flood, 2011).

It is important to remember that, at the time he first started making this content available, to read the Declaration of Independence, or any piece of text, one had to either buy a book or physically go to the library. Hart was the first person to think of electronics as a way to share texts and books openly and for free. Hart beat the invention of the World Wide Web by over 20 years, the invention of the e-reader by over 25, and the release of mainstream commercial books in e-format by over 30. It is rare for someone to be so completely ahead of his or her time.

Although Hart died in 2011, Project Gutenberg lives on. As of March 2014, Project Gutenberg has over 42,000 books available for free to download on any device. Over 100,000 free e-books are available online through partners of Project Gutenberg ("Project Gutenberg," n.d.).

"One thing about eBooks that most people haven't thought much is that eBooks are the very first thing that we're all able to have as much as we want other than air" ("Michael S. Hart," 2011).

Bibliography

Bensinger, G. (2013, January 4). The E-reader Revolution: Over Just as It Has Begun? *Wall Street Journal.* Retrieved from http://online.wsj.com/news/articles/SB10001424127887323874204578219834160573010.

Bradford, K. T. (2013, June 15). Paper rules: Why borrowing an e-book from your library is so difficult. *Digital Trends.* Retrieved April 1, 2014, from http://www.digitaltrends.com/mobile/e-book-library-lending-broken-difficult/.

Bradley, J., Fulton, B., Helm, M., & Pittner, K. A. (2011). Non-traditional book publishing. *First Monday, 16*(8). Retrieved from http://ojphi.org/htbin/cgiwrap/bin/ojs/index.php/fm/article/viewArticle/3353.

Buchanan, M. (2010, March 5). Official: iPad Launching Here April 3, Pre-Orders March 12. *Gizmodo.* Retrieved from http://gizmodo.com/5486444/official-ipad-launching-here-april-3-pre+orders-march-12.

Carmody, T. (2010, November 9). How E Ink's Triton Color Displays Work, In E-readers and Beyond | Gadget Lab. *WIRED.* Retrieved April 1, 2014, from http://www.wired.com/2010/11/how-e-inks-triton-color-displays-work-in-e-readers-and-beyond/

Catan, T., Trachtenberg, J. A., & Bray, C. (2012, April 12). U.S. Alleges E-book Scheme. *Wall Street Journal.* Retrieved from http://online.wsj.com/news/articles/SB10001424052702304444604577337573054615152.

Cramer, M. L. (2014, February 21). The State of Ebooks. *EContent Magazine.* Retrieved March 2, 2014, from http://www.econtentmag.com/Articles/Editorial/Feature/The-State-of-Ebooks-94237.htm.

Enderle, R. (2012, April 3). Anticipating the 4th-gen iPad. *TG Daily.* Retrieved April 9, 2012, from http://www.tgdaily.com/opinion-features/62504-anticipating-the-4th-gen-ipad.

Flood, A. (2011, September 8). Michael Hart, inventor of the ebook, dies aged 64. *The Guardian.* Retrieved from http://www.theguardian.com/books/2011/sep/08/michael-hart-inventor-ebook-dies.

Greenfield, J. (2013, October 30). Study: Ebook Growth Stagnating in 2013 | Digital Book World. Retrieved from http://www.digitalbookworld.com/2013/study-ebook-growth-stagnating-in-2013/.

Hart, M. (1992, August). The History and Philosophy of Project Gutenberg. *Project Gutenberg.* Retrieved April 9, 2012, from http://www.gutenberg.org/.

Hoffelder, N. (2011, October 9). Blast from the Past: Sony Data Discman DD-S35. *The Digital Reader.* Retrieved March 2, 2012, from http://www.the-digital-reader.com/2011/10/09/blast-from-the-past-sony-data-discman-dd-s35/.

Jensen, M. J. (2011, September 12). Michael Hart, 1947-2011, Defined the Landscape of Digital Publishing. *The Chronicle of Higher Education.* Retrieved from http://chronicle.com/article/Michael-Hart-Who-Defined-the/128953/.

Jones, N. (2014, February 4). School gives an iPad to every pupil. *New Zealand Herald.* Retrieved from http://www.nzherald.co.nz/nz/news/article.cfm?c_id=1&objectid=11196246.

Kaneshige, T. (2014, February 13). Will iPads in the Classroom Make the Grade for Students and Teachers? *CIO*. Retrieved March 2, 2014, from http://www.cio.com/article/748212/Will_iPads_in_the_Classroom_Make_the_Grade_for_Students_and_Teachers_.

Kendrick, J. (2012, March 14). Kindle Fire: Blurring the tablet and ereader markets. *ZDNet*. Retrieved from http://www.zdnet.com/blog/mobile-news/kindle-fire-blurring-the-tablet-and-ereader-markets/7148.

Kessenides, D. (2014, February 28). With E-book Appeal, Apple Sets Its Sights on the Supreme Court. *BusinessWeek: Technology*. Retrieved from http://www.businessweek.com/articles/2014-02-28/apples-e-book-war-moves-up-a-notch.

Kozlowski, M. (2010, May 17). A brief history of eBooks. *Good e-reader*. Retrieved April 9, 2012, from http://goodereader.com/blog/electronic-readers/a-brief-history-of-ebooks/.

Lee, K.-H., Guttenberg, N., & McCrary, V. (2002). Standardization aspects of eBook content formats. *Computer Standards & Interfaces*, *24*(3), 227–239. doi:10.1016/S0920-5489(02)00032-6.

Michael S. Hart. (2011, September). *Project Gutenberg*. Retrieved March 2, 2014, from http://www.gutenberg.org/wiki/Michael_S._Hart.

Murph, D. (2008, October 25). Oprah calls Kindle "her new favorite thing," gives everyone $50 off. *Engadget*. Retrieved March 2, 2014, from http://www.engadget.com/2008/10/25/oprah-calls-kindle-her-new-favorite-thing-gives-everyone-50/.

Patel, N. (2007, October 21). Kindle sells out in 5.5 hours. *Engadget*. Retrieved March 2, 2012, from http://www.engadget.com/2007/11/21/kindle-sells-out-in-two-days/.

Pressman, A. (2014, February 14). The book industry isn't dying, it's thriving with an ebook assist. *Yahoo Finance*. Retrieved March 2, 2014, from http://finance.yahoo.com/blogs/the-exchange/the-book-industry-isn-t-dying--it-s-thriving-with-an-ebook-assist-191025547.html.

Project Gutenberg. (n.d.). *Project Gutenberg*. Retrieved March 2, 2012, from http://www.gutenberg.org/wiki/Main_Page.

Rainie, L. (2014, January 16). E-Reading Rises as Device Ownership Jumps. *Pew Research Center's Internet & American Life Project*. Retrieved from http://www.pewinternet.org/2014/01/16/e-reading-rises-as-device-ownership-jumps/.

Rainie, L., Zickuhr, K., Purcell, K., Madden, M., & Brenner, J. (2012, April 4). The rise of e-reading. *Pew Internet*. Retrieved April 9, 2012, from http://libraries.pewinternet.org/2012/04/04/the-rise-of-e-reading/?src=prc-headline.

Reardon, M. (2011, December 6). An e-reader or tablet for Christmas? *CNET*. Retrieved April 9, 2012, from http://news.cnet.com/8301-30686_3-57336098-266/an-e-reader-or-tablet-for-christmas/.

Schofield, J. (2010, August 15). eBook DRM: Can't succeed with it, won't thrive without it. *Trusted Reviews*. Retrieved April 9, 2012, from http://www.trustedreviews.com/opinions/ebook-drm-can-t-succeed-with-it-won-t-thrive-without-it.

Souppouris, A. (2012, March 30). Ebooks, young readers stimulate publishing industry growth. *The Verge*. Retrieved from http://www.theverge.com/2012/3/30/2913366/ebook-sales-by-demographic-january-2012-aap.

15

Video Games

Brant Guillory, M.M.C.*

Why Study Video Games?

- 72% of American households play computer and video games (ESA, 2012).
- 55% of gamers play on mobile phones and other mobile devices (ESA, 2012).
- Video game hardware and games generate billions in revenue each year.
- New gaming devices and games are pushing the boundaries of creativity and technology.

Introduction

Video gaming is a multibillion dollar industry whose cultural penetration far belies its roots in entertaining mental exercises for overeager engineers. Video games include a variety of hardware and software, as well as multiple delivery and distribution models. Gamer culture has served as the subtext for successful media franchises, such as CBS's *The Big Bang Theory* and Disney's *Level Up*. Major video game conferences and events, such as E3 and Penny Arcade Expo (PAX) pull in tens of thousands of attendees.

In monetary terms, video games are easily competitive with the largest media. The opening day sales of *Call of Duty: Black Ops* exceeded $360 million worldwide, easily exceeding the largest-ever weekend movie gross through mid-2012 (*The Avengers*) and the first-day sales of the final *Harry Potter* book (TechWeb 2010; Thorsen, 2009). Additionally, videogame franchises are themselves becoming hot media properties, with such series as *Resident Evil, Bioshock, Fallout, HALO,* and *Mass Effect* spawning multiple sequels, websites, videos, downloadable games, and physical content ranging from action figures to comic books and coffee mugs.

Background

"Video games" as a catch-all term includes games with a visual (and usually audio) stimulus, played through a digitally-mediated system (see Figure 15.1). Video games are available as software for other digital systems (home computers, mobile phones, tablets), standalone systems (arcade cabinets), or software for gaming-specific systems (platforms). There have also been tentative forays into games delivered through set-top boxes and digital integration with offline games.

A video game system will have some form of display, a microprocessor, the game software, and some form of input device. The microprocessor may be shared with other functions in the device. Input devices have also evolved in sophistication from simple one-button joysticks or keyboards to replicas of aircraft cockpits and race cars. Recent controllers have integrated haptic feedback (enabling users to "feel" aspects of a game), as well as accelerometers that detect the movements of the controllers themselves. Finally, systems like Microsoft's Kinect enable games to be played using three-dimensional detection of the actions taken by a player's body. These systems include cameras used as input devices into the video game system.

Video gaming has advanced hand-in-hand with the increases in computing power over the past 50 years. Some might even argue that video games have pushed the boundaries of computer processors in

* Senior Consultant at Harnessed Electrons, and Editor-in-Chief of GrogNews.com, Raleigh, North Carolina

their quest for ever-sharper graphics and increased speed in gameplay. From their early creation on large mainframe computers, video games evolved through a variety of platforms, including standalone arcade-style machines, personal computers, and dedicated home gaming platforms.

As media properties, video games have shared characters, settings, and worlds with movies, novels, comic books, non-digital games, and television shows. In addition to a standalone form of entertainment, video games are often an expected facet of a marketing campaign for new major movie releases. Media licensing has become a two-way street, with video game characters and stories branching out into books and movies as well. As video gaming has spread throughout the world, the culture of video gaming has spawned over two dozen magazines and countless Web sites, as well as industry conventions, professional competitions, and a cottage industry in online "farming" of in-game items in massive multiplayer online role-playing games (MMORPGs).

Although some observers have divided the history of video games into seven, nine, or even 14 different phases, many of these can be collapsed into just a few broader eras, as illustrated in Figure 15.2, each containing a variety of significant milestones.

Figure 15.1
Video Game Genres

Sports

Space

Military

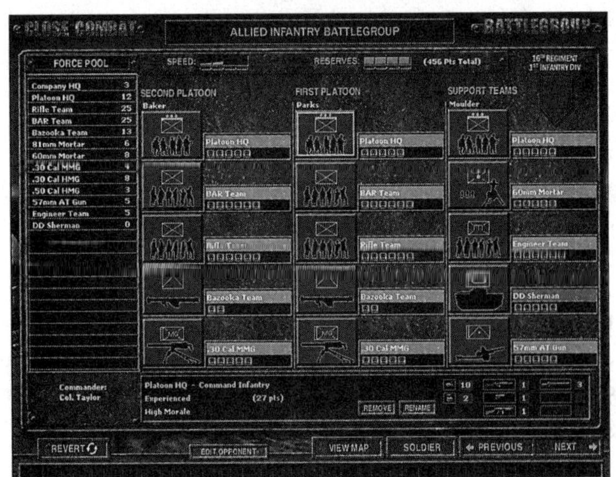

History

Source: Matrix Games *Background*

Figure 15.2
Video Game Chronology

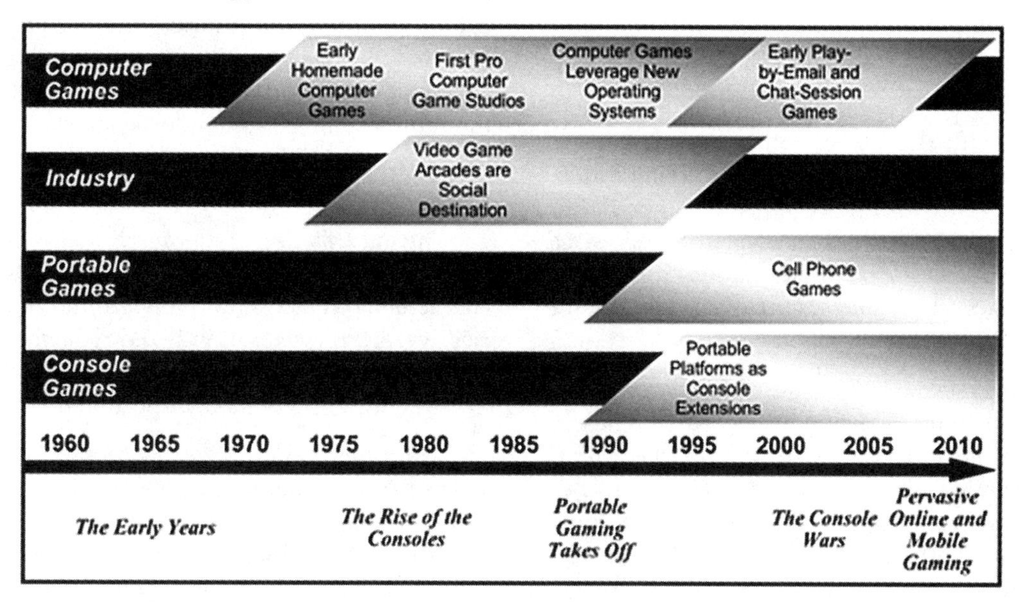

Source: Guillory (2008)

Most histories of video games focus on the hardware requirements for the games, which frequently drove where and how the games were played. However, it is equally possible to divide the history of games by the advances in software (and changes in the style of gameplay), the diffusion of games among the population (and the changes in the playing audience), or the increases in economic power wielded by video games, measured by the other industries overtaken through the years. Regardless of the chosen path, as the history of video games developed, however, the industry became increasingly fragmented into specialty niches.

Most industry observers describe the current generation of home gaming consoles as the seventh generation since the release of the first-generation Magnavox Odyssey. Handheld consoles are often said to be on their fourth generation.

No one has yet attempted to assign "generations" to computer gaming software, in large part because console "generations" are hardware-based and released in specific waves, while computer hardware is continually evolving, and major computer milestones are the releases of new operating systems (Windows 8, Mac OS X, etc).

The early years of video gaming were marked by small hobby programs, many developed on large university or corporate mainframes. Willy Higenbotham, a nuclear physicist with the Department of Energy, had experimented with a simple tennis game for which he had developed a rudimentary analog computer (Anderson, 1983).

A team of designers under Ralph Baer developed a variety of video game projects for the Department of Defense in the mid-1960s, eventually resulting in a hockey game, which started an ongoing tradition of leaving military sponsors underwhelmed (Hart, 1996). Baer also led another team that developed *Chase*, the first video game credited with the ability to display on a standard television set. In the early 1960s, *SpaceWar* was also popular among the graduate students at MIT, and inspired other work at the Pentagon.

Although many different treatises have been written arguing over the invention of the video game, it is still unclear how much, if at all, any of the early video game pioneers even knew of each others' work; it is completely unknown if they drew any inspiration from each other.

In the early 1970s, dedicated gaming consoles began to appear, starting with the Magnavox Odyssey in 1972. Built on switches, rather than a microprocessor, the Odyssey included "analog" components used in playing the video portions of the game, such as dice, play money, and plastic overlays for a common touchpad. The first home video game product built on a microprocessor was a home version of the popular coin-operated *Pong* game from Nolan Bushnell's Atari. Although it contained only the titular game hard-coded into the set, it was a popular product until the introduction of a console that could play multiple games by swapping out software (Hart, 1996).

The second generation of video gaming began around 1977 with the rise of consoles. This generation was marked by the integration of home video-gaming with other media licenses, whether tie-ins with popular movies such as *E.T. The Extra-Terrestrial* and *Raiders of the Lost Ark*, or adaptations of popular arcade games, such as *Pac-Man*, *Defender*, or *Missile Command*.

The Atari 2600 led the market for home video game sales, in which consumers would purchase a standard console and insert cartridges to play different games. While Colecovision and Intellivision (two other consoles) were popular in the market, nothing could compare with the market power wielded by Atari (which was eventually purchased by Warner Communications) from 1977 to 1982, during which an estimated $4 billion of Atari products were sold (Kent, 2001).

Atari's success also led to the formation of Activision, a software company founded by disgruntled Atari game programmers. Activision became the first major game studio that designed their games exclusively for other companies' consoles, thus separating the games and consoles for the first time. Business deals in the early 1980s also began to evolve that tied certain media licenses to specific platforms, establishing several precedents in exclusivity among game 'families' and the console systems.

After a brief downturn in the market from 1981 to 1984, mostly as a result of business blunders by Atari, home video game consoles began a resurgence. Triggered by the launch of the Sega Master System in the mid-1980s and the Nintendo Entertainment System (NES) shortly thereafter, home video game sales continued to climb for both the games and the hardware needed to play them.

The inclusion of 8-bit processors closed the gap between the performance of large standalone arcade machines and the smaller home consoles with multiple games, and it signaled the start of the decline of the video game arcade as a game-playing destination and focal point for the videogame subculture of teenagers in the US.

By 1987, the NES was the best-selling toy in the United States (Smith, 1999). The NES also continued the platform-specific media tie-ins with their repeated use of the character "Mario"—first made popular in the *Donkey Kong* arcade game—and leveraged that character into a burgeoning empire that today spans more than 50 titles. Nintendo's characters also crossed over between NES and traditional arcade games made by the company.

During this time, video games also began to appear in popular culture not as mere accessories to the characters, but as central plots around which the stories were built. *Tron* (1982), *War Games* (1983), and *The Last Starfighter* (1984) all brought video gaming into a central role in their respective movie plots.

Computer games were also developing alongside video game consoles. Catering to a smaller market, computer games were seen as an add-on to hardware already in the home, rather than the primary reason for purchasing a home computer system. However, the ability to write programs for home computers enabled consumers to also become game designers and share their creations with other computer users. Thus, a generation of school children grew up learning to program games on Commodore PET, Atari 800, and Apple II home computers.

The commercial success of the Commodore 64 in the mid-1980s gave game publishers a color system for their games, and the Apple Macintosh's point-and-click interface allowed designers to incorporate the standard system hardware into their game designs, without requiring add-on joysticks or other special controllers. Where console games were almost exclusively graphics-oriented, early computer games included a significant number of text-based adventure games, such as *Zork* and *Bard's Tale*, and a large number of military-themed board games converted for play on the computer by companies such as SSI (Falk, 2004).

In 1988, the first of several licensed games for *Dungeons & Dragons* appeared, and SSI's profile continued to grow. Other prominent early computer game publishers included Sierra, Broderbund, and Infocom.

Still, home computer game sales continued to lag behind console game sales, in large part because of the comparatively high cost and limited penetration of the hardware. Additionally, video games were still seen as an 'add-on' to a primary media channel (television, computer, etc.) and had not yet developed their own market channel through which to distribute titles. Thus computer games tended to be sold where other computer software was sold, and consoles (and their supporting games) were typically found in home electronics stores, alongside television sets.

With video games ensconced in U.S. and Japanese households and expanding worldwide, it was only a matter of time before portable consoles began to rival the home siblings in quality and sophistication, and thus began the third phase in the history of video games. Portable video games proliferated in the consumer marketplace beginning in the early 1980s. However, early handhelds were characterized by very rudimentary graphics used for one game in each handheld. In fact, "rudimentary" may even be generous in describing the graphics—the early Mattel handheld *Electronic Football* game starred several small red "blips" on a one-inch-by-three-inch screen in which the game player's avatar on the screen was distinguished only by the brightness of the blip.

Atari released the Lynx handheld game system in 1987. Despite its color graphics and relatively high-speed processor, tepid support from Atari and third-party developers resulted in its eventual demise.

In 1989, Nintendo released Game Boy, a portable system whose controls mimicked the NES. With its low cost and stable of well-known titles ported from the main NES, the Game Boy became a major force in video game sales (Stahl, 2003). Although technically inferior to the Lynx—black-and-white graphics, dull display, and a slower processor—the vast number of Nintendo titles for the Game Boy provided a major leg up on other handheld systems, as audiences were already familiar and comfortable with Nintendo as a game company.

Sega's Gamegear followed within a year. Like the Lynx before it, superior graphics were not enough to overcome Nintendo's catalog of software titles or the head start in the market the Game Boy already had. By the mid-1990s, most families that owned a home game console also owned a handheld, often from the same company.

Although the portable revolution had not (yet) migrated to computer gaming, it was hardware limitations, rather than game design, that prevented the integration of computer games into portable systems. The release of the Palm series of handheld computers (Hormby, 2007) gave game designers a new platform on which they could develop that was not tied to any particular company. This early step toward handheld computing would include early steps toward handheld computer gaming.

The third and fourth generations of video game history begin to overlap as the console wars included the handheld products of various console manufacturers, coinciding with the release of Windows 95 for Intel-powered PC computers. This gave game designers a variety of stable platforms on which to program their games. The console wars of the late 1990s have continued to today, with independent game design studios developing their products across a variety of platforms.

As Nintendo began to force Sega out of the console market in the mid-1990s, another consumer electronics giant, Sony, was preparing to enter the market. With the launch of the PlayStation in 1995, Sony plunged into the video game platform market.

Nintendo maintained a close hold on the titles it would approve for development on its system, attempting to position itself primarily as a "family" entertainment system. Sony developers, however, had the ability to pursue more mature content, and their stable of titles included several whose graphics, stories, and themes were clearly intended for the 30-year-old adults who began playing video games in 1980, rather than 13-year-old kids (Stahl, 2003).

Sony and Nintendo (and to a lesser extent, Sega) continued their game of one-upmanship with their improvements in hardware over the next several years. As the graphical processing power of consoles increased, and games gained greater notoriety in the press as they pushed the edges of storytelling and explicit graphics, the Interactive Digital Software Association (ISDA) finally caved to public pressure and instituted a rating system for games. This rating system sought to better inform consumers about the target ages for games, as well as the reasons for the ratings.

The ISDA became the ESA in 2004 (ESRB, 2009), and the ratings for video games have become a standard feature of every game, both console and computer. More granular than movie ratings, video games ratings break down into more than a half-dozen age categories and more than twenty content descriptors. Controversies have (predictably) ensued; unlike movies, video games can easily incorporate alternate, hidden, or add-on content which might alter their ratings. Among the most famous ratings controversies was the "Hot Coffee" content in *Grand Theft Auto*, in which downloaded patches opened up sexually explicit and violent content (BBC, 2005).

By early 2001, Sega admitted defeat in the hardware arena and focused instead on software. The next salvo in the platform wars was about to be launched by Microsoft, which debuted the Xbox in late 2001. With built-in networking and a large hard drive, Microsoft's Xbox began to blur the lines between computer video gaming and platform-based video gaming. Additionally, building their console on an Intel processor eased the transition for games from PC to Xbox, and many popular computer titles were easily moved onto the Xbox.

Around the same time, Sony entered the handheld arena to challenge Nintendo's Game Boy dominance with the PSP: PlayStation Portable. Capable of playing games as well as watching movies and (with an adapter) having online access, the PSP was intended to show the limitations of the Game Boy series with its greater number of features.

Although the platform wars continue today, every one of them supports networked gaming with high-speed data access, online accounts and multiplayer gaming, chat, and downloadable content to on-board hard drives.

With high-speed data networks proliferating throughout North America, Japan, Korea, Western Europe, and (to a lesser extent) China and Southeast Asia, online gameplay has become a major attraction to many video gamers, especially through MMORPGs. These pervasive worlds host shared versions of a variety of different games, including sports, military, and sci-fi and fantasy games. MMORPGs are most commonly accessed through computer platforms rather than game consoles. Since its launch in 2004, *World of Warcraft* has grown to exceed 10 million simultaneous subscribers at any one time, though their rate of subscriber turnover continues to be high.

MMORPGs have highly-developed in-game economies, and those economies have begun to spill over into the "real world" where websites and online classified listings offer game-world items, money, and characters for sale to players seeking an edge in the game but are reluctant to sacrifice the time to earn the rewards themselves. Fans' reactions have not been universally positive to these developments, and some have started petitions to ban such behavior from the games (Burstein, 2008). MMORPGs were among the first widespread systems that allowed a distributed user base to share a single game, rather than forcing a physical co-location on the participants to play together.

Wireless networking has also extended the ability to participate in online-based games to handhelds, both dedicated to gaming (Nintendo DS) and consumer-oriented (mobile devices and cell phones). Moreover, many software-specific companies have designed their online game servers such that the players' platforms are irrelevant, and thus gamers playing on an Xbox might compete against other gamers online who are using PCs.

The 2006 release of Nintendo's Wii game console drew a new audience by attracting large numbers of older users to the motion-based games enabled by the Wii's motion-sensitive remote. Not long after its release, the Wii began to appear on the evening network news as a new activity in senior citizens homes and in stories about children and grandparents sharing the game (Potter, 2008). Although graphically inferior to the 360 or PS3, the Wii developed an audience of players who had never tried video gaming before.

The availability of broadband connections has resulted in many software companies selling games online directly to the consumer (especially for computer gaming), with manuals and other play aids available as printable files for those players who wish to do so. Steam, from Valve Corporation, is a download site that boasts more than 65 million accounts (Steam, 2014) and delivers downloadable games directly to computer platforms.

Most computer game manufacturers sell virtually every title as a download directly to their customers either directly or through online stores. These sales are *not* simply mail-orders of physical copies, but actual direct-to-PC downloads. This direct-to-consumer sales route has reduced the dependence on local computer software stores for computer games; these stores have reacted by stocking more console games.

Similarly, with the ability of consoles to hook into in-home data networks (either by Ethernet or wirelessly), direct-to-console downloads are also growing in popularity, as storefronts such as the Xbox Live Arcade sell digital-only copies of games that are stored directly on the console's hard drive. This low-overhead market channel has allowed low-cost games to proliferate, as they eschew any physical costs (packaging, disks, etc.), transportation costs (no shipping), or retail support costs (no markup by a brick-and-mortar middleman).

Recent Developments

By some estimates, video games may be in their seventh, tenth, or twelfth generation. Those generations have been collapsed into five for this chapter: the early years, the rise of the consoles and computer games, portable gaming, the console wars, and pervasive online gaming.

Computer gaming roughly followed this same trajectory, although the introduction of portable computer gaming lags behind for hardware reasons. While the fourth generation described above is still ongoing, it seems as though the market has stabilized in that the three current major players in the console market released major new hardware in 2013.

Microsoft's Xbox One and Sony's Playstation 4 (PS4) spent the 2013 holiday shopping season dueling each other for consumer attention, and dollars. The Xbox One was released on November 22, 2014 in the US and retailed for $499. The Playstation 4 was released on November 15, 2014 and costs $399. Nintendo's Wii U was released a year earlier in November 2013, but their marketing focus was less on the technical specs of the system and more on the stable of characters in Nintendo's long-running game franchises.

Key among the marketing of the Xbox One and PS4 was the idea that the gaming console has morphed into a more robust media hub that can download (or stream) other non-game content as well as enabling gameplay. Netflix, a popular online video streaming service, is accessible on most consoles, enabling users to watch movies and television shows on-demand through their consoles. Additionally, the new consoles ship with DVD and Blu-ray players built in, which allows one device to serve multiple purposes when connected to a user's television.

The marketing buzz created by the Xbox One and PS4 was accompanied by a variety of game titles released specifically for the new consoles, hoping to piggy-back on the marketing buzz of the new hardware to capture consumer attention

Social Gaming—Your Friends Help You Play

With the proliferation of social networking sites over the past decade, from MySpace to Facebook to Google+ and others, software companies and game designers have developed a new twist on video gaming that not only encourages social interaction, but very nearly *requires* it for the players to fully experience the games.

Games such as *Farmville* and *Candy Crush Saga* on Facebook allow users to not only share their performance metrics online, they also allow users to invite their friends to join the game, and provide rewards to those recruiters. Additionally, players may help each other during the game by lending resources when online.

These sorts of persistent engagements have resulted in several effects. First, the lack of detailed commitment to learning rules, commands, or interfaces has resulted in very high rates of adoption of the games. Zynga Games, the largest of the social gaming developers/publishers, boasts over 65 million daily players. In comparison, *World of Warcraft* hovers at just over 10 million registered players at any one time, and Xbox Live has approximately 46 million users (Tassi, 2013). Second, many of the games have been designed with time-lapse effects that require the users to return to them after certain time intervals, resulting in continual engagement with the games.

Most importantly, though, is the business model through which many of these social games generate revenue. Players may play freely, but with time-lapse limitations, and certain in-game purchasing limits, attaining the highest-status rewards within the game (items for decorating a virtual restaurant, or virtual farm equipment, or just virtual cash for use with other online goods) can be extremely challenging. Social gaming companies offer these extensions within their games to players in exchange for real currency (Buckman, 2009). These monetary transactions have resulted in significant revenues for very small startups since 2005.

Within the game design industry, however, there have been significant criticisms of social games as merely a time-lapse collection of clicks, and many videogame creators have come down starkly on either side of this debate (Tanz, 2012).

Another variation on the social gaming theme is the delivery of certain rules-light games through mobile platforms. Games such as *Words With Friends* include mobile clients accessible through smartphones that allow players to interact with other live players, in real time, from their phones. These smartphone games, sold through a mobile phone app infrastructure that has been created and diffused since 2006, are marked by incredibly high sales volumes, allowing even very inexpensive games (as low as 99¢) to gross tens of millions of dollars.

Regulatory Environment—Continuing Legislative Efforts to "Protect the Children"

While the legislative landscape continues to treat video games as a child-oriented medium with an average player age of 10 years old, several major events cast new, negative lights on video games from a legislative perspective. Key among these was the school shooting at Sandy Hook Elementary School in Newtown, CT. Although the final report from the State's Attorney found no link between the shooter's behavior and violent video games (Sedesky, 2012), popular media and news commentators tried vociferously to draw links between playing of violent video games and lethal, violent behavior.

Laws targeting video games have continually been struck down in the courts, however, there are new post-Sandy Hook laws currently under consideration in several states that will doubtless face challenges should they pass. The last major legal cases to wind through the U.S. courts were each several years before Sandy Hook. In 2009, the Ninth Circuit Court upheld a ruling that a California law was too restrictive and violated free speech rights (Walters, 2010). The law restricted the sale of violent videogames to minors. In 2011, the Supreme Court of the United States also upheld this ruling (Savage, 2011).

Innovations in Mobile Gaming

With the increase in computing power available in handsets, mobile gaming has split along two lines. First, Nintendo and Sony both have handheld game platforms with wireless capability built in, allowing for head-to-head gameplay with other nearby systems, as well as shared gameplay through an Internet connection, where available. Platforms such as Nintendo's DS line and Sony's PSP and Vita machines are capable of establishing local networks for head-to-head gaming without any wireless service.

Apple's App Store, and the subsequent similar stores from Google (for the Android OS) and Microsoft (for the Windows Phone OS), have allowed the proliferation of huge numbers of "casual games" with simple, intuitive interfaces and playtimes measured in minutes rather than hours. Many app store games are also free to download and play, with in-game purchases to upgrade player capabilities or change the tempo of play to allow completion of in-game activities more quickly.

The continued improvement of mobile phone hardware (larger screens, greater resolution, faster processors, accelerometers) has made the devices far more attractive to game-makers. For consumers, the ability to have their games in-hand regardless of location, without an additional device, has proven quite popular. Sales of games for mobile phones have jumped dramatically over the past several years, with the loss of sales being felt primarily by the handheld "console" manufacturers—Sony and Nintendo (Farago, 2011).

In 2012 Sony introduced the PlayStation Vita. This handheld device incorporates a plethora of new technologies including 3G wireless and Wi-Fi (through AT&T), GPS, motion sensors, a front touch screen and rear touch pad, front and back cameras, and a stunning OLED display (Play-Station, 2012). It sounds like a fully equipped handheld gaming console, but these features could also describe a smartphone.

Game designers have now started leveraging the additional features of smartphones, such as cameras, barcode/QR code readers, GPS locators, and web connections to create "immersive reality" games (such as Niatic Labs' *Ingress*) that are designed to modify the players' perceptions of the ordinary world by providing location-specific challenges, geotagged competitions with nearby players, and real-world scavenger hunts for features that must be photographed or scanned by the game's software to unlock the next feature or challenge.

Figure 15.3

Annual 2013 Top 10 Games: New Physical Retail Only—Across all Platforms Including PC

Title (Platforms)	Rank	Publisher
Grand Theft Auto V (360, PS3)**	1	Take 2 Interactive (Corp)
Call Of Duty: Ghosts (360, PS3, XBO, PS4, PC, NWU)**	2	Activision Blizzard (Corp)
Madden NFL 25 (360, PS3, PS4, XBO)	3	Electronic Arts
Battlefield 4 (360, PS3, XBO, PS4, PC)	4	Electronic Arts
Assassin's Creed IV: Black Flag (360, PS3, PS4, XBO, NWU, PC)**	5	Ubisoft
NBA 2K14 (360, PS3, PS4, XBO, PC)**	6	Take 2 Interactive (Corp)
Call of Duty: Black Ops II (360, PS3, NWU, PC)**	7	Activision Blizzard (Corp)
Just Dance 2014 (WII, 360, NWU, XBO, PS3, PS4)**	8	Ubisoft
Minecraft (360)	9	Microsoft (Corp)
Disney Infinity (360, WII, PS3, NWU, 3DS)	10	Disney Interactive Studios

***(includes CE, GOTY editions, bundles, etc. but not those bundled with hardware)*

Source: NPD Group

Broadband-Enabled Downloads for Consoles as well as PCs

In a manner similar to the Steam download service, console manufacturers have integrated online participation and downloadable content through their consoles. Not only can networks of players compete cooperatively or head-to-head around the world, but broadband networking has enabled the push of content through storefronts including the Xbox Live Marketplace and the Nintendo Wii Virtual Console.

Xbox Live Marketplace includes not only games, but movies from major studios. Nintendo's Virtual Console has seen competition from WiiWare (Chan, 2010) for the download market on the Wii consoles.

Sony's PlayStation network is integrating connections to popular online destinations such as Facebook (Thorpe, 2009) and Netflix (Kennedy, 2009), further blurring the delivery lines between computers and other Internet-enabled devices. Additionally, online broadband delivery of video games have significantly reduced the physical overhead costs of PC game publishers, with major companies such as Firaxis releasing the popular *Civilization* franchise for download through Apple's App Store for Macintosh computers.

Current Status

After a half-decade of no new platforms from 2006-2012, all three major console manufacturers released new hardware in 2012 and 2013. Microsoft's Xbox One has sold 5 million units worldwide through mid 2014. Sony's PS4 has over 7 million, Wii U sales have been disappointing with the Xbox One outselling it even though the Xbox One has been on the market a shorter time than the Nintendo console. Because consoles are primarily dependent on their software to maintain customer interest, constant hardware upgrades may not be as necessary once the newest consoles leap ahead, and in fact might be considered detrimental to sales if the consoles are not backward-compatible with older games in the same product family. Computer-based games are not as dependent on regular hardware updates, and software continues to appear daily for computer-based videogamers.

In 2013, U.S. income from video games totaled slightly more than $15.4 billion, including almost $2 billion in rentals; this total is down from 2011's $17 million (Matthews, 2012; Sinclair, 2014). Given the multi-function nature of desktop and laptop computers, counting the hardware for computer game sales makes little sense. Console sales have typically fueled a significant end-of-year uptick in sales during the holiday seasons, and 2013 exceeded many expectations with the releases of Sony's and Microsoft's new entrants in the marketplace.

"Gamer parents" have continued to be a phenomenon of interest, and as yet an insufficient number of the media and legislative leadership have been replaced by longtime gamers. Frequently used to refute the argument that "video games are for kids," gamer parents are those game players that grew up with a game console in their households and are now raising their own children with consoles.

The average game player is 30 years old and has been playing for over 12 years (ESA, 2013). In fact, children who owned an Atari 2600 console in their home in the 1980s are now over 40 years old. Although legislative action has often been touted as a remedy for inhibiting access to video games that legislators feel is inappropriate, gamer parents have repeatedly noted that they are intimately familiar with video games and capable of making informed choices about their children's access to video games.

In addition, gamer parents tend to take the lead in game purchases for their households, thus making them a valuable target for the corporate marketing machines. In fact, 89% of all game players under age 18 note that their parents are present when they purchase or rent their games (ESA, 2013). By way of comparison to the stereotype of teenaged videogamers, a 10-year-old child who started playing videogames at home on an Atari 2600 console in 1982 is today a 42-year-old gamer with 30 years of videogaming experience.

As noted above, legislative action against video games continues in multiple venues. Not every legislative action is opposed by industry trade groups, however. The Entertainment Software Association has consistently supported measures designed to prohibit access to sexually explicit games by minors, as well as supporting legislation that increases access to ratings information for consumers (Walters, 2008). However, laws intended to severely limit games access to a large segment of the population have yet to stand up to judicial scrutiny.

Factors to Watch

The 2013 release of the Oculus Rift headset has generated a great deal of excitement in the video gaming community for its ability to fully immerse the gamer through the use of a pair of head-mounted projection screens. With its wide field of view (virtually identical to normal peripheral vision) and a head mount that completely blocks out the view of the 'real world,' the Oculus Rift headset is a level of immersion thus far not achieved by any other device.

It is no surprise that the initial list of games with Oculus Rift support are those played from a first-person point of view, such as shooters like *Half-Life 2* and *Left4Dead*. With a variety of new games in development for the headset, and Epic Games adding support for their Unreal Engine software development kit (on which many first-person games are built), the Oculus Rift headset is sure to spawn competing imitators as developers and manufacturers aim for greater player immersion in their games.

An economic factor to watch is the continual use of the crowd-funding websites like Kickstarter and IndieGoGo, as well as Steam's Greenlight program, to enable game designers and developers to raise funding for game development in advance of the release of the game. Building on a long-used preorder model from board game companies such as GMT Games, these crowdfunding sites reverse the process of "build game, then sell game" to "sell game, then build game." They allow creators to offer different "tiers" of product to customers in exchange for different levels of funding. Developers encourage support from customers by offering exclusive content to supporters who pre-order the games, or perhaps an earlier release to supporters than to the general public.

In 2013, Kickstarter (the most commercially successful crowdfunding site) passed the $1 billion threshold in pledges for all projects, with video games accounting for nearly 22% of the total (CrowdBox, 2014). The Oculus Rift headset's initial funding was through a Kickstarter campaign that raised more than $2 million (Kickstarter, 2012).

The ability to raise funding from prospective customers *before* the completion of the game development significantly reduces the economic risk to smaller game developers and allows games with a smaller commercial appeal to still find funding. The "pledge" system for preorders also allows game developers to judge the commercial viability of a project before committing an overwhelming amount of time towards it, reducing the likelihood that they might build a game for which there is no audience.

An interesting side note is Facebook's purchase of the Oculus VR, maker of the Rift, for $2 Billion in Spring 2014. Many Kickstarter supporters were angry for the "sell-out" especially since Facebook indicates intentions to use the device for more than gaming (Luckerson, 2014).

The release of the Xbox One and Microsoft's heavy focus on support of their new console hardware (as well as their tablet hardware) leaves most industry-watchers with the impression that Microsoft is unlikely to release any dedicated hand-held gaming-specific system for several years, at least. After years of misplaced rumors, it seems almost pointless to expect Microsoft to release a handheld game-specific device.

While the industry was watching Microsoft in anticipation of a yet-to-appear handheld device, Apple's iPhone has stormed forward to snag a significant share of the mobile gaming marketplace.

In fact while iOS and Android game sales have dominated more than half of the mobile gaming market, Nintendo's DS has lost over half of its market share (Farago, 2011). Moreover, the introduction of the iPad and iPad Mini have also contributed to the huge increase in iOS game sales, as the underlying OS, as well as the distribution channels, are shared with the iPhone. Tablet devices are now a significant factor in videogaming, especially for adaptations of popular board games and social games played across networks.

Microsoft's Kinect for Xbox 360 and the Xbox One has expanded on the motion-based gaming pioneered by Nintendo's Wii. Unlike the Wii and the PlayStation Move, which rely upon a remote control, Microsoft's Kinect relies on a camera that recognizes gestures, facial features, and body motion (Archibald, 2009). The ability to recognize motion will free game-players from at least part of the needed hardware for gameplay.

The growth in the numbers of women playing video games is expected to continue to accelerate. 45% of the game-playing public are women, a 7% increase from 2008, and adult women represent a greater share of the market (31%) than young males under 17 (19%) (ESA, 2013). The online titles favored by women are dependent not only on the continued diffusion of the software, but also on the continued diffusion of the high-speed Internet access needed to enable the online environment.

Government funding of new projects with video game developers will also continue as sponsors search for projects applicable to their specific fields. The U.S. Army's TCM-Gaming office, specifically designed to leverage video game technology for training purposes (Peck, 2007), has run into challenges with unifying the acquisition and development of game-based training tools across the Army enterprise.

Nevertheless, government-focused events, such as the Interservice/Industry Training, Simulation and Education Conference (I/ITSEC) bring in more than 20,000 attendees (IITSEC, 2012). With the continuing reduction in training budgets across the Department of Defense, military training organizations will continue to seek lower-cost alternatives to fielding large numbers of soldiers for maneuver exercises (Nichols, 2012). Many of these projects have also found commercial homes, such as *ARM-A*, the civilian version of the military simulator *VBS-2*.

Additionally, some videogame projects have migrated in the opposite direction, with games such as *Close Combat* being adopted by the US Marine Corps. However, controversy continues to boil around military-themed games that depict events that seem too similar to current operations (Suellentrop, 2010).

Despite years of video game legislation being rejected by the courts, legislators will continue to react to media coverage of parental concern about video game content, especially in the wake of the Sandy Hook school shooting. Overwhelming demographic data shows that video gamers are typically adults, and 25% of them are over age 50. Nonetheless, many news outlets and legislators continue to view video games as toys for kids and make no distinction in subject matter between mature-themed games and games clearly targeted at children. Legislative efforts are further complicated by legal precedents being established in cases about online distribution of content, which is increasingly relevant as Internet-enabled consoles are connected to broadband networks.

All three major console lines and many computer games allow for collaborative online play. Expect to see two developments in this area.

First, as game titles proliferate across platforms, expect to see more games capable of sharing an online game across those platforms, allowing a player on the Xbox to match up against an opponent on a PC system, as both players communicate through a common back-end server.

Second, many of these online systems, such as the Xbox Live, already allow voice conversations during the game through a voice over Internet protocol system. (VoIP is discussed in more detail in Chapter 21.) As more digital cameras are incorporated into consoles, either as an integrated component or an aftermarket peripheral, expect these services to start offering some form of videoconferencing, especially for players involved in games such as chess, poker, or other "tabletop" games being played on a digital system.

Video Games Visionary: Sidney "Sid" Meier

While many other groundbreakers blazed technical trails in video gaming, **Sid Meier** established one of the most successful video game franchises of all time with the initial release of the first *Civilization* in 1991. Already a successful designer with multiple hits to his credit (*Silent Service*, *Sid Meier's Pirates!*, *Red Storm Rising*, and *Railroad Tycoon*, among others), Meier's original game borrowed heavily from a board game of the same name, and launched the popular success of the "4X" strategy game – explore, expand, exploit, and exterminate – with a model that's been copied by hundreds of designers in the two decades since the original game was released.

Meier's *Civilization* franchise has spawned 5 major 'generations' of the game, multiple iterations and expansions for each successive version, hundreds of custom-built scenarios, and the ability to easily load user-created "mods" for custom content. Meier has famously coined the description of a video game as "a series of interesting decisions" and has been credited with a design philosophy as simple as "How do you make a good game? Get a game and remove all the parts that aren't fun."

In 30 years of game designing, Meier's hits have come almost non-stop. He has founded two different major computer game companies and sold tens of millions of copies of his games. His initial designs were flight simulators with MicroProse, one of the original PC gaming companies. After leaving MicroProse for Firaxis, Meier's name has continued to adorn the titles of his designs, including *Ace Patrol*, *Colonization*, *SimGolf*, *Alpha Centauri*, and *Railroads!*. He is an oft-sought speaker at industry conferences, and seen as one of – if not the most – successful computer game designer in the short history of video games.

Bibliography

Archibald, A. (2009). Project Natal 101. *Seattle Post-Intelligencer.* Retrieved from http://blog.seattlepi.com/digitaljoystick/archives/169993.asp.

Anderson, J. (1983). Who really invented the video game. *Creative Computing Video & Arcade Games 1* (1), 8. Retrieved from http://www.atarimagazines.com/cva/v1n1/inventedgames.php.

BBC. (2005). Hidden sex scenes hit GTA rating. *BBC News.* Retrieved from http://news.bbc.co.uk/2/hi/technology/4702737.stm.

Bruno, A. (2008). *Rock Band, Guitar Hero* drive digital song sales. *Reuters.* Retrieved from http://www.reuters.com/article/idUSN1934632220080120.

Buckman, R. (2009). Zynga's Gaming Gamble. *Forbes.* 16 November 2009. Retrieved from http://www.forbes.com/forbes/2009/1116/revolutionaries-technology-social-gaming-farmville-facebook-zynga.html.

Burstein, J. (2008). Video game fan asks court to ban real sloth and greed from *World of Warcraft. Boston Herald.* Retrieved from http://www.bostonherald.com/business/technology/general/view.bg?articleid= 1086549.

Carnoy, D. (2006). Nokia N-Gage QD. *CNET News.* Retrieved from http://reviews.cnet.com/cell-phones/nokia-n-gage-qd/4505-6454_7-30841888.html.

Chan, T. (2010). WiiWare market grows to nearly $60M USD in 2009. *Nintendo Life.* Retrieved from http://www.nintendolife.com/news/2010/02/wiiware_market_grows_to_nearly_usd60m_usd_in_2009.

CrowdboxTV (2014), *Crowdfunding Site Kickstarter Passes the $ 1 Billion Milestone,* retrieved from http://www.crowdbox.tv/crowdfunding-site-kickstarter-passes-the-1-billion-milestone/.

Entertainment Software Associaiton (2013). Essential Facts. Retrieved From http://www.theesa.com/facts/pdfs/esa_ef_2013.pdf.

Entertainment Software Association. (2012). Industry Facts. Retrieved from http://www.theesa.com/facts/index.asp.

Entertainment Software Rating Board. (2009). *Chronology of ESRB Events.* Retrieved from http://www.esrb.org/about/chronology.jsp.

Falk, H. (2004). *Gaming Obsession Throughout Computer History Association.* Retrieved from http://gotcha.classicgaming.gamespy.com.

Greenpeace. (2010). How the companies line up. *Guide to Greener Electronics*. Retrieved from http://www.greenpeace.org/international/en/campaigns/toxics/electronics/how-the-companies-line-up/.
Farago, P. (2011). Is it game over for Nintendo DS and Sony PSP?. *Flurry Blog*. Retrieved from http://blog.flurry.com/bid/77424/Is-it-Game-Over-for-Nintendo-DS-and-Sony-PSP.
Hart, S. (1996). A brief history of home video games. *Geekcomix*. Retrieved from http://geekcomix.com/vgh/.
Hinkle, D (2014) *NPD: US video game sales reach $15.39 billion in 2013*, retrieved from http://www.joystiq.com/2013/12/03/sony-ps4-sales-rise-to-2-1-million-700k-sold-in-europe-austr/.
Hormby, T. (2007). History of Handspring and the Treo (Part III). *Silicon User*. Retrieved from http://siliconuser.com/?q=node/19.
IITSEC. (2012). I/ITSEC statistics. IITSEC.org. Retrieved from http://www.iitsec.org/about/Pages/HighlightsFromLastIITSEC.aspx.
Kaiser, T. (2011). Activision kills Guitar Hero, confirms company layoffs. *Daily Tech*. Retrieved from http://www.dailytech.com/Activision+Kills+Guitar+Hero+Confirms+Company+Layoffs/article20879.htm.
Kennedy, S. (2009). Netflix Officially Coming to PS3. *1up.com*. Retrieved from http://www.1up.com/do/newsStory?cId=3176634.
Kent, S. (2001). *The ultimate history of video games: From Pong to Pokemon – The story behind the craze that touched our lives and changed the world*. New York: Patterson Press.
Kickstarter (2012), *Oculus Rift: Step Into the Game* retrieved from https://www.kickstarter.com/projects/1523379957/oculus-rift-step-into-the-game.
Kinsley, J (2014) *Report: Wii U sales reach 4.3 million*, retrieved from http://wiiudaily.com/2013/12/report-wii-u-sales-reach-4-3-million/.
Libby, B .(2009). Sustainability-themed computer games come to the classroom. *Edutopia*. Retrieved from http://www.edutopia.org/environment-sustainability-computer-games.
Linde, A. (2008). PC games 14% of 2007 retail games sales; World of Warcraft and Sims top PC sales charts. *ShackNews*. Retrieved from http://www.shacknews.com/onearticle.x/50939.
Luckerson, V. (2014). Facebook Buys Oculus Virtual Reality Company for $2 Billion. *Time*. Retrieved from http://time.com/37842/facebook-oculus-rift/.
Matthews, M. (2012). NPD: Behind the numbers of 2011. *Gamasutra*. Retrieved from http://gamasutra.com/view/news/39669/NPD_Behind_the_numbers_of_2011.php.
Nichols, P. (2012). Personal communication with the author.
NPD Group / Riley, D. (2012). Annual 2011 Top 10 Games. Personal correspondence with author.
Peck, M. (2007). Constructive progress. *TSJOnline.com*. Retrieved from http://www.tsjonline.com/ story.php?F=3115940.
Peckham, M. (2008). Did rare metallic ore fuel African "PlayStation War"? *PCWorld*. Retrieved from http://blogs.pcworld.com/gameon/archives/007340.html
PlayStation (2012). PlayStation Vita system features. Retrieved from http://us.playstation.com/psvita/features/.
Potter, N. (2008). Game on: A fourth of video game players are over 50. *ABC News*. Retrieved from http://abcnews.go.com/WN/Story?id=4132153.
Savage, D. (2011). Supreme Court strikes down California video game law. *Los Angeles Times*. Retrieved from http://articles.latimes.com/2011/jun/28/nation/la-na-0628-court-violent-video-20110628.
Schramm, M. (2008). EA Mobile prez: iPhone is hurting mobile game development. *TUAW.com*. Retrieved from http://www.tuaw.com/2008/01/08/ea-mobile-prez-iphone-is-hurting-mobile-game-development/.
Schreier, J (2013) *Sid Meier: Father of Civilization*, retrieved from http://kotaku.com/the-father-of-civilization-584568276.
Sedeskey, S (2012), Report of the State's Attorney for the Judicial District of Danbury on the Shootings at Sandy Hook Elementary School and 36 Yogananda Street, Newtown, Connecticut on December 14, 2012, retrieved from http://www.ct.gov/csao/lib/csao/Sandy_Hook_Final_Report.pdf.
Sinclair, B. (2014). US spent $15.39 billion on games in 2013 - NPD. *Games Industry International*. Retrieved from http://www.gamesindustry.biz/articles/2014-02-11-us-spent-USD15-39-billion-on-games-in-2013-npd.
Smith, B. (1999). Read about the following companies: Nintendo, Sego, Sony. *University of Florida Interactive Media Lab*. Retrieved from http://iml.jou.ufl.edu/projects/Fall99/SmithBrian/aboutcompany.html.
Stahl, T. (2003). Chronology of the history of videogames. *The History of Computing Project*. Retrieved from http://www.thocp.net/software/games/games.htm.
Steam (2014). Steam, the ultimate entertainment platform. Retrieved April 7, 2014 from http://www.valvesoftware.com//.
Stohr (2010). Violent video game law gets top U.S. court hearing. *Bloomberg BusinessWeek*. Retrieved from http://www.businessweek.com/news/2010-04-26/violent-video-game-law-gets-top-u-s-court-hearing-update1-.html.

Suellentrop, C. (2010). War games. *New York Times Magazine*. September 12, 2010. Pg 62.

Tabucki, H. (2010). Nintendo to make 3-D version of its DS handheld game. *New York Times*. March 23, 2010.

Tanz, J. (2012). The curse of cow clicker. *WIRED*. January 2012. Pg 98-101.

Tassi, P. (2013). Always On: Microsoft Xbox Live Subscriptions up to 46 Million, Will Never Be Free. *Forbes*. Retrieved from /sites/insertcoin/2013/04/20/always-on-microsoft-xbox-live-subscriptions-up-to-46m-will-never-be-free/.

Tassi, P (2014), *Microsoft Reveals 3 Million Xbox One Sales In 2013*, retrieved from http://www.forbes.com/sites/insertcoin/2014/01/06/microsoft-reveals-3-million-xbox-one-sales-in-2013/.

Techweb. (2010). COD Black Ops crosses $1 Billion Mark. *TECHWEB*. December 22, 2010. Retrieved from LexisNexis.

Thorpe, J. (2009). PlayStation 3 Firmware (v3.10) Update Preview. *Playstation.blog*. Retrieved from http://blog.eu.playstation.com/2009/11/17/playstation-3-firmware-v3-10-update-preview/.

Thorsen, T (2009). Modern Warfare 2 Sells 4.7 million in 24 hours. *GameSpot*. Retrieved from http://www.gamespot.com/news/6239789.html.

Troast, P. (2009). Vampire power check: Comparing the energy use of Xbox and Wii. *Energy Circle*. Retrieved from http://www.energycircle.com/blog/2009/12/29/vampire-power-check-comparing-energy-use-xbox-and-wii.

Vick, K. (2001). Vital ore fuels Congo's war. *Washington Post*. March 19, 2001, Pg A01.

Walters, L. (2008). Another one bites the dust. *GameCensorship.com*. Retrieved from http://www.gamecensorship.com/okruling.html.

Walters, L. (2010). *GameCensorship.com*. Retrieved from http://www.gamecensorship.com/legislation.htm.

16

Home Video

Steven J. Dick, Ph.D.*

Why Study Home Video?

- Television has been in 98% of American homes since 1980.
- Home video revenues exceed network television revenues.
- Despite the national recession, Americans continue to spend on home video.
- Home entertainment revenue topped $18.2 billion in 2013 led by digital delivery and Blu-ray sales.

Introduction

On June 12, 2009, full power television stations in the United States completed the transition to digital terrestrial broadcasting—the biggest change in television standards since the introduction of electronic television in 1945. Five years later, older tube televisions have virtually left the stores and standard definition has been relegated to a niche market.

Background

U.S. commercial television began in 1941 when the Federal Communications Commission (FCC) established a broadcasting standard. Over the years, different technologies have become more or less dominant in the delivery of home video. As this chapter considers each technology, issues of compatibility, quality, and ease of use determine success or failure. As a means of organization, technologies will be divided into three general means of reception including: by air, by conduit, and by hand.

Reception by Air

Television stations licensed in 1941 were crude, with low resolution pictures even compared to the 525-line, analog standard recently replaced. World War II stopped nearly all development, and only six stations were still broadcasting after the war.

The FCC completely revised the transmission system after the war with the introduction of electronic encoding. Yet, post-war confusion led to more delays as the FCC was inundated with new applications, and it was clear that the original very high frequency (VHF) band would not provide enough space.

After granting 107 licenses with 700 more to process, the FCC initiated a freeze on television applications in 1948 (Whitehouse, 1986). Initially, it was a short pause in processing, but the technical demands were daunting. Thus, from 1948 to 1952, there were only about 100 stations on the air nationally.

In 1952, the freeze was over, and the FCC formally accepted a plan to add ultra-high-frequency (UHF) television. Like VHF, UHF stations used six megahertz of bandwidth and encoded video using amplitude modulation (AM) and audio in frequency modulation (FM). However, because UHF transmitted on a higher frequency, requiring more electricity, existing television sets needed a second tuner for UHF, and new antennas were often needed.

This issue put UHF stations in a second-class status that was almost impossible to overcome. It was not until 1965 that the FCC issued a *final* all-channel receiver law, forcing television manufacturers to include a second tuner for UHF channels.

* Media Industry Analyst, Modern Media Barn (Youngsville, Louisiana).The author wishes to gratefully acknowledge the support of the University of Louisiana at Lafayette and Cecil J. Picard Center for Child Development and Lifelong Learning.

Initially, television was broadcast in black-and-white. In 1953, color was added to the existing (luminance) signal. This meant that color television transmissions were still compatible with black-and-white televisions. This ability to be compatible with previous technology is called *reverse compatibility* and can be a major advantage in the adoption of a new technology.

Reception by Conduit

A combination of factors, including the public's interest in television, the FCC freeze on new stations, and the introduction of UHF television, created a market to augment television delivery.

As discussed in Chapter 7, cable television's introduction in 1949 brought video into homes by wire conduit. At first, cable simply relayed over-the-air broadcast stations. In the 1970s, cable introduced a variety of new channels and expanded home video capability. An analysis of FCC data reveals that a total of 105 national and regional programming networks started operation prior to 1992. The *Cable Television Consumer Protection and Competition Act of 1992* stimulated growth in multichannel services. By 2002, the number of program networks more than tripled to 344 with the biggest increase (111) between 1997 and 1999 (McDowell & Dick, 2003).

From 1996 to 2005, the number of satellite-delivered programming networks increased from 145 to 565 (NCTA, 2007). Direct broadcast satelites (DBS) began to make serious inroads following the 1992 Cable Act. Then, the Telecommunications Act of 1996 also allowed telephone companies (telcos) to enter the market.

Today, fewer than 10% of U.S. households still receive television from traditional terrestrial broadcasting alone. Yet, traditional broadcast networks are still an essential part of the media landscape. The four top broadcast networks each still receive a weekly cumulative audience of 68% to 74% which is approximately double the top cable networks (Television Bureau of Adverting, 2013a, See Figure 16.1). A troubling trend, however, may be developing as the top program sources are all reaching a smaller cumulative audience from 2010 to 2011.

Reception by Hand

At first, television programs had to be either live or on film. In 1956, Ampex developed videotape technology that allowed recording of programs. These first videotape recorders were about the size of a double home refrigerator, used two inch wide videotape, and noisy vacuum systems held the tape in place. An hour long tape could weigh 30 pounds.

Figure 16.1

Weekly Cumulative Audience Reach for Top Broadcast and Cable Networks Change from 2010 to 2011

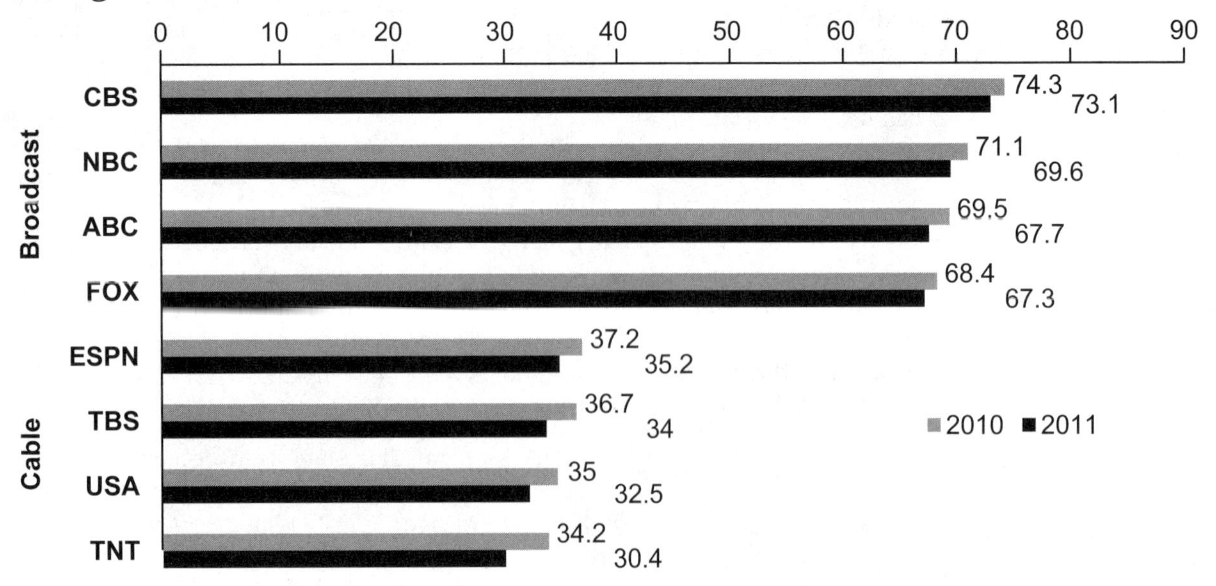

Source: TVB.org

1977 saw the introduction of practical home VCRs with two competing standards. Sony debuted the Betamax, and JVC the VHS. By 1982, a full-blown price and technology war existed between the two formats, and by 1986, 40% of U.S. homes had VCRs. Video distributers reluctantly distributed content in both formats until VHS eventually won the standard for the most homes and Betamax owners were left with incompatible machines.

VCRs gave consumers new power for on-demand programming through video rental stores or over-the-air recording. VCR penetration quickly grew from 10% in 1984 to 79% 10 years later. In 2006, the FCC estimated that 90% of television households had at least one VCR (FCC, 2006). However, VCRs were susceptible to damage by dirt, overuse, magnets, and physical shock. In the 1980s, the first videodiscs challenged VCRs.

Two main videodisc formats included RCA and MCA/Phillips. Both were based on disks the size of the long play (LP) audio disks—although the MCA/Phillips system used an optical recording system similar to today's digital videodisc or digital versatile disc (DVDs). Despite the higher quality signal, the discs were never accepted in the consumer market.

Based on the compact audio disc (CD), the DVD was introduced in 1997 as a mass storage device for all digital content. Bits are recorded in optical format within the plastic disc. Unlike earlier attempts to record video on CDs (called VCDs), the DVD had more than enough capacity to store an entire motion picture in analog television quality plus multiple language tracks, and bonus content. DVDs were disadvantaged by the lack of a record capability. However, they were smaller, lighter, and more durable than VHS tapes.

The introduction of a high-definition DVD player resulted in yet another format battle. In 2001, the Blu-ray format, supported by Sony, Hitachi, Pioneer, and six others, competed against the HD-DVD format supported by Toshiba, NEC, and Microsoft (HDDVD.org, 2003). Unlike the Beta/VHS battle, Blu-ray could hold significantly more content (50 GB compared to 30 GB for HD-DVD and 9 GB for standard DVDs). The initial cost for the players was around $500, but fell dramatically into the $250 range as the format war continued (Ault, 2008).

Both groups created exclusive deals with major film studios. Warner Brothers delivered a surprising blow at the 2008 Consumer Electronics Show by switching from HD-DVD to Sony's Blu-ray. Other studios, unhappy with multiple formats, soon followed suit. While Toshiba initially vowed to continue to fight, the company gave up within a month. Since 2003, consumers have shifted from VCR to digital formats (see Figure 16.2). 2007 marked the transition year between the dominance of the VCR and the DVD.

Figure 16.2

Technology Shifts in Home Video Penetration by Technology

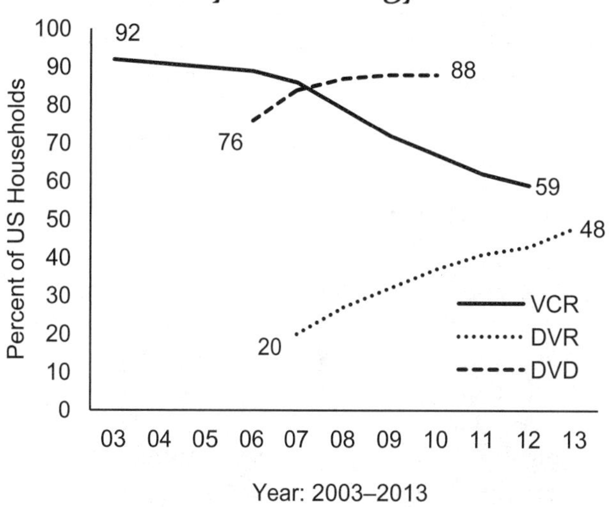

Source: TVB.org

The digital video recorder (DVR) was introduced in 1999. The heart of a DVR is a high-capacity hard drive capable of recording 100 or more hours of high definition video. Since it is a nonlinear medium, the DVR is able to record and playback at the same time. This ability gives the system the apparent ability to pause (even rewind) live television. Newer DVRs can provide on-demand programming and share content throughout the home while recording multiple programs at the same time.

DVR penetration grew dramatically between 2007 and 2013. However, as streaming and on demand options have matured (both of which are less complex and more reliable), DVR penetration has leveled off at about 48% of U.S. households (CAB, 2013)

Display

The television set has become a fixture in American homes. By 1975, televisions were seen in 95% of homes, and 66% viewed them as a "necessity" (Taylor, 2010). The television "set" was appropriately named because it included *tuner(s)* to interpret the incoming signals and a *monitor* to display the picture. Tuners have changed over the years to accommodate needs of consumers (e.g., UHF, VHF, cable-ready). Subprocessors later added interpreted signals for closed captioning, automatic color correction, and the V-chip (parental control).

The first type of television monitor was the cathode ray tube (CRT). The rectangular screen area of a CRT is covered with lines of phosphors that correspond to the picture elements (pixels) in the image. Color monitors use three streams of electrons, one for each color channel (red, blue, and green). The phosphors glow when struck by a stream of electrons sent from the back of the set. The greater the stream, the brighter the phosphor glows. The glowing phosphors combine to form an image.

The first U.S. color television standard was set by the National Television Standards Committee (NTSC), which called for 525 lines of video resolution with interlaced scanning. Interlacing means that the odd numbered video lines are transmitted first, and then the display transmits the even numbered lines. The whole process takes one-thirtieth of a second (30 frames of video per second). Interlaced lines ensured even brightness of the screen and a better feeling of motion (Hartwig, 2000).

While the 2009 economic downturn affected the media industry as much as any other industry, the transition to digital television broadcasting, combined with the exciting new digital displays, created a perceived need to buy. The Consumer Electronics Association (CEA) reported an industry-wide 7.7% decrease in revenue in 2008 (CEA, 2009). But HD displays grew from below 20% penetration in mid-2007 to 83% in 2013 (Nielsen, 2013).

Recent Developments

The transition to digital content caused consumers to make several changes in their home video environment. It is apparent that consumers chose to purchase a display to add rather than replace. The rush to meet the digital transition deadline corresponded with a growth in multi-set households which increased from 75% in 2003 to 84% in 2012 (TVB, 2012).

There is also evidence of shifts in the way people consume programming. First, since 1998, adults have increased average daily viewership by about one hour (see Figure 16.3). Viewing among women increased from about 4.5 to 5.5 hours per day. Men's viewing followed, moving from four to five hours per day. A separate study through the U.S. Department of Education indicated a drop in viewership on school nights until 2004 (Dick et.al., 2011, See Figure 16.4). In 1988, 65% of nine year olds reported watching at least three hours of television on school nights. The percent dropped to 49% in 2012. Similar drops in viewership were found at ages 13 and 17.

Figure 16.3

Daily Television Viewing for Adults

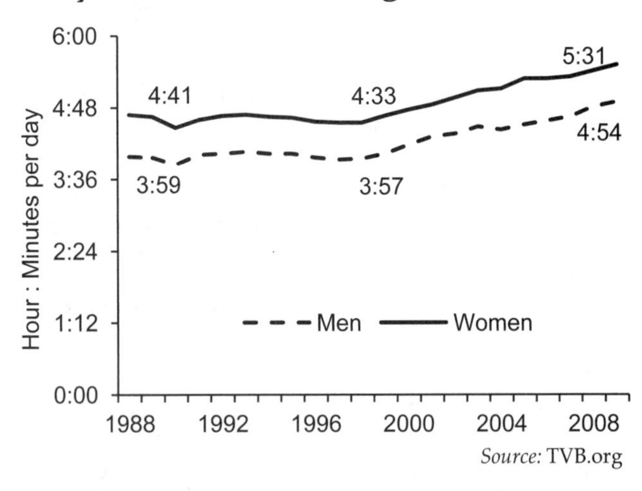

Source: TVB.org

As consumers were faced with new equipment choices, companies had to actively compete for a place in the home. Traditional cable television reached a penetration high of 71% in February 2001, and alternative multichannel services (mainly DBS) grew to 32% by November of 2013 with traditional cable penetration dropping to 59% (TVB, 2013b)

The additional competition has been answered with new technology centered on DVRs, on demand content, intelligent program guides, and mobile viewing including DirecTV's Genie, Dish's Hopper, and Cox's Contour. As a part of the package, premium program channels like HBO, Cinemax, Showtime, and Stars have streaming media channels for mobile viewers. While nearly 80% of all viewing is on live telecasts, time shifted viewing has increased—especially among 25-49 year old viewers (TVB, 2013b).

Figure 16.4

Percent of Students Watching At Least Three Hours of Television on School Nights 1984-2012

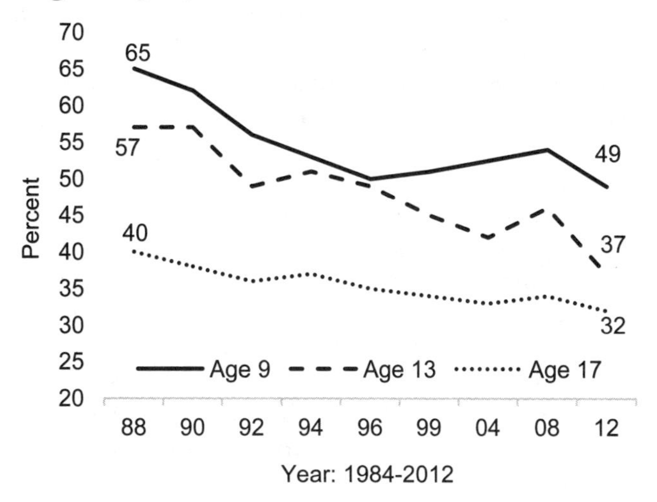

Source: National Center for Education Statistics

The line between Internet delivery and on demand programming began to blur as multi-channel program providers integrated broadband Internet delivery into their channel guides. Streaming to DVR was used to expand on-demand capacity. At the same time, companies including Hulu, Amazon, Google, and Netflix, expanded onto the home video screen through new interface boxes (Ruku), built in applications, and video game consoles.

While streaming media still only accounts for little more than an hour a week (Terrill, 2013), recent industry moves indicate an assumption of growth.

Annual digital sales surged to over $1 billion in 2013 (Chmielewski, 2014). In September 2011, Netflix angered many of its customers by splitting its DVD by mail and streaming media into two separate services—at the same time effectively doubling the cost. At first, the name of the mail based service was going to change to form two separate companies. Netflix reported a loss of 800,000 subscribers as a result of the move but has since rebounded from 24 to 39 million US subscribers (Lawler, 2013).

In November 2013, the dominant home video chain, Blockbuster, announced that it would close the last of its 300 corporate owned locations (Pappadems, 2013). Instead, the Blockbuster trademark will be a streaming media service. At the same time, Hulu reported 5 million paid subscribers and a billion dollars in revenue (Solsman, 2013).

Current Status

As the digital market has grown, consumers have more choices than ever when they select a new television. While most digital sets will at least attempt to produce a picture for all incoming signals, some will be better able to do it than others. The Consumer Electronics Association (CEA) suggests five steps in the decision process (CEA, 2007):

1) Select the right size.
2) Choose an aspect ratio.
3) Select your image quality.
4) Pick a display style.
5) Get the right connection.

Figure 16.5

Choosing the Correct Size Television

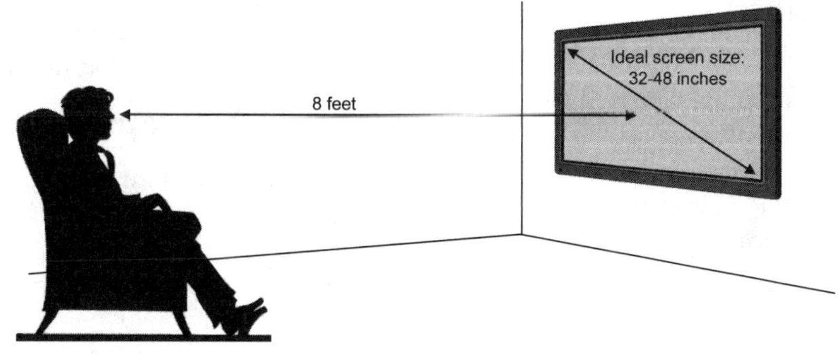

Selecting the right size set for a room is easy using this simple calculation: 1. Measure distance from TV to sitting position. 2. Divide by 2 and then by 3 to get ideal screen size range. The resulting numbers will be the ideal screen sizes. Example: Distance = 8 feet (or 96 inches), 96 / 2 = 48" set, 96 / 3 = 32" set, and Ideal set is 32" to 48".

Source: Technology Futures, Inc.

New sets are larger than ever, and it is easy to buy too large or small. Even with a high-definition set, if you sit too close, you see too much grain. If you sit too far away, you lose picture quality. The rule of thumb is to measure the distance from the picture to the seat. Divide that distance by three. The result is the smallest screen size for the room. Divide the distance by two and the result is the largest screen size for the room (see Figure 16.5).

Unlike analog, not all digital television signals are the same (See Chapter 6 for more on digital broadcast standards). The video monitor must be able to quickly interpret and display several incoming formats. Monitors display some video formats as transmitted—called native resolutions. All other signals are converted. If a monitor has only 720 lines of resolution, it cannot display 1,080 lines native, and the signal must be converted. The conversion process may reduce signal quality.

The two aspects of format that should concern the buyer the most are lines of resolution and aspect ratio. CEA has established three quality levels for lines of resolution:

- *Standard definition* uses 480 lines interlaced. It is most like analog television (525 lines), but contains the circuitry to convert higher-quality images down to the lower-resolution screen.
- *Enhanced definition* sets use 480p or higher. The image more smoothly presents high-definition content because of the progressive scan; it meets the quality of standard DVDs.
- *High-definition* pictures are at resolutions of 720p up to 1080p. They can display HD content and Blu-ray DVDs at full resolution.
- *Ultra High Definition:* A developing standard with sets/cameras in the market. It would require a 16 x 9 aspect ratio with a resolution of at least 2160p.

Aspect ratio is the relationship of screen width to the height. Standard definition sets have an aspect ratio of 4:3 or four inches wide for every three inches tall. The new widescreen format is 16:9. Most televisions will attempt to fill the screen with the picture, but the wrong image for the screen distorts or produces a "letterbox" effect.

In addition, picture size is a measure of the diagonal distance, and wider screens have proportionally longer diagonals. For example, a 55-inch diagonal set with a 4:3 aspect ratio has 1,452 square inches of picture space. The same 55-inch set with a 16:9 aspect ratio has only 1,296 square inches of picture space. Thus, it is misleading to compare the diagonal picture size on sets with different aspect ratios.

The major display styles are CRT, plasma, LCD, and projection. Although CRT screens tend to be smaller, for years they remained the most affordable choice for a bright picture and a wide viewing angle. LCD (liquid crystal display) and plasma are flat screen technologies. They take up less floor space and can be mounted like a picture on the wall. LCD displays tend to have a brighter image but a narrower viewing angle. Plasmas have a wider viewing angle but the shiny screens more easily reflect images from the room. Front or rear projection systems offer the best value for very wide images but use an expensive light bulb that must be replaced periodically. Projection systems are best in home theater installations.

Higher quality displays create video frames that bridge the motion between existing video frames. From an original 60 hertz frame rate, the system will create additional video frames to make the motion smoother and reduce perceived flicker. These sets are then marketed as 120 or 240 hertz sets. For presentation of motion pictures shot at 24 frames per second, the display can adapt to that frame rate or double it to 48 hertz—reducing flicker or a "video-like" image.

Factors to Watch

The future of home video will offer more choices. The two areas to watch for the coming years are continued development of monitors and more choices in delivery. The monitor market remains extremely competitive, and manufacturers are looking for an edge. Delivery systems are developing to reach the goal of anywhere, anytime, any-format video.

Monitor Development

Organic Light Emitting Diodes: The new OLED monitors produce a bright, crisp image without the need for backlighting. As a result, they are more energy-efficient and thinner than other flat screen monitors. Recent technology enhancements have resulted in increased dark and interframe resolution. New units coming on the market feature flexible screens.

Smart TVs and Game Console Integration: Some television receivers have added an array of interactive applications. In 2013, 76 million smart television were sold accounting for 1/3 of all units sold (Woods, 2014). In addition, cable and DBS boxes are likely to feature smart TV applications. New game consoles, like the Microsoft Xbox One have moved from DVD playback to serious video controller.

Time-shifting

Home video distributors may soon be facing new challenges and opportunities from time-shifted viewing. Currently, 86% of all viewing occurs on the same day as the original broadcast but that percent can vary wildly (TVB, 2013b). People in the top 25 markets are more likely to time-shift programming (24%) and certain programs are far more likely to be viewed later. For example, NBC's *Parenthood* topped out at 52% time-shifted in the TVB study.

In local markets, the effect of time shifting can be dramatic. TVB found one episode of *Fringe* in Cleveland/Akron was 91% time shifted. Comedy and Reality programs are the most likely to benefit from later viewing. Broadcast networks are also more likely to be time-shifted with 44% (18-49) time-shifting broadcast networks compared to 24% of advertising supported cable programs. The possible under-reporting has prompted Nielsen to include streaming in audience estimates.

Conclusion

Options are just too enticing as a new generation of home video technologies comes of age. The industry is readying for an age of multiple platforms from the very large home theater to the small mobile device. Which devices and markets will become a success is up to consumers and producers alike. The consumer must accept the platform, and the producer must create a business model that will work for everyone.

Home Video Visionaries: Chuck Lorre & Newton Minow

Chuck Lorre. Chuck Lorre began his career as a musician and song writer—traveling the country with a band. He moved on to television by writing scripts for cartoons for shows like Teenage Mutant Ninja Turtles (Internet Movie Database, 2014).

Lorre finally landed a job as a staff writer on the show My Two Dads based on a script written on speculation. His move to situation comedies (sitcoms) was highly successful as it launched him on a path to write and produce show such as Roseanne, Grace Under Fire, and Dharma and Greg. More recently has been responsible for Mom, Mike and Molly, Two and a Half Man and The Big Bang Theory.

For his effort, Lorre has earned two Emmy's, a star on Hollywood Walk of Fame and numerous other awards. Lorre drew attention and scorn of Charlie Sheen when Lorre fired Sheen from the popular show, Two and a Half Men. Still, Lorre has carved out an important niche in his brand of sitcoms highlighting common people dealing with their own personal problems.

The much maligned sitcom genre is not the most popular with only 11 percent of the prime time audience and 15 percent of the advertising revenue (Marketing Charts, 2012). Yet, the genre can be economical with cheaper sound stage only productions and limited cast size. Successful programs do very well in post network syndication. In the 2012-13 syndication season, Lorre's Big Bang Theory claimed six of the top ten highest viewed program slots (Television Advertising Bureau, 2014). The remaining four went to ESPN NFL syndicated content with much higher production costs.

Newton Minow. Soon after becoming head of the Federal Communications Commission (FCC), Newton Minow issued a challenge in a speech before the National Association of Broadcasters. In his now famous speech, he said.

When television is good ... nothing is better. But when television is bad, nothing is worse. I invite each of you to sit down in front of your television set when your station goes on the air and stay there, for a day, without a book, without a magazine, without a newspaper, without a profit and loss sheet or a rating book to distract you. Keep your eyes glued to that set until the station signs off. I can assure you that what you will observe is a vast wasteland.

Minow's "Vast Wasteland" speech, delivered May 9, 1961, echoes as a continual challenge to broadcasters even today. At its core, the speech sought a balance between the profit concerns of the fledgling industry and the goal of public service. Broadcasting, at that time, was charged with operating in the public interest, convenience, and necessity.

The reasoning for more intense regulation was at least partially based on the scarcity of available channels. In today's multi-channel environment, one passage begs an interesting question today.

I do not accept the idea that the present over-all programming is aimed accurately at the public taste. The ratings tell us only that some people have their television sets turned on and of that number, so many are tuned to one channel and so many to another. They don't tell us what the public might watch if they were offered half-a-dozen additional choices.

Increased channel capacity reduces or eliminates the scarcity rational. The 1996 change in telecommunication regulation recognized the change in the market as a result of the technology. Is television still a "vast wasteland"? If so, does it matter? Does the industry still bear a responsibility to the public to serve the public? How might the answers to these questions affect the expectations of insist success?

Bibliography

Ault, S. (2008, January 21). The format war cost home entertainment in 2007. *Video Business*. Retrieved from http://www.videobusiness.com.

Cable Adverting Bureau. (2013). Cable Nation: Fast Forward to DVR Facts. Retrieved from http://www.thecab.tv/main/bm~doc/cablenation-timeshifting-report.pdf.

Chmielewski, D.C. (2014, January 7).Digital video sales' rise breathes new life into home entertainment. *Los Angeles Times*. Retrieved from http://www.latimes.com/business/la-fi-ces-digital-sales-20140103,0,5970703.story#axzz2povFQDF0.

Consumer Electronics Association. (2007). *DTV—Consumer buying guide*. Retrieved from http://www.ce.org/Press/CEA_Pubs/1507.asp.

Consumer Electronics Association. (2009). Retrieved from http://www.ce.org/Research/Sales_Stats/default.asp.

Dick, S.J., Davie, W.R. & Miguez, B.B. (2011). "Adding depth to the relationship between reading skills and television viewing: an analysis of national assessment of educational progress (NAEP) Reading Scores" *Association for Education in Journalism and Mass Communication National Conference*, St Louis, MO.

Federal Communications Commission. (2006). Annual assessment of the status of competition in the market for the delivery of video programming. Retrieved from http://www.fcc.gov/mb/csrptpg.html.

Hartwig, R. (2000). *Basic TV technology: Digital and Analog*, 3rd edition. Boston: Focal Press.

HDDVD.org. (2003). The different formats. Retrieved from http://www.hddvd.org/hddvd/ difformatsblueray.php.

Internet Movie Database. (n.d.). Chuck Lorre. Retrieved from http://www.imdb.com/name/nm0521143/

Lawler, E. (2013, July 7) Netflix: We Got It Right! *Huffington Post*. Retrieved from http://www.huffingtonpost.com/ed-lawler/netflix-we-got-it-right_b_3530819.html.

McDowell, W & Dick, S (2003). Has Lead-in Lost its Punch? *The International Journal on Media Management, Vol. 5,* no. IV. pp. 285-293.

Nielsen (2013), Consumer Electronics Ownership Blasts off in 2013. Retrieved from http://www.nielsen.com/us/en/newswire/2013/consumer-electronics-ownership-blasts-off-in-2013.html.

Pappadems, A. (2013, November 7). Blockbuster Video 1985-2013. *Hollywood Prospectus*. Retrieved from http://grantland.com/hollywood-prospectus/blockbuster-video-1985-2013/.

Solsman, J. (2013, December 18). Hulu in 2013. *CNet.com*. Retrieved from http://news.cnet.com/8301-1023_3-57616003-93/hulu-in-2013-sales-up-nearly-half-to-$1b-with-5m-paid-users/.

Taylor, P. (2010, August 19). The fading glory of the television and telephone. *Pew Social and Demographic Trends*. Retrieved from http://www.pewsocialtrends.org/2010/08/19/the-fading-glory-of-the-television-and-telephone/.

Television Bureau of Adverting. (2012, December). TV basics. Retrieved from from http://www.tvb.org/media/file/TV_Basics.pdf.

Television Bureau of Adverting. (2013a, November). Media trends track. Retrieved from http://tvb.org/nav/build_frameset.aspx.

Television Bureau of Adverting. (2013b, November). Time shifted viewing: A local phenomena. Retrieved from http://www.tvb.org/media/file/LPM_Time_Shifted_Viewing.pdf.

Terrill, D. (2013, December 3). A Look Across Media: The Cross-Platform Rreport Q3 2013. Retrieved from Nielsen.com: http://www.diary.tvratings.com/us/en/reports/2013/a-look-across-media-the-cross-platform-report-q3-2013.html.

Whitehouse, G. (1986). *Understanding the new technologies of mass media*. Englewood Cliffs, NJ: Prentice-Hall.

Woods, B. (2014, January). Smart TVs accounted for One Third of Flat Panel TV Shipments in 2013. The Next Web.

Digital Audio

Ted Carlin, Ph.D.*

Why Study Digital Audio?

- There are more than 500 licensed music services worldwide enabling users to download, stream, or access music legally.
- It is estimated that more than 28 million people worldwide now pay for a music subscription, up from 20 million in 2012 and just eight million in 2010.
- Music fans' growing appetite for subscription and streaming services helped drive trade revenue growth in most major music markets in 2013, with overall digital revenues growing 4.3 percent and Europe's music market expanding for the first time in more than a decade.
- Revenues from music streaming and subscription services leapt 51.3 percent globally, crossing the $1 billion threshold for the first time.

Introduction

Producers, consumers, and advertisers alike are gravitating to the digital domain, which offers ever-expanding media content options via a growing variety of devices. For example, Randall Roberts's *Los Angeles Times* review of Van Halen's latest album, *Different Kind of Truth*, effectively describes this digital evolution and the format choices today's music listeners have:

> In the 28 years since Roth recorded a full album with Van Halen, the landscape has completely changed. When the band's original lineup last released a record, home taping was "killing" music and the question was whether to buy "1984" on LP or cassette, or borrow a friend's copy and tape over Foreigner "4."
>
> Now the dilemma isn't just should you spend money on the CD ($14.99 list price) or a digital copy (also—frustratingly—$14.99). It's also, how much are you willing to commit to buying in? Will a few dropped bucks on a handful of the best individual tracks suffice? Or will "A Different Kind of Truth" be the perfect Spotify streaming album, not good enough to pay hard money for but worth a mouse-click when you've got a spare few minutes? Or should you just ask your computery friend to Sendspace you a pirated copy?
>
> —Roberts, 2012

Just as in 1984, in the end, the consumer gets to make a choice—to buy, to rent, to steal—but today the choice is immediate and with fewer people in the way. Today's digital audio technology drives this choice by removing the traditional retail barriers between consumer and content, and enabling consumers to interact with content in new, direct and personal ways. Through various software and hardware tools, consumers and content providers become intertwined in the discovery process, co-creating playlists, sharing recommendations, enabling file and information flow, and ultimately, facilitating an acquisition decision.

* Professor, Department of Communication/Journalism, Shippensburg University (Shippensburg, Pennsylvania).

Background

Analog Versus Digital

Analog means "similar" or "a copy." An analog audio signal is an electronic copy of an original audio signal as found in nature, with a continually varying signal. Analog copies of any original sound suffer some degree of signal degradation, called generational loss, and signal strength lessens and noise increases for each successive copy. However, in the digital domain, this noise and signal degradation can be eliminated (Watkinson, 2013).

Audio is digitized by sampling the amplitude (strength) of a waveform from a capturing device (typically a microphone) using an analog to digital converter (see Figure 18.1). Depending on the equipment used to make this digital copy, various samples of the original sound wave are taken at given intervals using a specified sampling rate (i.e., 32 kHz for broadcast digital audio; 44.1 kHz for CD and MP3; 192.4 kHz for Blu-ray) to create a discrete digital wave (Alten, 2014). These samples are then quantized as binary numbers at a specific bit level (16-bit, 32-bit, etc.); the higher the bit level, the higher the quality of the digital reproduction.

Figure 18.1
Analog Versus Digital Recording

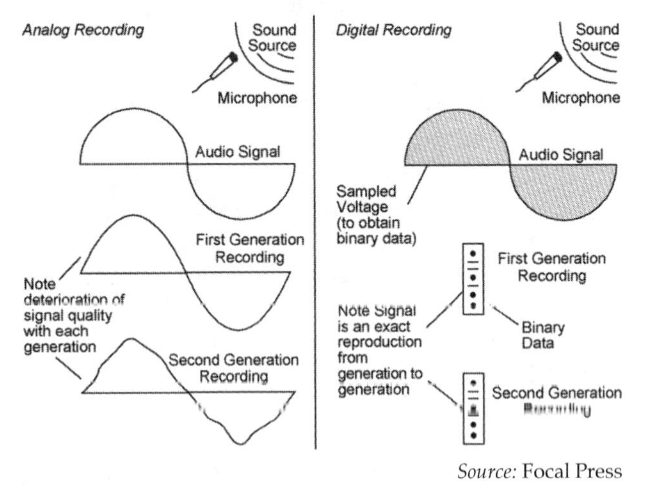

Source: Focal Press

File Formats and Codecs

Once the signal is digitized, sampled, and quantized, the digital signal can be subjected to further processing—in the form of audio data compression—to reduce the size of the digitized audio file even further. This is similar to "zipping" a text file to more easily send the digital file across the Internet and to store more digital files on devices.

Audio compression algorithms are typically referred to as *audio codecs*. As with other specific forms of digital data compression, there exist many "lossless" and "lossy" formulas to achieve the compression effect (Alten, 2014). Lossless compression works by encoding repetitive pieces of information with symbols and equations that take up less space, but provide all the information needed to reconstruct an exact copy of the original. Lossy compression works by discarding unnecessary and redundant information (sounds that most people cannot hear) and then applying lossless compression techniques for further size reduction. With lossy compression, there is always some loss of fidelity that becomes more noticeable as the compression ratio is increased.

Unfortunately for consumers, there is no one standard audio codec, and Table 18.1 provides a list of the major codecs in use in 2014. However, there is one codec that has emerged as the clear leader—MP3. Before MP3 came onto the digital audio scene in the 1990s, computer users were recording, downloading, and playing high-quality sound files using an uncompressed codec called .WAV. The trouble with .WAV files, however, is their enormous size. A two-minute song recorded in CD-quality sound uses about 20 MB of a hard drive in the .WAV format. That means a 10-song CD would take more than 200 MB of disk space.

The file-size problem for music downloads has changed thanks to the efforts of the Moving Picture Experts Group (MPEG), a consortium that develops open standards for digital audio and video compression. Since the development of MPEG, engineers have been refining the standard to squeeze high-quality audio into ever-smaller packages. MP3—short for MPEG 1 Audio Layer 3—can take music from a CD and shrink it by a factor of 12 (Digigram, 2014).

Digital Audio Technologies

Compact Disc

In the audio industry, nothing has revolutionized the way we listen to recorded music like the compact disc (CD). Originally, engineers developed the CD solely for its improvement in sound quality over LPs and analog cassettes. After the introduction of the CD player, consumers became aware of the quick random-access characteristic of the optical disc system. In addition, the size of the 12-cm (about five-inch) disc was easy to handle compared with the LP. The longer lifetime of both the medium and the player strongly supported the acceptance of the CD format.

The next target of development was the rewritable CD. Sony and Philips jointly developed this system and made it a technical reality in 1989. Two different recordable CD systems were established. One is the write-once CD named CD-R, and the other is the re-writable CD named CD-RW (Alten, 2014).

Digital Media Player

To play MP3s and other digital audio files, a computer-based or digital portable media player (PMP) is needed. Hundreds of portable media players are available, from those with 512 MB flash drives to massive 250 GB hard drive-based models. They are sold by manufacturers such as Apple, Archos, Cowon, Creative, and Sony. The amount of available disc space is most relative to the way a person uses the portable player, either as a song selector or song shuffler. Song selectors tend to store all of their music on players with larger drives and select individual songs or playlists as desired, whereas song shufflers are more likely to load a group of songs on a smaller drive and let the player shuffle through selections at random.

Peer-to-Peer (P2P) File Sharing

In a P2P network, the "peers" are computer systems which are connected to each other via the Internet. Digital files can be shared directly between computers on the network without the need of a central server. In other words, each computer on a P2P network becomes a file server as well as a client.

Table 17.1

Popular Audio Codecs

Codec	Developer	Date	Compression Type
AAC	MPEG	2002	lossy
AACplus or HE-ACC	MPEG	2003	lossy
AIFF	Electronic Arts/Apple	1985	uncompressed
ALAC	Apple	2004	lossless
AU	Sun	1992	uncompressed
Cook	RealNetworks	2003	lossy
FLAC	Xiph.org	2003	lossless
LAME	Cheng/Taylor	1998	lossy
Monkey's Audio	M. Ashland	2002	lossless
MP3	Thomson/Fraunhofer	1992	lossy
MP3 Pro	Thomson/Fraunhofer	2001	lossy
MP3 Surround	Thomson/Fraunhofer	2004	lossy
Musepak	Buschmann/Klemm	1997	lossy
SDII (Sound Designer II)	Digidesign	1997	lossless
SHN (Shorten)	T.Robinson/W. Stielau	1993	lossless
Speex	Xiph.org	2003	lossy
Vorbis (Ogg Vorbis)	C. Montgomery/Xiph.org	2002	lossy
TTA (True Audio)	Alexander Djourik	2007	lossless
VoxWare MetaSound	A. Penry	2004	lossy
WavPack	D. Bryant	1998	lossless
WMA (Windows Media Audio)	Microsoft	2000	lossy
WMA Lossless	Microsoft	2003	lossless

Source: T. Carlin

The only requirements for a computer to join a P2P network are an Internet connection and P2P software. These programs connect to a P2P network, for example, BitTorrent, which allows the computer to access thousands of other computers on the network. Once connected to the network, the P2P client software allows you to search this computer "swarm" for torrent files located on other people's computers. The torrent files are "chunks" of the desired song or movie that are located and assembled on your computer to make one complete file.

So, rather than getting the whole file from one central server, the file is assembled, or "leeched," from pieces located on various clients within the swarm. Meanwhile, other users on the network can search for files on your computer, but typically only within a single folder that you have designated to share (Strufe, 2014). While P2P networking makes file sharing convenient and quick, is also has led to software piracy and illegal music downloads, as discussed below.

Podcasting

Podcasting is the distribution of audio or video files, such as radio programs or music videos, over the Internet using "really simple syndication" (RSS) for listening on mobile devices and personal computers (Podcast Alley, 2014). A podcast is basically a Web feed of audio or video files placed on the Internet for anyone to download or subscribe to. Podcasters' websites may also offer direct download of their files, but the subscription feed of automatically delivered new content is what distinguishes a podcast from a simple download or real-time streaming. A quick search of the Internet will uncover a myriad of content that is available—and accumulating daily.

Copyright Legislation, Cases and Actions

With this ever-increasing ability to create and distribute an indefinite number of exact copies of an original sound wave through digital reproduction comes the incumbent responsibility to prevent unauthorized copies of copyrighted audio productions and safeguard the earnings of performers, writers, and producers. Before taking a closer look at the various digital audio technologies in use, a brief examination of important legislative efforts and resulting industry initiatives involving this issue of digital audio reproduction is warranted.

Audio Home Recording Act of 1992

The *Audio Home Recording Act (AHRA) of 1992* exempts consumers from lawsuits for copyright violations when they record music for private, noncommercial use and eases access to advanced digital audio recording technologies. The law also provides for the payment of modest royalties to music creators and copyright owners, and mandates the inclusion of the Serial Copying Management Systems (SCMS) in all consumer digital audio recorders to limit multigenerational audio copying (i.e., making copies of copies). This legislation also applies to all future digital recording technologies, so Congress is not forced to revisit the issue as each new product becomes available (*AHRA*, 1992).

The Digital Performance Right in Sound Recordings Act of 1995

This law (*Digital Performance Right in Sound Recordings Act*, 1996) allows copyright owners of sound recordings to authorize certain digital transmissions of their works, including interactive digital audio transmissions, and to be compensated for others. This right covers, for example, interactive services, digital cable audio services, satellite music services, commercial online music providers, and future forms of electronic delivery.

No Electronic Theft Law (NET Act) of 1997

The *No Electronic Theft Law* states that sound recording infringements (including by digital means) can be criminally prosecuted even where no monetary profit or commercial gain is derived from the infringing activity. Punishment in such instances includes up to three years in prison and/or $250,000 in fines (*NET Act*, 1997).

Digital Millennium Copyright Act (DMCA) of 1998

The main goal of the DMCA was to make the necessary changes in U.S. copyright law to allow the United States to join two new World Intellectual Property Organization (WIPO) treaties that update international copyright standards for the Internet.

The DMCA amended copyright law to provide for the efficient licensing of sound recordings for Webcasters and digital subscription audio services via cable and satellite. In this regard, the DMCA:

- Made it a crime to circumvent anti-piracy measures [i.e., digital rights management (DRM) technology] built into most commercial software. DRM technology controls the trading and monitoring of a digital work. Apple's Fairplay and Blu-ray's Advanced Access Content System (AACS) are two prominent DRM technologies.

- Outlawed the manufacture, sale, or distribution of code-cracking devices used to illegally copy software.
- In general, limited Internet service providers from copyright infringement liability for simply transmitting information over the Internet. Service providers, however, are expected to remove material from users' websites that appear to constitute copyright infringement.
- Provided for a simplified licensing and compensation system for digital performances of sound recordings on the Internet, cable, and satellite systems. This licensing system is similar to ASCAP and BMI compulsory licensing for music used on radio and television stations (U.S. Copyright Office, 2014).

The U.S. Copyright Office designated a nonprofit organization, SoundExchange, to administer the performance right royalties arising from digital distribution via subscription services. Once rates and terms are set, SoundExchange collects, administers, and distributes the performance right royalties due from the licensees to the record companies and artists (SoundExchange.com, 2014).

MGM Studios, Inc. v. Grokster, Ltd.

Although there has been a constant debate in all three of these copyright areas, it is the copyright protection pertaining to Internet music downloading that has led to the most contentious arguments. On June 27, 2005, the U.S. Supreme Court ruled unanimously against peer-to-peer (P2P) file-sharing service providers Grokster and Streamcast Networks. The landmark decision in *MGM Studios, Inc. v. Grokster, Ltd.* (545 U. S. 125 S. Ct. 2764, 2005) was a victory for the media industry and a blow to P2P companies. At the same time, the decision let stand the main substance of the Supreme Court's landmark "Betamax" ruling (*Sony Corp. of America v. Universal City Studios*, 1984), which preserved a technologist's ability to innovate without fear of legal action for copyright infringement, and which the media industry sought to overturn.

The Supreme Court found that technology providers should be held liable for infringement if they *actively promote* their wares as infringement tools. Writing for the court, Justice David Souter stated, "We hold that one who distributes a device with the object of promoting its use to infringe copyright, as shown by clear expression or other affirmative steps taken to foster infringement, is liable for the resulting acts of infringement by third parties" (*MGM Studios, Inc. v. Grokster, Ltd.*, 2005, p. 1).

The decision created a new theory of secondary infringement liability based on "intent to induce" infringement, which now extends the existing theory of contributory liability (knowingly aiding and abetting infringement). This inducement concept is derived from an analogous and established theory in patent law (LII, 2014). Again, technology inventors must promote the illegal use of the product to be found liable:

> *...mere knowledge of infringing potential or of actual infringing uses would not be enough here to subject a [technology] distributor to liability.... The inducement rule, instead, premises liability on purposeful, culpable expression and conduct, and thus does nothing to compromise legitimate commerce or discourage innovation having a lawful promise.*
>
> —MGM v. Grokster (2005, p. 19)

Overall, this decision does not mean that P2P services have shut down altogether—as of mid-2014, the top 10 P2P services were uTorrent, Tixati, Transmission, Vuze, BitComet, ABC, Deluge, Frostwire, TurboBT and iMesh (P2PON!, 2014). However, those P2P services that have been targeted by Recording Industry Association of America (RIAA) and the Department of Justice (DOJ) over the last decade, including Morpheus, LimeWire, and Pirate Bay, have been shut down or forced to alter their practices.

DRM-Free Music

Although buying practices are rapidly changing, much of the music sold in the world is still sold on compact discs. CDs have no encryption. They are DRM-free and can play on any CD player, including computer drives. CDs also provide high-quality digital content that can easily be ripped to a digital file, copied, or shared (legal issues notwithstanding) at the discretion of the buyer.

Digital downloads, in contrast, are accessible online at any time, making their purchase convenient. Often sold as singles and not full albums, they are economical as well. That convenience comes at the cost of quality and, especially, portability. Smaller digital files appropriate for downloads mean that the purchaser gets the music with lesser sound quality. Because the major recording labels insisted (until 2007) that downloadable music be encrypted with DRM, and there was no universal, open standard for that encryption, the music could only play on devices capable of decrypting the specific DRM encryption the music was encoded with (as opposed to universally on any digital media player).

One of the biggest developments in digital audio technologies was the 2008 announcement, by the Big Four record companies, Amazon.com, Apple, and others, to offer DRM-free music tracks over the Internet (Holahan, 2008). For years, DRM was simply a way to protect the rights of the artists and record labels whose music was being illegally distributed. Most in the music industry believed that the technology was a necessary evil that needed to be put in place for content creators to make sure that they were being fairly compensated for their work. However, as mentioned earlier in the chapter, DRM wraps music tracks in a copy-protection code that is not only restrictive but also confusing for many potential users. Very few consumers wanted to purchase music on one service using a specific DRM, only to find out later that they cannot play it on this digital media player, that computer, or this operating system.

However, DRM-free does not mean consumers are free from having to purchase the music. And, unlike the mid-2000s, where providers offered music tracks (or albums) for a single price, the 2010s have ushered in the era of variable pricing and limited-time discounts by many providers, including Amazon and iTunes. For example, rather than static track pricing of 99-cents each, services now offer tracks at $0.99 or $1.29 depending on the age and popularity of the track. To further entice consumers, limited-time promotional offers—such as release date discounts or holiday specials—have become a standard marketing approach as well.

An excellent resource for learning about all of the legal DRM-free music options available globally on the Internet is Pro-music. This non-profit organization is comprised of international musicians, performers, managers, artists, record companies, and retailers across the music industry. The group's website (www.pro-music.org) houses an engaging, searchable, map-based database of digital music information (Pro-music, 2014).

U.S. Department of Justice Operations & RIAA Lawsuits

The DOJ's Computer Crimes and Intellectual Property Section (CCIPS) is responsible for coordinating and investigating violations of the *U.S. Copyright Act*, the *Digital Millennium Copyright Act 2014*, and other cyber laws discussed earlier in this chapter. The CCIPS, along with the Federal Bureau of Investigation's (FBI's) Cyber Division and numerous state Attorneys General, have been actively monitoring and investigating illegal operations associated with digital audio and copyright violations since 2004.

And, in an increased effort to implement the Obama administration's commitment to the importance of the Internet to U.S. innovation, prosperity, education, and political and cultural life, the DOJ created a new Task Force on Intellectual Property to crack down on the growing number of domestic and international intellectual property crimes (DOJ, 2014a). The task force works closely with the President's Office of the Intellectual Property Enforcement Coordinator (IPEC) and the Commerce Department's newly created Internet Policy Task Force.

The RIAA and several of its member record companies provide assistance to the DOJ and FBI in their investigations. The legal issues surrounding digital audio are not limited to file sharing networks. The RIAA assists authorities in identifying music pirates and shutting down their operations.

Based on the *Digital Millennium Copyright Act*'s expedited subpoena provision, the RIAA sends out information subpoenas as part of an effort to track and shut down repeat offenders and to deter those hiding behind the perceived anonymity of the Internet. Information subpoenas require the Internet service provider (ISP) providing access to or hosting a particular site to provide contact information for the site operator. Once the site operator is identified, the RIAA takes steps to prevent repeat infringement. Such steps range from a warning e-mail to litigation against the site operator. The RIAA then uses that information to send notice to the operator that the site must be removed. Finally, the RIAA requires the individual to pay an amount designated to help defray the costs of the subpoena process (RIAA, 2014).

Recent Developments

Digital Audio Technologies

Wireless Digital Media

In addition to mobile smartphone users who can find many of the cloud, playlist, and subscription services described below via downloaded apps or websites, the major wireless phone companies continue to offer music to data-equipped mobile phone subscribers.

AT&T eMusic Mobile. AT&T subscribers can download music, from about seven million tracks, directly to their phone with eMusic Mobile. Users can purchase five songs for $7.49/month and receive a free MP3 copy online (eMusic, 2014b).

Verizon Wireless Music Partners. In lieu of V Cast, which was Verizon's in-house solution for music and video for the past several years, Verizon now promotes customer choice via two partner third party applications—Rhapsody and Slacker. Customers can subscribe to these streaming services or select another provider of their choice (Verizon Wireless, 2014).

Sprint Music Plus. Sprint Music Plus allows users to create ringtones, music playlists, search for music by artist, title or keyword, and organize full tracks/albums by artist, genre and custom playlists using the music library manager. Tracks are available from 99¢ to $1.29 (Sprint, 2014).

Subscription-Based Services

With the evolution of track-based DRM-free Internet and wireless music options, astonishing growth continues in the U.S. subscription arena from four major services in 2010 (eMusic, Napster, Rhapsody and Zune) to over 25 in mid-2014. Some of the major ones are discussed here.

Beats Music. Beats Electronics purchased MOG and has deployed the streaming service as a foundation to build Beats Music. (MOG shut down on April 15th.) While the MOG infrastructure is evident in Beats Music, there is a completely new user interface and process for curating music. There is no free listening, ad-supported option. Subscribers can access the 20 million songs catalog for $9.99/month. AT&T Wireless customers can subscribe to a five-member family plan for $14.99/ month (Beats Music, 2014).

eMusic. eMusic started as the first online music store in 1998 and was the first to offer DRM-free tracks in 2007. It has licensing deals with all of the major record labels, but is known for its large number of contracts with independent labels and its diverse music catalog. eMusic is also one of the more expensive services, with the Basic plan of 24 downloads per month offered for $11.99/month (eMusic, 2014a).

Google Play All Access Music. The 2013-launched service is integrated with the music section of the Google Play store, as well as the subscriber's existing digital music collection. After a 30-day free trial, subscribers can join for $9.99/month. The Listen Now feature offers "smart" recommendations for similar music, similar to Pandora and YouTube (Google play, 2014).

Grooveshark. Part streaming music service, part peer-to-peer network, Grooveshark is one of the most unique streaming music sites on the Web. Grooveshark lets its users upload MP3 files and stream them to others, engage in social networking, and receive music recommendations. It is also being sued by the major record labels for copyright infringement (see below). Grooveshark is an ad-supported service that offers free music by enabling users to post their own tracks to the site and then share them with other users.

In addition to its free, ad-supported tier, Grooveshark offers a $9/month ad-free "Anywhere" level. Once a user finds a song or artist, a playlist can be started. Grooveshark will then recommend songs based on the selection, and the user can favorite songs, play the songs, add them to a queue or playlist, or embed them as a widget on a website. As with most playlist sites, users can also share Grooveshark musical selections via social media (Grooveshark, 2014).

Pandora. Built on the Music Genome Project, a patented mathematical algorithm that scans over 400 musical attributes of songs selected by users (like rhythm, tempo, syncopation, key tonality, vocal harmonies, etc.), Pandora creates customized "stations" based on similarly-matched artists, songs, and styles to provide a free streaming playlist music service. Pandora also features a premium option: Pandora One, a $3.99/month subscription that removes all visual and audio ads, increases bits-per-second to 192k, and removes the 12 skips per day limit (although you're still limited to six skips per station per hour).

Rdio. This online streaming service features three levels of subscription service, all without advertising: a free, but restricted-play-per-month Web-only service, an unlimited Web-only service ($4.99/month), and an unlimited web and mobile streaming service ($9.99/month). Family unlimited plans can also be selected. Over 20 million songs are available from a variety of labels (rdio.com, 2014).

Rhapsody. Rhapsody, which purchased Napster from Best Buy in October 2011, offers a 30-day free-trial and then its subscription service, Rhapsody Premier, for $9.99 per month. The service allows subscribers to listen to all available music (over 16 million tracks) on a variety of media playback devices. (Rhapsody, 2014).

Sony Music Unlimited. This service offers two subscription plans, an Access Plan ($4.99/month) and a Premium Plan ($9.99/month). The Access Plan allows subscribers to listen to over 20 million tracks only on a computer or PlayStation console. The Premium Plan adds mobile phones, tablets and smart TVs (Sony Entertainment Network, 2014).

Spotify. Coming to the U.S. from Sweden, Spotify is a streaming music service similar to Pandora and Rdio. Spotify's free account is advertiser-supported, and allows playback on mobile devices via shuffle mode or Spotify's "ready-made" playlists. The Spotify Premium $9.99/month option removes the ads, has unlimited desktop streaming, increases the audio bitrate, and enables unfettered mobile access (Spotify, 2014).

Xbox Music. From Microsoft, it replaced Zune Music Pass in October 2012. For no cost, via an ad-supported interface, users can stream songs and play them on a Windows 8.1 tablet or PC, Xbox One, Xbox 360, or the web. Users can also purchase the $9.99/month Xbox Music Pass subscription to listen to ad-free music on these and non-Microsoft devices (Xbox Music Pass, 2014).

Free Playlist & Webcaster Services

Blip.fm. A blip is a combination of a song and a short message that accompanies it. Blip centers on the user as DJ and the individual songs that are blipped. The DJ would search for a track he or she wanted to share, write a Twitter-style note about the selection, and "blip" it for listeners. Then, depending on user-selected settings, the blip could be broadcast to social sites such as Twitter or Tumblr. DJs can reply to other DJs and can give out tokens of respect called "props." As a user starts to blip a bit more and add favorite friends as DJs, the blip-generated playlist starts to expand. Users can customize a few settings, but essentially every time DJs blip, their songs will be populated in the user's mix for a potentially never-ending assortment of songs (Blip.fm, 2014).

Digster. Digster is a service that publishes playlists for use in Spotify and iTunes. Playlists are compiled by a team of Universal Music Group editors, can be imported into iTunes and Spotify, and can be listened to for free on Spotify Free, which is funded by ads. Ads can be disabled by paying a monthly fee. Every track and playlist on Digster is linked to a corresponding track on iTunes, so users can buy a playlist or a selection of tracks with just a few clicks (Digster, 2014).

last.fm. last.fm is a free global music playlist service that exposes its community of over 40 million users worldwide to new and relevant music through its proprietary "scrobbling" technology. Scrobbling allows the audience to track the music they play on last.fm, Radio.com, and on more than 600 music applications. Last.fm then uses the collective intelligence to recommend the songs and artists that power its personalized radio stations. Related music will populate the station with the ability to love, ban, and skip tracks as appropriate. To play content on a mobile device, a $3/month subscription is required (last.fm, 2014).

Rara. A British-based music subscription service that streams out playlists created by Rara's programmers from its 22 million tracks for web listening ($4.99/month) or listening on any device ($9.99). Rara markets itself as the subscription service for non-techies and features an imaged-based website/app to engage these users (Rara, 2014).

Slacker Personal Radio. Slacker is an interesting playlist music service. On the one hand, the free version—like Pandora—lets users listen to pre-made stations or create stations based on artists, albums, or tracks they choose. When users create such a station, it is populated with tracks that are supposed to be harmonious with the original selection. If users purchase a Radio Plus ($3.99/month) or Premium Radio subscription ($9.99 per month)—as with such subscription services as Rdio, Rhapsody, and Spotify—

they additionally have the ability to play songs, albums, and single-artist stations on demand, as well as cache albums and playlists on portable devices (Slacker, 2014).

Songza. Songza builds playlists based on whatever activity the user is currently engaged in (working, making dinner, driving) or his/her mood (gloomy, happy, romantic), and does so without interruptions and for free, In 2013, a Club Songza ad-free subscription account, at 99¢/week, was made available (Songza, 2014).

SoundCloud. SoundCloud.com is a free basic service that allows sound creators, from home users to musical artists and composers, to instantly record audio on the site or via mobile applications and share them publicly or privately. The service operates much in the same way as Twitter, where registered users can follow other users and receive notification of new activity in their stream.

Listening to music via SoundCloud provides a different kind of visual experience. For example, each track is represented by a waveform with member comments layered on top. Comments are tied to specific points in the song, and will be displayed accordingly for future listeners to read as they listen to the song. SoundCloud offers additional features to users with paid subscriptions. Such users are given more hosting space and may distribute their tracks or recordings to more groups and users, create sets of recordings, and more thoroughly track the statistics for each of their tracks (SoundCloud, 2014).

8tracks. This is a free, ad-supported Internet radio service where users do the music selecting. Members pick at least eight songs (hence the name) from their own collections, upload them, and share them as handcrafted mixes that seem like a happy throwback from the cassette era. Users can find mixes created by Facebook friends or simply follow other members who share similar tastes (8tracks, 2014). And it's legal—8tracks takes advantage of a provision in the *Digital Millennium Copyright Act*, which gives webcasters a compulsory license to operate as a "non-interactive Webcaster." This means that 8tracks can stream any uploaded music as long as 8tracks operates, like it currently does, as a radio station and pays the Sound Exchange royalty fees. An ad-free option, 8tracksPlus, is available for $4.16/month

Cloud Music Services

The four major cloud service providers offer similar products with a similar mission: Allow users to buy new music and access existing libraries from multiple devices, via streaming from the cloud.

Amazon Cloud Drive with Cloud Player. The service is free for up to 5GB of uploads. As with Google Music, Amazon.com MP3 purchases do not count against this cap. Unlimited space for all music, regardless of purchase location, is included with all paid storage plans of 20GB and up. Like Google Music, all music is streamed and does not require download to a device in order to listen. Amazon provides a mobile app for Android and an iPad-friendly web view of its library (Amazon, 2014).

Google Play. Google Play is free to all users. This includes access on the web, desktop, and mobile devices. Google does impose a library limit of 20,000 songs that a user may upload. However, purchased songs do not count against this cap. All music is streamed and does not require download to a device in order to listen. Users can combine an All Access paid subscription with this service (Google, 2014).

iCloud with iTunes Match. This service is available from Apple for $24.99 a year. For that $24.99, users get the ability to upload 25,000 songs plus any iTunes purchases. Apple uses iTunes as its central music listening hub. Users can authorize up to 10 devices for use with iTunes Match—five can be computers with iTunes—to stream or download songs. Apple's mobile solution is the most limited in terms of device access. Playlist syncing and file access is only available on an iOS device (Apple, 2014).

Ubuntu One with Music Streaming. Users can stream all the music saved in their Ubuntu One personal cloud via the web or mobile devices. With a $3.99/month Music Streaming subscription, users get 20 GB of cloud storage to store all files, photos, and music (Ubuntu One, 2014).

Cyberlocker Services

The "bad guy on the block" in illegal file sharing is the cyberlocker service (legal issues are discussed in the next section). Cyberlockers are online data hosting services that provide remote storage space within a secure cloud-based storage system. They can be accessed globally over the Internet and are often called online storage or cloud storage services. Megaupload, RapidShare, Hotfile, 4Shared, and Mediafire are the most popular cyberlockers. The storage capacity provided by these cyberlockers varies depending on the price. Services normally offer some storage and downloading services for free, and then charge for premium accounts that include more storage capacity and faster file transfers (Wyman, 2012).

Because of the anonymity and restricted access they provide, cyberlockers are criticized for being used as a safe haven for piracy. Why? Cyberlocker users get an IP address (or URL), and then can share it with whom they choose. Unlike P2P programs such as BitTorrent clients, where the IP addresses of those sharing infringing files can easily be collected from the network using search crawlers, authorities are not able to track IP addresses of those who download infringing works through cyberlocker services. This makes it difficult to forward RIAA notices of infringement to such users. And, just as important, cyberlocker services, in order to stay within DMCA guidelines, do not provide search engines on their sites to allow users to search for what other users have uploaded there. A number of third party search engines, like FilesTube, do provide this solution (Torrentfreak.com, 2013).

Copyright Legislation, Cases & Actions

Copyright Alert System. Begun in July 2012 and first implemented in February 2013, the RIAA is relying on partnerships with major ISPs, including AT&T, Cablevision, Comcast, Time Warner Cable, and Verizon, to track down pirated online content. In this private industry "graduated response" enforcement system, the RIAA alerts an ISP that a customer appears to be file sharing illegally. The ISP will then notify the person that he or she appears to be file sharing by sending a "copyright alert." If the behavior by the customer does not change, up to four more alerts are sent—six total alerts. If the customer ignores these, then the ISP may choose to use "mitigation measures" to limit, suspend, or terminate the person's service. Monitoring will be via a third party company, which will watch BitTorrent networks and other public networks, and gather IP addresses. Those who think that the ISP is in the wrong, can call for an independent review, at the cost of $35, before a mitigation measure is initiated (Center for Copyright Information, 2014).

The Pirate Bay. In January 2014, Sweden's Court of Appeals lifted a block that was in place to prevent Swedish ISPs from connecting to The Pirate Bay, the global BitTorrent service. The Court ruled the site blocking was ineffective and a "disproportional" hardship for the ISPs. Evidence in the case actually showed that content piracy substantially increased during the blockade (RT.com, 2014). Previously, in November 2010, the Swedish appeals court upheld the copyright infringement convictions of the site's owners, who then switched the service to a "trackerless" file, magnet link-based service to avoid future lawsuits (Olanoff, 2012).

Megaupload. The DOJ describes cyberlocker Megaupload.com as a massive criminal operation in which the main website acted as a conduit to other sites where pirated content—including books, movies, and music—was easily available. In carrying out the arrests, the DOJ worked with law enforcement agencies in at least eight other countries. As of mid-2014, Kim Dotcom, the owner of the service, is appealing the charges and fighting extradition to the U.S. from New Zealand (Reuters, 2014).

Seven individuals and two corporations were indicted by a Virginia grand jury with running an international organized criminal enterprise allegedly responsible for massive worldwide online piracy of numerous types of copyrighted works. Through Megaupload.com and other related sites, these alleged activities generated more than $175 million in criminal proceeds and caused more than half a billion dollars in harm to copyright owners.

The DOJ describes this case as "among the largest criminal copyright cases ever brought by the United States and directly targets the misuse of a public content storage and distribution site to commit and facilitate intellectual property crime" (DOJ, 2012b, p. 1). In addition, Megaupload allegedly offered financial incentives and a rewards program for users to upload popular copyrighted content and drive web traffic to the site via third party linking sites and search engines.

Grooveshark. Universal Music Group filed a lawsuit against Grooveshark.com on November 18, 2011 because, unlike Pandora and Spotify, the site did not obtain licenses to stream Universal content. Since January 2012, three other record labels (EMI, Sony/ATV, and Warner Music) joined the lawsuit. Ordinarily, websites are protected by the DMCA safe harbor provision which ensures that sites are not liable for copyright violations by their users. In order to preserve this legal shield, the site owners must respond to take-down notices provided by copyright owners. They must also ensure they are not directly controlling and profiting from the infringement. In Grooveshark's case, however, the facts provided by Universal suggest the company may have forfeited the safe harbors (Peoples, 2012).

In August 2013, Grooveshark signed an out-of-court agreement with EMI Music Publishing and announced an agreement with Sony/ATV, though legal battles with Universal Music Group and Warner Music Group continued (Sandoval, 2013).

Current Status

CEA Sales Figures

Details on the current state of the digital audio marketplace from the *2013 Digital America* report included:

- Digital music downloads surpassed CD sales for the first time in 2012.
- Nearly 1.4 billion digital music singles and 117 million digital albums were downloaded in 2012.
- Revenues from digital music files and royalties topped $4 billion in 2012.
- Nearly half (45%) of U.S. households owned an MP3 player as of January 2013.
- Total home audio sales in 2013 outpaced portable audio sales in factory-level dollar volume for the first time in nine years as home audio component sales continue to rise and portable MP3 sales continue their dramatic decline.
- Sales of MP3 players, which peaked in 2010, have been falling at double-digit rates ever since, thanks to market saturation and a rise in smartphone adoption, many of which double as MP3 players.
- In 2014, sales of MP3 players were expected to reach $4 billion, down 22 percent from 2012. (CEA, 2013).

Nielsen SoundScan Data

Nielsen SoundScan, the sales source for the Billboard music charts, tracks sales of music and music video products throughout the United States and Canada. Sales data is collected weekly from over 14,000 outlets, including brick and mortar merchants, online retailers, and performance venues. SoundScan data from 2013 was compiled in the *Nielsen Company Year-End Music Industry Report.* Some of the report's most interesting findings:

- For the first time since the iTunes store opened its doors, the U.S. music industry finished the year with a decrease in digital music sales.
- Overall for 2013, digital track sales fell 5.7% from 1.34 billion units to 1.26 billion units while digital album sales fell 0.1% to 117.6 million units from the previous year's total of 117.7 million.
- The CD declined 14.5% to 165.4 million units, down from 193.4 million in the prior year, while vinyl continued its ascension rising to 6 million units from the 4.55 million the format tallied in 2012.
- That means vinyl is now 2% of album sales in the U.S; digital albums comprise 40.6% CDs comprise 57.2% and cassettes and DVDs comprise 0.2%
- The top selling track in 2013 was Robin Thicke's *Blurred Lines,* which scanned nearly 6.5 million units, followed by Macklemore & Ryan Lewis' *Thrift Shop,* with 6.1 million units, and Imagine Dragon's *Radioactive* with 5.5 million units.
- In 2013, 106 songs hit the million unit mark, versus 108 titles that achieved that feat in 2012.
- Despite the decline in digital album sales, download stores like iTunes gained market share growing to 40.6% of U.S album sales, while mass merchants like Target and Walmart saw sales drop 16.3% (Christman, 2014).

IFPI Digital Music Report

The International Federation of the Phonographic Industry (IFPI) is the organization that represents the interests of the recording industry worldwide. IFPI is based in London and represents more than 1,400 record companies, large and small, in 75 different countries. IFPI produces an annual report on the state of international digital music, and the *Digital Music Report 2014* also provides some interesting information on the current state of the global digital music industry:

- Despite positive trends in most markets, overall global music trade revenues fell by 3.9% to $15.0 billion in 2013. The result was heavily influenced by a 16.7% fall in Japan, which accounts for more than a fifth of global revenues.

- The digital market has continued to diversify with revenues from subscription services growing by 51.3%, passing the $1 billion mark for the first time. Global revenues from subscription and advertising-supported streams now account for 27% of digital revenues, up from 14% in 2011.
- There are some 450 licensed digital music services internationally, including global services such as Spotify (which expanded into 38 new markets in 2013), Deezer, Google Play, and regional services such as Muve in the US and Asia's KKBOX.
- Physical format sales still account for a major proportion of industry revenues in many major markets. They accounted for more than half (51.4%) of all global revenues in 2013, compared to 56% in 2012.
- Performance rights income to record companies crossed the $1 billion threshold for the first time in 2013 to hit $1.1 billion. This was an increase of 19%, more than double the growth rate in 2012, accounting for 7.4% of total record industry revenue.
- Internet service providers have a demonstrable effect on reducing copyright infringement, when required to act. European countries where ISPs are required by courts to block access to infringing sites saw Bit Torrent usage fall by 11% during 2013. Countries without the block saw Torrent usage rise by 15% over the same period. (IFPI, 2014).

Factors to Watch

By the end of 2014, the factors to watch in digital audio will be:

Subscription-based music streaming. The explosion of cloud-based music streaming services has been rapid and diverse. This trend will continue as the current providers, as well as startups, vie for consumers in the always-dynamic mobile phone arena, where smartphones will be the driving force. ABI Research forecasts that by 2016, streaming cloud-based services will become a more important form of access to music than owning albums, songs, or tracks. According to the ABI, this shift will primarily be driven by the growing use of mobile handsets, especially smartphones, as listening devices. ABI Research believes that number in the U.S. will exceed 161 million subscribers in 2016, a growth rate of nearly 95% per year (ABI Research, 2011).

In-car music enhancements. Digital audio has made its way into vehicles via various dashboard solutions, including Ford's SYNC and Chrysler's Uconnect, have begun to drive the sale of cars: Luxury cars are marketing themselves based on the enhanced digital music experience in the car, and now more than 100 vehicle models have digital radio in the dashboard that include apps like Pandora, Slacker, and iHeart Radio. Expect these systems to become standard features in even more 2015 vehicle models.

Illegal file sharing. The four factors to watch are:

1) The impact of the pending Megaupload case on future lawsuits and interventions by government and copyright holders into cyberlocker activities.
2) The increased involvement of ISPs, as mentioned earlier, in the newly implemented Copyright Alert System in the U.S. and similar initiatives abroad.
3) The potential involvement of search engines, such as Google and Yahoo, to work with the music community to meaningfully prevent illegal sources of music from appearing in top search results.
4) In June 2013, the President's Office of the Intellectual Property Enforcement Coordinator (IPEC) released its *2013 Joint Strategic Plan on Intellectual Property Enforcement* to provide ongoing efforts to grow the legitimate digital marketplace while maintaining a proper balance between copyright protections, free speech, and innovation (IPEC, 2013). Will groups targeted in the Report, like online payment processors (Visa, PayPal), data storage services (Dropbox, cyberlockers), and domain name registrars (GoGaddy, 1&1) adopt voluntary infringement initiatives?

Expect to see continued innovation, marketing, and debate in the next few years. Which technologies and companies survive will largely depend on the evolving choices made by consumers, the courts, and the continued growth and experimentation with wireless, mobile, and Internet streaming technology. Chris Ely, CEA's manager of industry relations, sees a bright, but competitive future:

Technology allows consumers to access almost any content they desire instantaneously on Internet-connected devices. The rise of mobile broadband has resulted in the emergence of connected devices that are able to stream content directly from the Internet, and services that allow consumers to store and access content without the need of a hard drive. Digital media consumption will continue to grow as the number of connected devices and services for accessing content improves and expands. Manufacturers, content providers, aggregators, and service providers must work together to ensure the content customers want is accessible through different devices

—Consumer Electronics Association, 2012

Digital Audio Visionary: Karlheinz Brandenburg

In 1982, **Karlheinz Brandenburg,** "Father of the MP3," was a math and electronics PhD student at the University of Erlangen-Nuremberg in Germany, when one of his professors instructed him to work on a problem that had yet to be solved: how to use a mathematical algorithm to transmit music over a digital ISDN phone line without spoiling the musical fidelity of the piece. It took several years, and much collaboration, but Brandenburg was able to solve this dilemma, help create the resulting MP3 file format, and launch the digital audio industries we have today.

Early audio encoding processes of the 1970s were designed to filter an audio signal into layers of sound which could each be saved or discarded depending on relative significance of the information. But these early systems were considered very structured and inflexible (Ganz and Rose, 2011). So, Brandenberg proposed a new system that would replicate the limitations of human hearing. In 1988, as part of the newly formed Motion Picture Experts Group (MPEG), Brandenburg's musical test-case was not a German classic from Beethoven or Mozart, but an acapella version of folk-singer Suzanne Vega's pop song *Tom's Diner*. The nuances of Vega's voice on the acapella track meant that Brandenburg's algorithm had to be very precise at picking out which parts of the sound could be discarded without ruining the transmitted file or the listener's musical experience. The result, in 1992, was the MP3 format which became the basis of our current audio compression systems.

So what does that mean for consumers and the music industry? The hallmark of the MP3 is that its algorithm lets users massively shrink (by a factor of 10 at least) the amount of data needed to represent the original audio recording, while still allowing the track to sound authentic to most listeners. This means that while the MP3 almost always sounds just as good as the older recording formats that were in play, the MP3 is also easier to store and to transmit—hence, today, you can carry a whole music library around on your portable player or your smartphone.

Bibliography

8tracks. (2014, March 18). What is 8tracks? Retrieved from http://8tracks.com.

ABI Research. (2011, March 17). Mobile cloud-based music streaming services will be mainstream by 2016. Retrieved from http://www.abiresearch.com/press/ 3640.

Alten, S. (2014). Audio in media, 10th edition. Belmont, CA: Wadsworth.

Amazon. (2014, March 18). Introducing Amazon Cloud Drive. Retrieved from https://www.amazon.com/clouddrive/learnmore.

Apple. (2014, March 18). iTunes match. Retrieved from http://www.apple.com/itunes/itunes-match/.

Audio Home Recording Act of 1992, 17 U.S.C. §§ 1001–10 (1992).

Beats Music. (2014, March 18). One service. Two ways to pay. Retrieved from https://beatsmusic.com/pricing.

Blip.fm (2014, March 18). FAQ. Retrieved from http://blog.blip.fm/faq/.

Center for Copyright Information. (2014, March 18). Copyright Alert System FAQ. Retrieved from http://www.copyrightinformation.org/resources-faq/copyright-alert-system-faqs/.

Consumer Electronics Association. (2013). *Digital America 2013: Audio Overview*. Retrieved from http://content.ce.org/PDF/2013DigitalAmerica_abridged.pdf.

Department of Justice. (2014a, March 18). Intellectual Property Task Force. Retrieved from http://www.justice.gov/dag/iptaskforce/.

Department of Justice. (2012b, January 19). Justice Department charges leaders of Megaupload with widespread online copyright infringement. Retrieved from http://www.justice.gov/opa/pr/2012/January/12-crm-074.html.

Digital Performance Right in Sound Recordings Act, Pub. L. No. 104-39, 109 Stat. 336 (1996).
Digigram. (2014, March 18). About world standard ISO/MPEG audio. Retrieved from http://www.digigram.com/support/library.htm?o=getinfo&ref_key=282.
Digster. (2014, March 18). FAQ. Retrieved from http://digster.fm/support/faq/.
eMusic. (2014a, March 18). Plans & Pricing. Retrieved from http://www.emusic.com/info/plans-pricing/.
eMusic. (2014b, March 18). eMusic Mobile. Retrieved from http://www.emusic.com/promo/mobile/index.html?fref=150606.
Ganz, J. and Rose J. (2011, March 23). The MP3: A History of Innovation and Betrayal. Retrieved from http://www.npr.org/blogs/therecord/2011/03/23/134622940/the-mp3-a-history-of-innovation-and-betrayal.
Google. (2014, March 18). Unlimited music from Google. Retrieved from https://play.google.com/about/music/allaccess/#/.
Grooveshark.com. (2014, March 18). About Grooveshark. Retrieved from http://grooveshark.com/#!/about.
Holahan, C. (2008, January 4). Sony BMG plans to drop DRM. *Business Week*. Retrieved from http://www.businessweek.com/technology/content/jan2008/tc2008013_398775.htm.
IFPI. (2014, March 18). *IFPI digital music report 2014*. Retrieved from http://www.ifpi.org/downloads/Digital-Music-Report-2014.pdf.
IPEC. (2013, June). *2013 Joint Strategic Plan on Intellectual Property Enforcement*. Retrieved from http://www.whitehouse.gov/sites/default/files/omb/IPEC/2013-us-ipec-joint-strategic-plan.pdf.
Last.fm. (2014, March 18). About last.fm. Retrieved from http://www.last.fm/about.
Legal Information Institute (LII). (2014, March 18). *Contributory Infringement*. Retrieved from: http://www.law.cornell.edu/wex/contributory_infringement.
Metro-Goldwyn-Mayer Studios, Inc., et al. v. Grokster, Ltd., et al. (2005, June 27). 545 U. S. 125 S. Ct. 2764.
No Electronic Theft (NET) Act, Pub. L. No. 105-147, 111 Stat. 2678 (1997).
Olanoff, D. (2012, February 28). The Pirate Bay makes official switch to magnet links. Retrieved from http://thenextweb.com/insider/2012/02/28/as-promised-the-pirate-bay-officially-drops-torrent-files-for-magnet-links/.
P2PON!. (2018, March 18). Top 10 most popular P2P file sharing clients of 2012/13. Retrieved from http://www.p2pon.com/top-10-most-popular-p2p-file-sharing-clients-of-20112012/.
Peoples, G. (2012, March 1). Grooveshark files for dismissal of copyright-infringement case, blasts major labels' lawsuit. Retrieved from http://www.billboard.biz/bbbiz/industry/legal-and-management/grooveshark-files-for-dismissal-of-copyright-1006331352.story.
Podcast Alley. (2014, March 18). What is a podcast? Retrieved from http://www.podcastalley.com/what_is_a_podcast.php.
Pro-music. (2014, March 18). Find music services. Retrieved from http://www.pro-music.org/legal-music-services.php.
Rara. (2014, March 18). Frequently asked questions. Retrieved from https://www.rara.com/static/site/US/en/faq.html.
Rdio. (2014, March 18). Millions of songs, for free or a small fee. Retrieved from http://www.rdio.com/.
Recording Industry Association of America. (2014, March 18). About copyright notices. Retrieved from http://riaa.com/toolsforparents.php?content_selector=resources-music-copyright-notices.
Reuters. (2014, February 18). *NZ court rules Megaupload warrant legal, dealing blow to Dotcom*. Retrieved march 18, 2014 from http://www.reuters.com/article/2014/02/19/us-newzealand-megaupload-warrant-idUSBREA1I02F20140219.
Rhapsody. (2014, March 18). Pricing. Retrieved from http://www.rhapsody.com/discover/pricing.html.
Roberts, R. (2012, February 4). *Album Review: Van Halen's 'A Different Kind of Truth.'* Retrieved from http://latimesblogs.latimes.com/music_blog/2012/02/ album-review-van-halens-a-different-kind-of-truth.html.
RT.com. (2014, January 29). *Dutch court rules in favor of unblocking Pirate Bay as ban 'ineffective'*. Retrieved March 18, 2014 from http://rt.com/news/court-unblock-pirate-bay-308/.
Sandoval, R. (2013, August 6). *Grooveshark settles EMI Publishing lawsuit, still faces uncertain future*. Retrieved from http://www.theverge.com/policy/2013/8/6/4592346/grooveshark-settles-emi-publishing-lawsuit-still-faces-uncertain.
Slacker. (2014, March 18). About Slacker. Retrieved from http://www.slacker.com/about;jsessionid=6836082A58F3BA277F0E77AEBF8CA147.
Songza. (2014, March 18). Frequently asked questions. Retrieved from https://songza.desk.com/.
Sony Corp. of America v. Universal City Studios, Inc., 464 U.S. 417. (1984). Retrieved March 18, 2014 from http://caselaw.lp.findlaw.com/scripts/getcase.pl?navby=CASE&court=US&vol=464&page=417.
Sony Entertainment Network. (2014). Music Unlimited. Retrieved from http://www.sonyentertainmentnetwork.com/music-unlimited/internet-radio/.

SoundCloud. (2014, March 18). Be heard everywhere. Retrieved from https://soundcloud.com/creators.
SoundExchange.com. (2014, March 18). FAQ. Retrieved from http://www.soundexchange.com/about/.
Spotify. (2014, March 18). Music for everyone. Retrieved from https://www.spotify.com/us/.
Sprint. (2014, March 18). Learn more about Sprint Music Plus. Retrieved from http://support.sprint.com/support/article/Learn_more_about_Sprint_Music_Plus/6949ad4c-49f4-4cad-83ae-7934a6105322?ECID=vanity:sprintmusic.
Strufe, T, (2014, March 18). Peer-to-Peer Networks: Lecture. Retrieved from http://www.p2p.tu-darmstadt.de/teaching/winter-term-20112012/p2p-networks-lecture/.
Torrentfreak.com. (2013, December 1). *Filestube search engine first to smash 10 million DMCA notice barrier*. Retrieved from http://torrentfreak.com/filestube-search-engine-first-site-to-smash-10-million-dmca-notice-barrier-131201/.
Ubuntu. (2014, March 18). Retrieved from https://one.ubuntu.com/services/music/.
U.S. Copyright Office. (2014, March 18). *The Digital Millennium Copyright Act of 1998: U.S. Copyright Office summary*. Retrieved from http://www.copyright.gov/legislation/dmca.pdf.
Verizon Wireless. (2014, March 18). Music & tones. Retrieved from http://www.verizonwireless.com/wcms/consumer/products/music-tones.html.
Watkinson, J. (2013). *The art of digital audio, Kindle edition*. London: Focal Press.
Wyman, B. (2012, January 20). So long, and thanks for all the pirated movies. Retrieved from http://www.slate.com/articles/business/technology/2012/01/megaupload_ shut-down_what_the_site_s_departure_means_for_other_traffic_hogging_cyberlockers_.html.
Xbox Music Pass. (2014). Xbox Music Pass. Retrieved from http://www.xbox.com/en-US/music/music-pass.

18

Digital Imaging & Photography

Michael Scott Sheerin, M.S.*

Why Study Digital Imaging and Photography?

- Digital imaging and photography is the most popular and ubiquitous visual medium in the world, and its growth has accelerated at a pace never seen before.
- Digital imaging and photography offers a way to place our world in context, as well as a way to document our daily lives.
- Digital imaging and photography, coupled with social media apps, make it easier to be creative, and make it easier to share our creations with others in real time.

Big Brother is watching you.

–from Orwell's 1984

But first, let me take a Selfie.

–The Chainsmokers, 2014

Introduction

George Orwell's classic fictional novel *Nineteen Eighty-Four* deals with a loss of civil liberties, based on government's mass surveillance techniques. This practice was seen in a negative light by the people of Oceania (the mythical totalitarian state in the book).

Flash forward to today. It's not just big brother watching you anymore; it's your big sister, your cousin, a spouse, friends, and acquaintances. It's everyone around you viewing the world through the prism of a lens, including yourself, as evidenced by the plethora of self shot camera phone "selfies." The practice is so common that "selfie" was Oxford Dictionary's Word of the Year in 2013 (Oxford Dictionary, 2014).

This digital imagery of our lives is captured and posted to social networks like Instagram (40 million uploaded/day), Flickr (4.5 million photos uploaded/day), or Facebook (300 million photos uploaded/day), for everyone to see (Donegan, 2014). According to a study conducted by SelfieCity, led by Dr. Lev Manovich of CUNY, approximately 4% of all images posted on Instagram are selfies. Taken from a larger sample of images, the study analyzed 3,200 selfies from five major cities: New York, Moscow, Berlin, Bangkok, and Sao Paulo.

This opt-in social surveillance via digital self-imaging is especially popular with the Millennial Plus, or Generation Y, group (18-35 year olds), as the average age of these selfie takers is 27.3, with the vast majority being women (SelfieCity, 2014). It seems Orwell's imagined surveillance takes place in reality with complete compliance by the subjects a vast majority of the time, as a typical Millennial probably poses for a picture more often in a month than a Baby Boomer does in a decade (purely anecdotal observation)! As Fred Ritchin, professor of Photography

* Associate Professor, School of Journalism and Mass Communications, Florida International University (Miami, Florida)

and Imaging at New York University's Tisch School of the Arts, points out, "we are obsessed with ourselves" (Brook, 2011).

The digital image has allowed the photo industry to fully converge with the computer and cell phone industry, thus changing the way we utilize and change our images "post shutter-release." These digital images are not the same as the photographs of yesteryear. Those analog photos were continuous tone images, while the digital image is made up of discreet pixels, ultimately malleable to a degree that becomes easier with each new version of photo-manipulating software. And unlike the discovery of photography, which happened when no one alive today was around, this sea change in the industry has happened right in front of us; in fact, we are all participating in it—we are all "Big Brother." This chapter will look at some of the hardware and software inventions that continue to make capturing and sharing quality images easier, as well as the implications on society, as billions of digital images enter into all our media.

Background

Digital images of any sort, from family photographs to medical X-rays to geo-satellite images, can be treated as data. This ability to take, scan, manipulate, disseminate, or store images in a digital format has spawned major changes in the communication technology industry. From the photojournalist in the newsroom to the magazine layout artist to the vacationing tourist posting to Facebook via Instagram, digital imaging has changed media and how we view images.

The ability to manipulate digital images has grown exponentially with the addition of imaging software, and has become increasingly easier to do. Don't like your facial skin tones on that selfie you just took? A simple swipe of the finger on the FaceTune app lets you "hide imperfections, whiten teeth, and make your subjects look like they just stepped out of a salon" (Gil, 2014). Want to make your images look like they were shot with a 35 mm film camera? Use Mextures, an iPhone and iPad app. "Created from actual 35mm film scans, the textures in this app give you everything from grainy overlays to light leaks you can combine together" (C/Net, 2013).

Looking back, history tells us photo-manipulation dates back to the film period that Mextures attempts to capture. Images have been manipulated as far back as 1906, when a photograph taken of the San Francisco Earthquake was said to be altered as much as 30% according to forensic image analyst George Reid. A 1984 National Geographic cover photo of the Great Pyramids of Giza shows the two pyramids closer together than they actually are, as they were "moved" to fit the vertical layout of the magazine (Pictures that lie, 2012). In fact, repercussions stemming from the ease with which digital photographs can be manipulated caused the National Press Photographers Association (NPPA), in 1991, to update their code of ethics to encompass digital imaging factors (NPPA, 2014). Here is a brief look at how the captured, and now malleable, digital image got to this point.

The first photograph ever taken is credited to Joseph Niepce, and it turned out to be quite pedestrian in scope. Using a technique he derived from experimenting with the newly-invented lithograph process, Niepce was able to capture the view from outside his Saint-Loup-de-Varennes country house in 1826 in a camera obscura (Harry Ransom Center, University of Texas at Austin, 2014). The capture of this image involved an eight-hour exposure of sunlight onto bitumen of Judea, a type of asphalt. Niepce named this process heliography, which is Greek for sun writing (Lester, 2006).

The next 150 years included significant innovation in photography. Outdated image capture processes kept giving way to better ones, from the daguerreotype (sometimes considered the first photographic process) developed by Niepce's business associate Louis Daguerre, to the calotype (William Talbot), wet-collodion (Frederick Archer), gelatin-bromide dry plate (Dr. Richard Maddox), and the now slowly disappearing continuous-tone panchromatic black-and-white and autochromatic color negative films. Additionally, exposure time has gone from Niepce's eight-hour exposure to 1/500th of a second or less.

Cameras themselves did not change that much after the early 1900s until digital photography came along. Kodak was the first to produce a prototype digital camera in 1975. The camera had a resolution of .01 megapixels and was the size of a toaster (Kodak, 2012). In 1981, Sony announced a still video camera called the MAVICA, which stands for magnetic video camera (Carter, 2012a). It was not until nine years later, in 1990, that the first digital still camera (DSC) was introduced. Called the Dycam (manufactured by a company called Dycam), it captured images in monochromatic grayscale and had a resolution that was lower than most video cameras of the time. It sold for a little less than $1,000 and had the ability to hold 32 images in its internal memory chip (Aaland, 1992).

In 1994, Apple released the Quick Take 100, the first mass-market color DSC. The Quick Take had a resolution of 640 × 480, equivalent to a NTSC TV image, and sold for $749 (PCMAG.com, 2012). Complete with an internal flash and a fixed focus 50mm lens, the camera could store eight 640 × 480 color images on an internal memory chip and could transfer images to a computer via a serial cable. Other mass-market DSCs released around this time were the Kodak DC-40 in 1995 for $995 (Carter, 2012b) and the Sony Cyber-Shot DSC-F1 in 1996 for $500 (Carter, 2012c).

The DSCs and the digital single lens reflex (DSLR) cameras work in much the same way as a traditional still camera. The lens and the shutter allow light into the camera based on the aperture and exposure time, respectively. The difference is that the light reacts with an image sensor, usually a charge-coupled device (CCD) sensor, a complementary metal oxide semiconductor (CMOS) sensor, or the newer, live MOS sensor. When light hits the sensor, it causes an electrical charge.

The size of this sensor and the number of picture elements (pixels) found on it determine the resolution, or quality, of the captured image. The number of thousands of pixels on any given sensor is referred to as the megapixels. The sensors themselves can be different sizes. A common size for a sensor is 18 × 13.5mm (a 4:3 ratio), now referred to as the Four Thirds System (Four Thirds, 2012). In this system, the sensor area is approximately 25% of the area of exposure found in a traditional 35mm camera. Many of the sensors found in DSLRs are full frame 35mm in size (Canon, 2014).

The pixel, also known in digital photography as a photosite, can only record light in shades of gray, not color. In order to produce color images, each photosite is covered with a series of red, green, and blue filters, a technology derived from the broadcast industry. Each filter lets specific wavelengths of light pass through, according to the color of the filter, blocking the rest. Based on a process of mathematical interpolations, each pixel is then assigned a color. Because this is done for millions of pixels at one time, it requires a great deal of computer processing. The image processor in a DSC must "interpolate, preview, capture, compress, filter, store, transfer, and display the image" in a very short period of time (Curtin, 2012).

This image processing hardware and software is not exclusive to DSCs or DSLRs, as this technology has continued to improve the image capture capacities of the camera phone. Starting with the first mobile camera that could take digital images, the Sharp J-phone released in Japan in 2000, the lens, sensor quality, and processing power have made the camera phone the go-to-camera for most of the images captured today (Hill, 2013). (To prove this point: If you had to take an image of this page right now, what camera do you have with you?) An example of the improved image capture technology can be found on the Lumia 1020, perhaps the best camera phone on the market as of mid-2014, with its 41-megapixel resolution capabilities (Van Camp, 2014).

Recent Developments

The digital imaging industry, which is midway into the third, and last, decade of Saffo's 30-year rule, is an industry reaching full maturity. But that doesn't imply that it is not a dynamic industry. Camera hardware and software improvements continue to evolve, as do apps for the distribution and sharing of images across many electronic platforms. However, these technical developments pale in comparison to the holistic way the industry has changed over the past few years. This sea change is directly linked to the rise in use of the camera phone, due in part to the image capture and distribution technology advancements in these smart phones.

These technological advancements usually happen in one of two ways. The first is by incremental improvements, illustrated by the way camera sensors continue to increase resolution capacity, as well as "better high ISO performance, improved autofocus, and sharper lenses" (Cicala, 2014). The second is called disruptive innovation, and this is the category that this smart phone-as-camera trend falls under.

In addition to the ubiquitous use of today's camera phones for capturing images, there is another factor that contributes to this disruptive innovation, and that is the rise of social media. In the past, photography represented a window to the world, and photographers were seen as voyeurs, outsiders documenting the world for all of us to see. Today, with the image capture ability that a camera phone puts in the hands of millions, photographers are no longer *them*—outsiders and talented specialists. Rather, they are *us*—insiders that actively participate in world events and everyday living, and, thanks to social media sites such as Instagram, we share this human condition in a way that was never seen before. In *Bending the Frame*, Fred Ritchin writes, "The photograph, no longer automatically thought of as a trace of a visible reality, increasingly manifested individuals' desires for certain types of reality. And rather than a system that denies interconnectedness, the digital environment emphasizes the possibility of linkages throughout—from one image to another" (Ritchin, 2013).

Of course, any disruption innovation comes with its naysayers. When Damon Winter won Pictures of the Year International's (POYI) third place award for a photo he captured using his iPhone and the app Hipstamatic, photojournalist Chip Litherland blogged, "What we knew as photojournalism at its purest form is over and POYI just killed it" (Winter, 2011). To be fair, Litherland was not so upset with the use of the camera phone, but rather the use of the app that "changes what was there" (Litherland, 2011).

This argument, involving technologic advances in the medium, has played out before as photography evolved, starting with George Eastman's mass production of the dry plate film in 1878 (previous to this innovation, photographers made their own wet-collodion plates). Lewis Carroll, noted early photographer and author, upon examination of this dry plate film, stated, "Here comes the rabble" (Cicala, 2014). Soon after, film became the capture method of choice in photography. "No one likes change," freelance photographer Andrew Lamberson says. "People who shot large format hated on the people who shot medium format, who in turn hated on the people who shot 35mm, who in turn hated on people who shot digital" (McHugh, 2013).

But based on the sheer number of users, including a growing number of professionals, camera phone photography is not going away. Its growing number of award shows, such as the Mobile Photography Awards, the iPhone Photography Awards, as well as its inclusion in the aforementioned POYI, only helps this image-capture method gain credibility. This credibility is seeping into other aspects of digital imaging. CEO Matt Munson of Twenty20, an Instagram stock photo service, states that the stock photo industry is also being disrupted. "Flash forward five to 10 years from now, and the traditional tight control over commercial photography and the ways people find photos for articles, or to use in books, or for t-shirts, or greeting cards… is going to be totally different" (McHugh, 2013).

Recent legal precedent, as seen in the *Morel v. AFP* copyright verdict, will also help insure that digital images will continue to be shared over social media networks, without fear of the loss of copyright control. The Federal Court ruling stated that Morel, a photojournalist who posted images of the 2010 Haitian earthquake, did not forfeit any copyrights by posting to TwitPic (an image sharing app for Twitter) (Walker, 2013).

The posting of images on social media networks, first started in 2005 with the advent of Flickr, and turned into a gusher by Facebook and Instagram, with an estimated 350 million images posted per day, raises other issues beyond copyright law. How do all these images about the human condition play a role in citizen journalism, or in cultural research? As Ariella Azoulay, a theorist of photography and visual culture, writes, "We cannot deny anymore that photography is not about the world, and therefore outside it, but is rather part of the world" (Azoulay & Thompson, 2014). Documenting a revolution, or civil strife is "becoming more of a live activity and less about retroactive interpretation of objects," says Hito Steyerl, filmmaker and author. "The event is being made across different platforms and networks, as a stack of actions, images, and feedback loops, travels from cloud to cobblestone" (Keenan & Steyerl, 2014).

This abundance of images has also played a role in the emerging field of social media documentation, which allows us to study our culture, not from the viewpoint of one, but from an aggregate production of millions. Meta-data stored on our shared images can be tagged, sorted, and indexed, allowing us to study complex patterns of life that are actively documented by these images. As Manovich states, "The photo-universe created by hundreds of millions of people might be considered a mega-documentary, without a script or director" (Manovich, 2014).

The disruptive innovation brought upon the field has shaken some of the giants of the industry on the manufacturing and production side as well, as there are some dire predictions about the future of some of these companies. Recent speculation hinted that "Nikon—yes, Nikon—might be gone in five years" (Elrich, 2013), as falling demand and shrinking revenue threaten all major DSC and DSLR manufacturing companies. But the sheer increase in users means that every camera phone photographer is a potential customer, so opportunity exists. This is signified by the projection that the overall digital photography industry will grow from a market value of $68.4 billion in 2011 to a projected $82.5 billion in 2016 (BCC Research, 2012). Here are some new technology advancements that are predicted to help with this market growth, starting first with the hardware side.

New mirrorless/compact interchangeable lens cameras (CILC), such as the Panasonic Lumix GF6, the Sony Alpha NEX-6, and the Samsung NX300, are recent additions to the compact camera line-up. The removal of the mirror has reduced the camera body size to be closer to a camera phone, while still being able to use high-end lenses. Many of these cameras also have touch-screen menu technology, the ability to shoot HD video, and built in Wi-Fi capabilities. Sensors found on these cameras range from an APS-C sensor to Micro-Four Thirds MOS sensors, though some models, such as the new Sony Alpha 7R, come with a full frame CMOS sensor (the 7R boosts a 36.4 MP sensor, giving it a resolution equivalent to the best DSLRs). The overall benefit of the CILCs is that you get a more compact body with the functionality and connectivity of a camera phone, but with much higher image-capture quality, based on the interchangeable lens system and sensors.

As evidenced by both the CLIC Sony Alpha 7R and the Lumia 1020 camera phone, digital image resolution continues to improve. In their book *Converging Media*, Pavlik & McIntosh wrote, "Unlike the evolution of other types of media, the history of photography is one of declining quality" (Pavlik & McIntosh, 2004). Images shot with large format cameras in the 1880s were superior to the digital images of only a decade ago. But this statement is no longer accurate, as we now push past the resolution of those film cameras.

Roger N. Clark concludes in his study of digital still versus film resolution, that the resolution obtained from an image sensor of 10 megapixels or more is equal to or greater than the resolution of 35mm film (Clark, 2014). Even camera phones are now topping this resolution threshold. But the reality is that unless we are making large, gallery-quality prints, we already have all the resolution we need to share quality images on our electronic devices. The screen resolution on the new UltraHD television is only 8 MP. And all that is needed for the iPad with Retina Display is 3 MP (Pitt, 2013). Camera-phone screens require even less resolution in order for an image to display as sharp and clear.

As much as the number of megapixels helps determine the resolution and quality of the digital image, the sensor is the most important factor in determining image quality. A 10 MP camera phone with a sensor area of 1/1.6 does not produce the same quality images as a 10 MP DSLR with a full frame (24 x 36 mm) sensor (see Figure 18.1). New developments in sensor technology include the ASP-C sensor, whose size falls between the smaller Micro-Four Third sensors and full frame DSLR sensors. Sigma's new Quattro camera sports a new, three-layer sensor design. With an area of 23.5 x 15.7 mm, the Quattro sensor captures luminance information on its top layer, with about five MP of color information recorded on each of the two lower layers. This camera produces images with "incredible resolution, precise gradation, gorgeous color, breathtaking realism with a 3D feel, and a 29-megapixel-equivalent ultrahigh resolution" (Worthington, 2014).

Figure 18.1
Comparison of Image Sensors

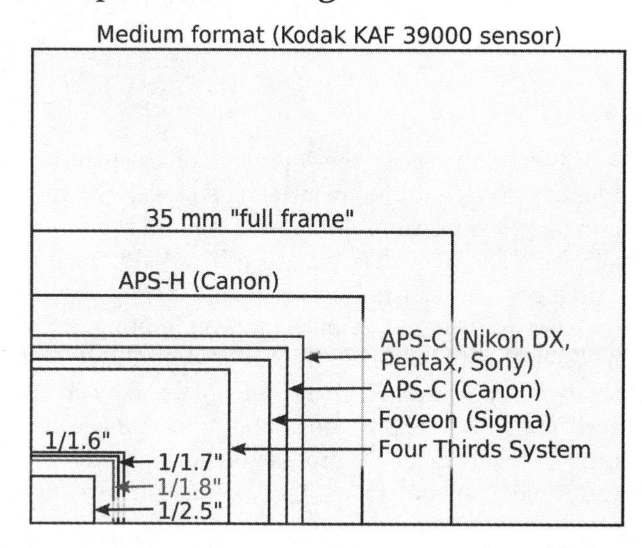

Source: Moxfyre

Imagine being able to refocus a picture after taking it. This is possible with cameras from Lytro. Using light field data, the Lytro captures data like any other digital camera such as color but it also collects data on where light is coming from. This data allows the image to be refocused after it is captured. Lytro calls these images—living pictures. These images cannot only be refocused on the camera LCD screen but also on the screens of mobile devices and computers. So a user can post a living picture on Facebook and friends can refocus the image. The original Lytro camera was released in 2011 and costs $199 for a 16 Gb model. Looking like something between a telescope and a stick of butter, the Lytro was marketed to the young hip crowd looking for an unusual and fun consumer camera. See Figure 18.2. In 2014 Lytro announced the Illium camera—a high-end camera with a 40 Megaray sensor. Think of it as the DSLR while the original Lytro is a point and shoot. The Illium will cost $1,499 and ships in August 2014 (Lytro, 2014).

Figure 18.2

Lytro Camera

Source: Lytro

Looking at recent developments on post-shutter release software, three main items emerge. The first is the growth in use of image editing and workflow software. The two main players in this field are Apple's Aperture (Mac only), the "first to come out with an easy to use image management/editing suite aimed at the digital photographer" (Moss, 2014), and Adobe's Lightroom (both PC and Mac), now in its third release. Both programs' ability to manage images, enable slide show production, and process images, have made them the go-to software for professional photographers.

One of the items that both of these software solutions are geared to work with are the RAW camera files. A RAW file "contains all the pixel information captured by the camera's sensors. The RAW file has not been compressed or processed in any way" (PC Mag, 2014). Photographers with cameras that can shoot RAW capture multiple images (usually three) that have different exposures. Using High Dynamic Range (HDR) apps, found in Aperture, Lightroom 3, Photomatix and Photoshop CC, among others, one can "blend the photos together and create a single image comprised of the most focused, well-lit, and colorful parts of the scene" (Prindle, 2013). This process gives images a more life-like, 3D feel, equivalent to what we see with our eyes. HDR photography is on the upswing and doesn't look to be going away soon.

The third item that is a recent addition to the photographer's toolbox is software that allows one to control the digital image once disseminated. Due to the abundance of shared images, controlling one's output seems like a good idea, as these shared images stick around the web and can come back to haunt you. Software apps such as Snapchat (average user age: 18) (Colao, 2014) and Wickr allow the user to program the length of time one can view an image before it self-destructs. Security issues arose with Snapchat in 2014, as its site was hacked, possibly lowering user confidence in its ability to make a sent image disappear after a limited pre-set time of 10 seconds or less (Snapchat, 2014). Wickr, on the other hand, was built by cryptographers and promises to be more robust on the security side. According to Nico Sell, long time organizer of the DEF CON Hacking Conference, "Snapchat is an app for kids. Wickr is an app for spies that kids will use" (Love, 2014). But, remember, those who receive the images can always take a screenshot.

Current Status

Ritchin pointed out earlier in the chapter that we have an obsession with capturing our own image. His point converges with another factor that has occurred in the digital imaging industry. As the ubiquitous nature of the smart phone continues to expand, the number of mobile subscriptions has, for the first time in consumer technology history, matched the world population. And this growth does not look like it is slowing, evidenced by Hong Kong's +% mobile penetration per capita (Ahonen, 2013). Of these mobile phones in use, 83% of them are camera phones, equaling roughly 4.4 billion digital cameras. It has been estimated that "more than 90% of all humans who have ever taken a picture, have only done so on a camera phone, not a stand-alone digital or film-based 'traditional' camera" (Ahonen, 2013).

An estimated 880 billion digital images will be taken in 2014 (Horaczek, 2013), meaning that about 25% of all photos ever taken (estimated at 3.8 trillion) will be snapped this year. A popular equation making the rounds in the digital photography blogosphere is that we currently take more pictures in two minutes than were taken in all of the 1800's (Schwartz, 2012). The overwhelming vast majority of these images will not be printed, but will instead "jump to screen," as we view, transfer, manipulate, and post these images onto high-definition televisions (HDTV), UltraHD televisions, computer screens, tablets, and mobile phones via web pages and social media sites. Because of the Wi-Fi capabilities of our smartphones, we send images via email, post them in collaborative virtual worlds, or view them on other smartphones, tablets, and handheld devices, including DSCs and DSLRs. As noted, some images will only "jump to screen" for a limited time, due to the increased use of photo apps such as Snapchat, with its 26 million active users sending 350 million photos per day (Colao, 2013) and Wickr, with one million apps downloaded to date (Eddy, 2014).

Social media networks have played a huge role in the growing use of camera phones (and vice versa). For instance, the four most popular cameras for Flickr users are Apple's iPhone5, iPhone4s, iPhone5s, and iPhone4. Coming in at number five is the first non-camera phone, the Canon EOS Rebel T3i, though Canon is the most popular brand of camera used by Flickr users overall (Camera Finder, 2014). Instagram, the all-around social sharing site for digital images, boasts 150 million active users. Most IGers are using camera phones to shoot and share their images, as Instagram etiquette frowns on the posting of images captured on DSLRs.

Paul Worthington, PMA Magazine senior editor, sums up the popularity of the camera phone saying, "Phones lead the capture business. The camera is still just a capture device—but photography is everything else too—capture, share, enhance, and display. The smartphone is the first full photography device that allows you to do all of this. With a smartphone, everyone has a complete photography device with them all the time" (Kruger, 2014). Add to that these demographics of the camera phone owner—79% of the 18-29 age group and 67% of the 30-49 age group own one, 69% of college grads own one, and 76% of those that make over $75,000, own at least one (Pew Research, 2013). But the increased use of the camera phone as our go-to image-capture device has changed the outlook for other digital photography devices and forced many manufacturers to rethink their business plans. Chris Ely, manager of industry analysis at Consumer Electronics Association (CEA), looks for digital camera revenues to drop by double-digits through 2016 (CEA, 2013). Leading this precipitous drop are the point and shoot DSCs, whose number of units shipped in 2013 dropped 41.4%. Shipments of DSLRs and CILCs, both interchangeable lens cameras with better sensors, also fell 15% (CIPA, 2014).

In order to stem this decreasing revenue, DSLR manufacturers are trying to match the Wi-Fi, onboard editing, and photo-sharing capacity of the camera phone. Because the resolution of most camera phones is more than adequate for sharing on all electronic platforms, without the aforementioned capabilities, the only true selling point of DSLRs is the interchangeable lenses. The increased resolution and picture quality of DSLRs actually hinders their marketability, as sending a large RAW file over Wi-Fi in less than a second is still not an easy operation, even with Eye-Fi cards, wireless transmitters, and mobile hotspots. The hope here is that the market will grow as camera-phone photography, and its billions of users, acts as a gateway for those who want higher quality image-capture capabilities.

Figure 18.3

Fuji's Ad Campaign Promoting Photo Printing

Source: Fujifilm

One bright spot, as far as economic growth in the photo industry is concerned, is seen in the printing of custom products such as photo books. This category has risen steadily since 2007 in the US, as a steady annual rate of 20% was recorded through 2012 (Johnson, 2013). It is projected to continue to grow, but at a slower rate, as estimates for 2014 indicate a 9% growth rate from 2011 (PMA, 2011). The lower growth rate is due to the increased purchase of digital photo frames, but overall, the online photo printing market has still taken in $2 billion in revenue since 2008 (IBISWorld, 2013).

Perhaps due to the decreasing rate of growth in the online printing market, the Photo Marketing Associations of the United Kingdom and Australia got together with FujiFilm and launched the *Print It or Lose It* campaign in 2013 (see Figure 18.3). The campaign is "an industry-wide, global marketing effort that addresses the dangers of losing your photos due to accidental deletion, hard drive or memory card corruption, lost or damaged camera phones, or even natural disasters" (PMA, 2013), otherwise known as the dark side of digital imaging.

Factors to Watch

- *Moore's Law (discussed in Chapter 12) remains in play,* as storage capacity continues to expand. Supplanting flash memory cards with 128 Gb of capacity that reached the market in 2012, were the newer 256 Gb cards in 2014, as demand for highly compact, robust memory cards with maximum storage capacity continues. One challenge to this market is the addition of high capacity built-in memory found on some newer portable devices, decreasing the need for additional flash memory cards.
- Will Wi-Fi capability and on-camera image processing tools find their way onto lower end DSLRs? Nikon's D5300, released in October, 2013, adds to the short list of DSLRs (Canon's EOS 70D, for one) that have this capability, but at a high price point. Many CILCs, like the Sony NEX-6, offered this capability in late 2012, as the need for this option became apparent in the shrinking DSC market. This is the one gap between camera phones and DSLRs that needs to close in order to see more photographers make the jump to DSLR ownership. The Mobi Wireless Memory card (up to 32 Gb) can automatically transfer images from a DSLR to a camera phone without a wireless network, but this is a bridge solution. The lack of image processing tools, other than the histogram tool found on all DSLRs, remains an issue as well.
- *Increased use in the newer all-in-one Super Lenses.* The ability to go from wide angle to telephoto using one lens and still get high quality images is an advantage DSLR and CILC users have over camera phones (unless they have add-on lens kits attached). Image quality on these Super Zooms has increased, as has lens speed, and the fact that one doesn't have to constantly change lenses lessens the chance of dust entering the sensor area.
- The increased use of tablets in digital photography. With the same capability to capture images as with a camera phone (save the more conspicuous nature of tablets' size), the image processing tools available, and the screen size of tablets are a definite advantage over the camera phone. One can edit with hundreds of apps already available, share, and easily manage one's photo library without the need of a laptop or desktop computer.
- *Faithful recordings of our past will blur with manipulated recordings.* How will we differentiate between faithful recordings and those images that have been manipulated to represent a "fantasy" world? Where do we draw the line between manipulated image and photojournalism? Is adding a filter (say, Instagram's Toaster) to a digital image on Instagram (not accepted in many photojournalism circles) different from changing a digital image captured in color to a black and white image (accepted in all photojournalism circles)? How is it different? How will this manipulation change the way we record our past, and how will it change the role of the digital image in society.

Digital Imaging & Photography Visionaries: Kevin Systrom & Mike Kreiger

With the October, 2010 launch of Instagram, the world's first social media network built specifically around the sharing of digital images captured by camera phones, the world of digital photography changed forever. Founders **Kevin Systrom**, CEO of the company, and **Mike Kreiger** were two Stanford graduates who saw the burgeoning boom of camera phone shutterbugs, along with the increased popularity in social media networks, and developed Instagram to bridge the two movements. In a matter of a few months, the photo app turned photo and video social media network had a million users. Less than a year later, the 150 millionth photograph was uploaded, and they haven't looked back (Instagram, 2014).

The two-man team had been developing different parts of the app for years prior to its launch. Systrom, the one with a passion for photography, worked on filters, based on looking at Flickr images and noticing how photos had changed in appearance over the years (the frame, once a popular add-on in the early period of Instagram, worked as a great marketing tool for the company, signifying an "Instagram" image). And he practices what he preaches. With over a million followers, Systrom's Instagrams tend to be stoic landscape images and meticulously framed food and beverage shots. Kreiger, on the other hand, concentrated more on the software development side, having focused on human-computer interaction in his studies. After discarding one idea on what the app should be, he focused on the user interface and how the interactions would work and then wrote the basic app in a week, all before any code was written. Next, a decision was made to design the app for mobile. This was perhaps their best idea, as that path led directly to the initial success of Instagram (Hamburger, 2013).

The marketing potential of Instagram became very obvious early on, and the addition of 15 second video to the mix (though beaten to the punch by Vine by a few months) only increased this potential. In April of 2012, Facebook bought Instagram, eventually paying $1 billion for a company that still had no revenue stream at the time.

At a recent Q & A on the UC Berkeley campus conducted after the acquisition, Systrom offered some advice to the students, in terms of the importance of the people behind the technology. "People don't realize when you start a startup, you're not just starting a product—you're starting a company," Systrom said. "And companies are people, and people are the hardest part of the startup, meaning finding the right ones, retaining the great ones" (Handler, 2013). Look for the company to try to monetize its popularity in the very near future. Systrom gives his vision for the future of Instagram, with undertones of Marshall McLuhan's "medium is the message" theory, "Well, we will be everywhere—on every platform, on every kind of phone and tablet and on wearables, which will be a core component of sharing. I want Instagram to be the place I learn about the world" (Kiss, 2013).

Bibliography

Aaland, M. (1992). *Digital photography*. Avalon Books, CA: Random House.

Ahonen, T. (2013). The Annual Mobile Industry Numbers and Stats Blog. *Communities Dominate Brands*. Re-trieved February 22, 2014 from http://communities-dominate.blogs.com/brands/2013/03/the-annual-mobile-industry-numbers-and-stats-blog-yep-this-year-we-will-hit-the-mobile-moment.html.

Azoulay, A. & Thompson, N. (2014) Photography and Its Citizens. *Aperture*. 214, 52-57.

BCC Research. (2012). Digital Photography: Global Markets. *Market Forecasting*. Retrieved February 26, 2014 from http://www.bccresearch.com/market-research/information-technology/digital-photography-global-markets-ift030c.html.

Brook, P. (2011). Raw Meet: Fred Ritchin Redefines Digital Photography. *Wired*. Retrieved February 22, 2014 from http://www.wired.com/rawfile/2011/09/fred-ritchin/all/1.

C/Net. (2013). An excellent interface for arty photos. *C/Net Reviews*. Retrieved February 22, 2014 from http://reviews.cnet.com/software/mextures-ios/4505-3513_7-35782639.html.

Camera Finder. (2014). Most Popular Cameras in the Flickr Community. *Flickr*. Retrieved February 27, 2014 from https://www.flickr.com/cameras.

Canon. (2014). Technology Used in Digital SLR Cameras. *Canon Global*. Retrieved February 22, 2014 from http://www.canon.com/technology/canon_tech/explanation/35mm.html.

Carter, R. L. (2012a). *DigiCam History Dot Com*. Retrieved February 22, 2014 from http://www.digicamhistory.com/1980_1983.html.

Carter, R. L. (2012b). *DigiCam History Dot Com*. Retrieved February 22, 2014 from http://www.digicamhistory.com/1995%20D-Z.html.

Carter, R. L. (2012c). *DigiCam History Dot Com*. Retrieved February 22, 2014 from http://www.digicamhistory.com/1996%20S-Z.html.

CEA. (2013). Camera Industry Pivots as Smartphones Target Point-and-Shoots. *Digital America* 2013. Retrieved February 27, 2014 from http://www.ce.org/i3/Features/2013/Digital-America/Camera-Industry-Pivots-as-Smartphones-Target-Point.aspx.

Cicala, R. (2014) Disruption and Innovation. *PetaPixel*. Retrieved February 26, 2014 from http://petapixel.com/2014/02/11/disruption-innovation/.

CIPA. (2014). 2014 Outlook on the Shipment Forecast by Product-Type Concerning Cameras and Related Goods. *Press Release*. Retrieved February 27, 2014 from http://www.cipa.jp/documents/e/PRESSRELEASE20140203_e.pdf.

Clark, R. N. (2014). Film versus digital information. *Clark Vision*. Retrieved February 26, 2014 from http://clarkvision.com/articles/how_many_megapixels/index.html.

Colao, J.J. (2013). Pew Study Suggests That Snapchat Has 26 Million U.S. Users. *Forbes*. Retrieved February 22, 2014 from http://www.forbes.com/sites/jjcolao/2013/10/28/pew-study-suggests-snapchat-has-26-million-u-s-users/.

Colao, J.J. (2014). The Inside Story Of Snapchat: The World's Hottest App Or A $3 Billion Disappearing Act? *Forbes*. Retrieved February 27, 2014 from http://www.forbes.com/sites/jjcolao/2014/01/06/the-inside-story-of-snapchat-the-worlds-hottest-app-or-a-3-billion-disappearing-act/.

Curtin, D. (2012). How a digital camera works. Retrieved February 25, 2012 from http://www.shortcourses.com/guide/guide1-3.html.

Donegan, T.J. (2014). Smartphone Cameras are taking over. *USA Today*. Retrieved Febraury 22, 2014 from http://www.usatoday.com/story/tech/2013/06/06/reviewed-smartphones-replace-point-and-shoots/2373375/.

Eddy, M. (2014). Wickr (for iPhone). *PCMag*. Retrieved February 27, 2014 from http://www.pcmag.com/article2/0,2817,2430025,00.asp.

Elrich, D. (2013). Camerapocalypse: How photography's titans will survive smartphones. *Digital Trends*. Re-trieved Febraury 22, 2014 from http://www.digitaltrends.com/photography/camera-companies-risk-going-bust/.

Four Thirds. (2012). Overview. *Four Thirds*: Standard. Retrieved February 22, 2014 from http://www.four-thirds.org/en/fourthirds/whitepaper.html.

Gil, L. (2014). The 10 best photography apps of 2013. Retrieved February 22, 2014 from http://www.idownloadblog.com/2013/12/03/best-photography-apps-2013/.

Hamburger, E. (2013). Sketching Instagram: co-founder Mike Krieger reveals the photo app's humble beginnings. *The Verge*. Retrieved February 28, 2014 from http://www.theverge.com/2013/5/13/4296760/sketching-instagram-co-founder-mike-krieger-reveals-apps-humble-beginnings.

Handler, M. (2013). Instagram founders offer stories, advice on campus. *The Daily Californian*. Retrieved February 28, 2014 from http://www.dailycal.org/2013/09/10/instagram-founders-offer-stories-advice-on-campus/.

Harry Ransom Center-The University of Texas at Austin. (2014). The First Photograph. Exhibitions. Retrieved February 22, 2014 from http://www.hrc.utexas.edu/exhibitions /permanent/firstphotograph/#top/.

Hill, S. (2013). From J-Phone to Lumia 1020: A complete history of the camera phone. *Digital Trends*. Retrieved February 22, 2014 from http://www.digitaltrends.com/mobile/camera-phone-history/.

Horaczek, S. (2013). How Many Photos Are Uploaded to The Internet Every Minute? *PopPhoto*. Retrieved February 22, 2014 from http://www.popphoto.com/news/2013/05/how-many-photos-are-uploaded-to-internet-every-minute.

IbisWORLD. (2013). Online Photo Printing in the US: Market Research Report. Retrieved February 27, 2014 from http://www.ibisworld.com/industry/online-photo-printing.html.

Instagram. (2014). Our Story. *Instagram*. Retrieved February 27, 2014 from http://instagram.com/press/#.

Johnson, H. (2013). Digital Photo Printing: 10 Years After. *PetaPixel*. Retrieved February 27, 2014 from http://petapixel.com/2013/07/25/digital-photo-printing-10-years-after/.

Keenan, T. & Steyerl, H. (2014). What is a Document? *Aperture*. 214, 58-64.

Kiss, J. (2013). Kevin Systrom, Instagram's man of vision, now eyes up world domination. *The Guardian/The Observer*. Retrieved February 28, 2014 from http://www.theguardian.com/technology/2013/oct/11/instagram-kevin-systrom-world-domination.

Kodak. (2012a). Milestones-chronology: 1960-1975. *Our Company*. Retrieved February 22, 2014 from http://www.kodak.com/ek/US/en/Our_Company/History_of_Kodak/Milestones_-_chronology/1960-1979.htm.

Kruger, J. (2014). Top imaging technology trends revealed at 2014 PMA@CES session. *Newsline*. Retrieved February 27, 2014 from http://pmanewsline.com/2014/01/10/top-imaging-technology-trends-revealed-at-2014-pmaces-session/#.Uw-JWM5AdD0.

Love, D. (2014). Wickr Is The Messaging App You'll Turn To If Snapchat Screws Up Again. *Business Insider*. Re-trieved February 27, 2014 from http://www.businessinsider.com/wickr-an-alternative-to-snapchat-2014-2.
Lester, P. (2006). Visual communication: Images with messages. Belmont, CA: Wadsworth.
Litherland, C. (2011). there's an app for photojournalism. Retrieved February 22, 2014 from http://www.chiplitherland.com/blog/2011/02/09/theres-an-app-for-photojournalism/.
Manovich, L. (2014). Watching the World. *Aperture*. 214, 48-51.
McHugh, M. (2013). Photographers tussle over whether 'pro Instagrammers' are visionaries or hacks. *Digital Trends*. Retrieved February 26, 2014 from http://www.digitaltrends.com/social-media/are-professional-instagrammers-photographic-visionaries-or-just-hacks/.
Moss, K. (2014). 10 Recent Photography Trends. *Digital Photography Daily*. Retrieved February 27, 2014 from http://digitalphotographydaily.com/10-recent-photography-trends/.
National Press Photographers Association. (2014). NPPA Code of Ethics. Retrieved February 22, 2014 from https://nppa.org/node/5145.
Oxford Dictionary. (2014). The Oxford Dictionaries Word of the Year 2013 is... Retrieved Febraury 22, 2014 from http://blog.oxforddictionaries.com/2013/11/word-of-the-year-2013-winner/.
Pavlik, J. & McIntosh, S. (2004). Converging media: An introduction to mass communications. Pearson Education, MA: Allyn & Bacon.
PCMAG.com. (2012). 21 Great Technologies That Failed. *Features*. Retrieved February 22, 2014 from http://www.pcmag.com/article2/0,2817,2325943,00.asp.
PCMAG.com. (2014). Definition of:RAW file. Encyclopedia. Retrieved February 27, 2014 from http://www.pcmag.com/encyclopedia/term/50204/raw-file.
PEW Research. (2013). Mobile Technology Fact Sheet. *PEW Research Internet Project*. Retrieved February 27, 2014 from http://www.pewinternet.org/fact-sheets/mobile-technology-fact-sheet/.
Photo Marketing Association Marketing Research Department (PMA). (2011). *Photo Industry 2011: Review and Forecast*. Jackson, MI: PMA.
PMA. (2013). What's New. *PMA Australia*. Retrieved February 27, 2014 from http://www.pmai.org/content.aspx?id=106.
Pictures that lie. (2012). *C/NET News*. Retrieved February 22, 2014 from http://news.cnet.com/2300-1026_3-6033210-1.html.
Pitt, B. (2013). How many megapixels do you need? *Digital Photography Review*. Retrieved February 26, 2014 from http://connect.dpreview.com/post/1313669123/how-many-megapixels.
Prindle, D. (2013). HDR as easy as 1, 2, 3: A beginner's guide to High Dynamic Range photography. *How To. Digital Trends*. Retrieved February 27, 2014 from http://www.digitaltrends.com/photography/what-is-hdr-beginners-guide-to-high-dynamic-range-photography/.
Ritchin, F. (2013). Bending the Frame: Photojournalism, Documentary, and the Citizen. New York, NY: Aperture Foundation.
SelfieCity. (2014). Findings. Retrieved February 22, 2014 from http://selfiecity.net/#dataset.
Schwarz, H. (2012). How Many Photos Have Been Taken Ever? *BuzzFeed*. Retrieved Febraury 22, 2014 from http://www.buzzfeed.com/hunterschwarz/how-many-photos-have-been-taken-ever-6zgv.
Snapchat. (2014). What does my teen need to know about using Snapchat? Parents. Retrieved Febraury 22, 2014 from http://www.snapchat.com/static_files/parents.pdf.
Van Camp, J. (2014). Nokia Lumia 1020 Review. *Digital Trends*. Retrieved Febraury 22, 2014 from http://www.digitaltrends.com/cell-phone-reviews/nokia-lumia-1020-review/.
Walker, D. (2013). Morel v. AFP Copyright Verdict: Defense Strategy to Devalue Photos and Vilify Photographer Backfires. *Photo District News*. Retrieved Febraury 26, 2014 from http://www.pdnonline.com/news/Morel-v-AFP-Copyrig-9598.shtml.
Winter, D. (2011). Through My Eye, Not Hipstamatic's. Lens. *The New York Times*. Retrieved Febraury 26, 2014 from http://lens.blogs.nytimes.com/2011/02/11/through-my-eye-not-hipstamatics/?_php=true&_type=blogs&_r=0.
Worthington, P. (2014). Sigma debuts Quattro camera, new sensor design. PMA *Newsline*. Retrieved Febraury 26, 2014 from http://pmanewsline.com/2014/02/10/sigma-debuts-quattro-camera-new-sensor-design/#.Uw5rNc5AdD3.

19

eHealth

Heidi D. Campbell, Ph.D.*

Why study eHealth?

- Between 70-75% of U.S. adults look online for health information (Aitken, 2014).
- eHealth or digital health funding surpassed $1.9 billion in 2013, growing 39% from the previous year and more than doubling since 2011 (Tecco, 2014a).
- The number of hospitals and office-based physicians who have adopted eHealth records has nearly tripled since 2010 (RWJF, 2013).
- Mobile healthcare and medical app downloads are expected to reach 142 million by 2016 (Cox, 2012).

Introduction

When was the last time you were at the doctor, the emergency room, or other healthcare provider? When you were there, how many encounters with technology did you have? Did you fill out patient information online? Did the nurse check your blood pressure and temperature with a digital device? Did the doctor use a computer or tablet in the exam room? Did the prescription you received go directly to the pharmacy via electronic delivery? When you left the office, did you think about the diagnosis and treatment and seek further clarification or information online or on a mobile app? This is eHealth—the use of electronic information and communication technologies in the field of health care (Cashen, Dykes, & Gerber, 2004).

eHealth affects everyone and is one of the fastest growing areas of innovations in communication technology. The proliferation of eHealth, sometimes known as digital health, has been driven by a convergence of digital technologies. Throughout this textbook you have read about the impacts of the digital revolution and how that has affected every aspect of our lives. The rapidly decreasing size of computer technology in the form of mobile phones and tablets, near ubiquitous Internet access and increasing bandwidth speeds, the connectedness of social media, and cloud computing have created the perfect storm for eHealth to burst on the scene and fundamentally change health care as we once knew it (Topol, 2013).

Essentially, eHealth is empowering the consumer to play a bigger role in managing his or her own health. Patients now have access to their own electronic health records (EHR) and can access information about every common illness known to man. This shifts the balance of healthcare and creates a way for individuals to take control of their own health focusing on personalized health plans and prevention.

Definitions

- eHealth—the use of electronic information and communication technologies in the field of health care.
- EHR—Electronic Health Records are electronic versions of patients' medical history maintained by the health care provider.
- mHealth—or mobile health is the use of mobile technologies such as cellphones and tablets for

*Associate Professor, Dept. of Journalism and Mass Communication, Bob Jones University (Greenville, South Carolina)

health communication and delivery of health information.

- Health eGames—electronic games used to promote health and wellness.
- Gamification—the application of different aspects of game playing to encourage engagement with a product or service.

Background

Convergence of digital technologies has given health communication increased traction over the past 30 years; however, the history of medicine is rich with information sharing from the earliest days of recorded history. It is the sharing of information that is critical to the growth of healthcare in this digital age.

The first recorded evidence of healthcare predated the use of paper, pens, or even books. The first evidence of health care was chronicled in cave paintings dating back thousands of years in what is now France (Hall, 2014). The cave art depicted the use of plants to treat ailments. Early evidence of surgeries and the use of anatomy in diagnosis were found in Egypt around 2250 BC (Woods & Woods, 2000). These drawings were some of the first forms of health communication because they documented health conditions and treatments for future generations of physicians.

It wasn't until the 19th century that the field of medicine advanced exponentially with developing sciences and growing knowledge of chemistry, anatomy, and physiology. The mass dissemination of information can be credited to innovations in mass communication, which delivered vital medical information to doctors and health care workers. Medical information was communicated through the use of telegraphs, printed journals, books, photography, and telecommunications (Kreps, Bonaguro, & Query, 2003).

Computers were used in healthcare since the early 1970s, serving as an administrative tool, storing patient information and medical practitioner records. Developments in surgical and diagnostic instruments continued, but it wasn't until 1995 that the information floodgates were opened to provide individuals with access to a pool of knowledge about every aspect of healthcare (Kreps et al, 2003).

Innovations in communication technology also created concerns regarding the protection of patient information on the Internet. In 1996, the Health Insurance Portability & Accountability Act (HIPAA) propelled the healthcare industry to adopt new ways of safeguarding the privacy of health records. HIPAA gave more control of information to the consumer, but because of the regulated safeguards, the healthcare industry has been slow to adopt technologies that would give patients ready access to their own medical records. The days of limited access to personal health information are quickly coming to a close with newer technologies emerging that are designed to facilitate the flow of information between the healthcare provider and the healthcare consumer.

Recent Developments

mHealth

- mHealth companies generated $6.2 billion in revenue in 2013, a figure that's expected to more than triple by 2018 (Freedman, 2014).
- 52% of smartphone users have looked up health or medical information and 19% of smartphone users have health apps (Fox & Duggan, 2012).

Technologies in eHealth have been featured at major technology shows worldwide for the past few years as an introduction to a growing industry. That growth has been exponential as new innovations have been adopted in virtually every area of the health care industry. The rapid growth can be attributed in part to ubiquitous mobile phone technologies, which connect users to health information (Steinhubl, Muse & Topol, 2013).

mHealth, or mobile health, is the use of mobile technologies such as cellphones and tablets for health communication and delivery of health information. Gone are the days when cellphones were used to simply make a call; now phones are multifunctional tools able to perform as photo and video devices, word processors, electronic organizers, and now even ECGs (electrocardiograms) or thermometers to monitor your health. There are more than 4.5 billion unique mobile phone users on the planet, with one in four mobile phones in use today being smartphones, and that number is expected to reach 50% penetration by 2017 (eMarketer, 2014). The use of smartphones in mHealth has also expanded healthcare services by providing access to medical information, hotlines for medical advice, and visually connecting with surgeons up to 5,000 miles away.

Mobile technologies for eHealth are expanding through software application and the use of mobile apps. By 2015, smartphone apps will help the eHealth industry connect to more than 500 million users (Jahns, 2013). With more than one million apps available in the iTunes store alone, smartphones can carry out diverse eHealth functions from monitoring glucose to monitoring heart rhythms with a mobile ECG (Costello, 2013).

More than 4.4 billion users are expected to use mobile apps on their smartphones by the year 2017, and the healthcare and business sectors are quickly jumping on board by engaging individuals through new eHealth apps (Protio, 2013). The number of mobile apps for health purposes has nearly tripled since 2009 (Fox & Duggan, 2012). Currently there are 97,000 mHealth applications in major app stores but that number growing daily. Traditional healthcare providers are also joining the mobile app market, and the business model is broadening to include healthcare services, and advertising and drug sale revenues (Jahns, 2013).

Corporations and makers of mHealth apps have found that their success comes from consistent engagement with the use of gamification, which is the application of game playing to encourage engagement with a product or service. Gamification uses challenges, points, badges, and leaderboards to encourage continuous participation (Weintraub, 2012). Many healthcare organizations are using this type of engagement to keep employees and customers engaged in healthy lifestyles, including diet and exercise. The use of gamification in mobile apps incentivizes healthy choices, encourages accountability, and provides a higher level of satisfaction to users.

Health e-Games

- Health e-Games is a $6.6 billion market (Donner, Goldstein, & Loughran, 2012).

In 2013, global revenue from video gaming reached $68 billion and is projected to grow to $96 billion by 2018 (DFC, 2014). That is nearly $20 billion in global growth in just five years. Innovations in cloud gaming, mobile gaming, and social media gaming have been the driving forces behind the growth of this industry, and the trend towards eHealth and health e-games is expected to significantly impact the gaming industry in the near future (Donner et al., 2012). Major video game companies such as Nintendo, Sony, and Microsoft have developed consoles, games, and accessories that not only provide entertainment enjoyment, but also improve overall quality of life (QOL) by engaging consumers with health literacy and education to help them take control of their health through entertaining and interactive games (Nintendo, 2013).

Health e-games can be classified into four key areas: *Exergames* or games that involve physical movement and interactivity; *health and nutrition* used in weight management and meal planning; *brain fitness* meant to improve cognitive function; and *condition management* used to manage chronic health issues such as diabetes, asthma, or cancer. The exergames category alone is a $6.7 billion segment of the video game market and dominates the health e-game sector (Donner et al., 2012). Games using dance pads, motion tracking, balance boards, and other types of controllers are categorized under exergames and range from dancing to snowboarding to golf.

Probably the biggest indicator that health e-games will be a key player in the future of eHealth is the fact that Nintendo is entering the market full force. Nintendo is not new to health gaming, having produced health and fitness accessories and games for years, including the Wii Fit, which has sold nearly 23 million units worldwide and is used for consumer fitness as well as physical therapy in hospitals and clinics worldwide (Nintendo, 2013).

The next generation of this product is the Wii Fit U, which incorporates the Wii accessories including the Wii U consoles, Wii Balance Boards, and the wearable Fit Meter. The game features some of the same activities from the original Wii Fit like yoga and strength training, as well as 19 new activities that incorporate the system's innovative second-screen controller, the Wii U GamePad. The game incorporates the Fit Meter, which allows people to wirelessly track daily activity, from calories burned to steps taken, including elevation, and compare accomplishments with other users online (Nintendo, 2014).

Nintendo has also partnered with companies like Bayer to engage diabetic patients in gaming. Bayer's Didget™ is the only glucose monitor designed specifically for children. The unit connects to Nintendo gaming systems and rewards children for consistent testing of their blood sugar (Bayer, 2014).

Consumers can expect to see more of these kinds of partnerships and innovations in eHealth in the future. In a letter to shareholders in February of 2014, Nintendo's president, Satoru Iwata, announced that the company plans to redefine itself as a health-oriented entertainment company in the coming decade, focusing its efforts in the expanding digital health sector (Iwata, 2014).

> *"We have set 'health' as the theme for our first step and we will try to use our strength as an entertainment company to create unique approaches that expand this business ...We have decided to redefine entertainment as something that improves people's quality of life ('QOL') in enjoyable ways and expand our business areas ..."*
>
> *—Iwata in a letter to shareholders and investors on February, 2014.*

Major health insurance companies are using health e-games in their wellness programs to encourage group participation in healthy activities, such as diet and exercise, while also lowering costs resulting from claims. Companies such as Aetna, CIGNA, Inland Empire Health Plan, Humana, and Kaiser Permanente offer e-games that encourage healthy eating and increased physical activity (Donner et al., 2012). Recent research conducted by Towers Watson and the National Business Group on Health found that 60% of employers planned to include e-games as part of their health initiatives for employees by the end of 2013 (Watson, 2013).

UnitedHealth Group is one of the leaders in the insurance industry using gaming to encourage healthier living among their insured. The company has partnered with Kanami, producer of the *DanceDanceRevolution* games to create the *Activate for Kids* exergame program. *DanceDanceRevolution* "Classroom Edition" is an interactive video game that combines dance movements with fast music and stimulating visuals. The goal of the exergame program is to help fight childhood obesity and encourage physical activity and healthy eating for all children, and the program seems to be working. A recent study conducted on the program found that overweight children involved in a 16-week weight loss program that used the *DanceDanceRevolution* exergame lost two and a half times their body mass index compared to those children who only dieted. UnitedHealth Group introduced their "Exergame" in classrooms in California, Florida, Georgia, and Texas; however, the program is expected to be adopted in other states as well. Watch for expanded growth in the health e-game market over the next five years (Trost, Sundal, Foster, Lent, & Vojta, 2014).

Wearables

In 2014, the big news at the digital health summit was the rapid expansion of wearable health devices. Wearables include clothing, wristbands, sensor rings, computerized glasses, and dermal devices equipped with sensors that wirelessly collect and transmit health data over extended periods of time with minimal lifestyle disruptions.

There are companies innovating clothing and textiles with wireless sensors for sportswear and health monitoring. Cityzen Sciences (2014) has developed a smart shirt with enhanced micro sensors equipped with mobile connectivity that enables users to monitor temperature, heart rate, and physical activity. The shirt is also equipped with a global positioning system (GPS) and is used primarily by runners.

While there are sports and fitness applications for smart clothing, the Chronious project in Europe has developing a t-shirt designed specifically for patients suffering from chronic obstructive pulmonary disease and chronic kidney disease. The t-shirt is outfitted with sensors to monitor heart, respiratory, and physical activity. The sensors connect wirelessly to electronic devices in the home, including blood pressure monitors, glucometers, and air quality sensors which then transmit the data to "intelligent data processing software" equipped with algorithms specifically designed to create unique therapies for the patient. This technology has been in use for a few years in Europe and is being tested on patients with other chronic diseases (Bellos, 2013).

Pedometers and Wristbands

The most widely used wearable device for health and fitness is the pedometer, which measures the number of steps an individual has taken. The pedometer has undergone several levels of innovation and has evolved into a more powerful digital device, worn as a clip on a belt or as a necklace or wristband.

Wristbands dominate the wearable fitness devices, and many of the major sports companies have designed wristbands that wirelessly communicate with smartphones to measure steps, physical activity, sleep patterns, and calories burned. The Fitbit Flex accounted for 67% of the total units sold in this category in 2013, and the Fitbit mobile app is also the No. 1 app in the Apple iTunes store in the Health & Fitness category (NPD Group, 2014b; Dolan, 2014). Other health and fitness wristbands available include Nike's FuelBand, Adidas' miCoach, and Up by Jawbone which all work similarly by recording physical data and connecting wirelessly to a smartphone that has the designated app loaded (Business Wire, 2014).

Figure 19.1
Fitbit Flex

Source: Fitbit

Major cellular companies, such as Samsung and Sony, have developed smart watches that have mobile connectivity and also serve as a fitness companion monitoring heart rate and physical activity. The Samsung Fit bracelet and Sony Smart-Band allow users to log events and photographs taken during the day as well as track physical activity including sleep cycles (Gibbs, 2014).

Developers are working on advanced sensors that will take eHealth wearables, such as wristbands, to a new level. Smart Monitor's *Smart Watch* is a device with mobile connectivity that alerts the user of impending convulsions, tremors, or seizures. The watch connects with a smartphone and alerts users when repetitive motion is detected and then connects with chosen emergency contacts with alerts and GPS coordinates of where the individual Smart Watch user is located. Other data collected by the Smart Watch is then compiled for review by the user and can be shared with a healthcare provider (Smart Monitor, 2014). This type of technology is in its infancy, but is gaining ground for users with these chronic conditions.

Google Glass

You may remember Lt. Commander Geordi La Forge from Star Trek Next Generation who was blind but received his first visual VISOR in the year 2340. The computerized device worn like glasses allowed him to see. While the creators of Star Trek thought of this technology as being futuristic in the late 1980s, the innovation of a wearable computer is here. Google Glass looks much like a pair of glasses and allows users to see data images, scheduling calendars, surf the web, take photos or video, and allow communication with various Internet-based apps and services, mostly provided by Google (Albrecht et al., 2014). Doctors have been some of the early adopters of Google Glass and are using the technology for remote monitoring of patients, viewing lab reports as they come available, and live streaming medical procedures to medical students. Hospitals are traditionally slow to adopt new communications technologies so this type of wearable device is not likely to be widely used in healthcare in the near future (Glauser, 2013).

Figure 19.2
Google Glass

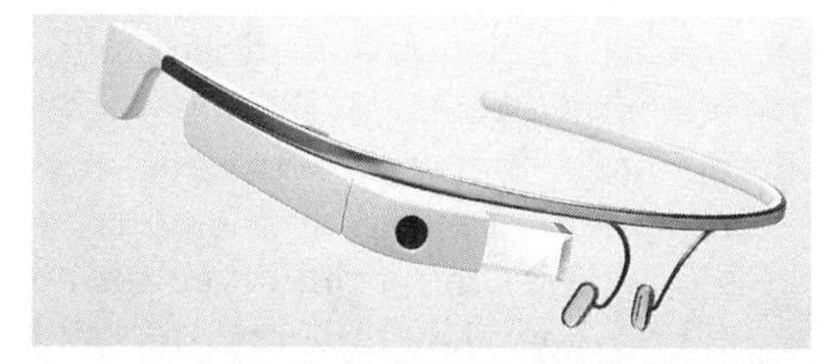

Source: Google

Dermal Devices

While wearables such as wristbands have definitely moved into the forefront of eHealth hardware technology, the next innovation in wearables is found in dermal devices, which are attached directly to the body allowing freedom of movement and more precise health monitoring (Ahlberg, 2011). Scientists from the University of Illinois developed a chip tattoo that is applied directly on the skin using electronic components for sensing, communication interfaces, and medical diagnostics. The chip is applied much like a temporary tattoo and is comfortable to wear, unlike other types of probes and sensors (Kim et al., 2011). This type of device connects to a remote device to collect data ranging from ECG type monitoring to brainwave activity. Researchers are working to take this technology to a new level by developing sensors that measure glucose in diabetic patients.

Figure 19.3
Dermal Tattoo

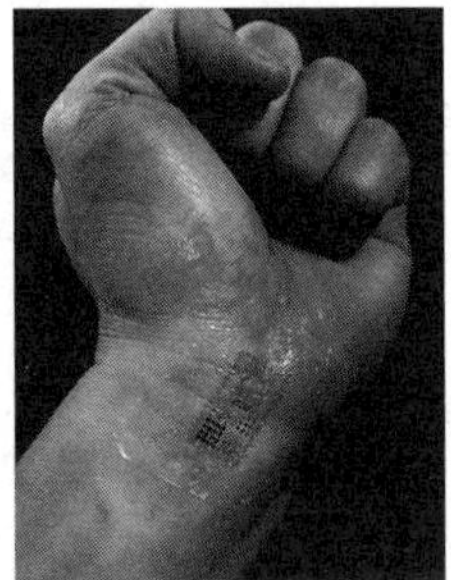

Source: John Rogers

There are 25.8 million people in the United States who have diabetes (ADA, 2011). New technologies are emerging that will give more freedom to diabetes patients through monitoring and glucose delivery systems. Google is testing a prototype for a contact lens that will continuously measure glucose in tears using a wireless chip and miniaturized glucose sensor. Google is hoping this smart contact lens will be a more efficient way for people with diabetes to manage their disease (Klonoff, 2014).

Dexcom (2014), producers of diabetic products, has also produced a continuous glucose monitor device implanted just below the skin that checks blood sugar every 5 minutes. The sensor sends data to a cellphone and alerts users of spikes or dips in blood sugar allowing the user to monitor their glucose levels more closely. The data can then be downloaded for historical comparison and for health care professionals to analyze (Comstock, 2014).

Developing Countries

The World Health Organization reports that there is a global shortage of 7.2 million health care workers but in the next 20 years, that number is expected to double (GHWA & WHO, 2013). Advances in communication technology are helping in this area, and some believe eHealth can help. eHealth expands access to remote areas of the globe that never dreamed of having modern medical care.

Technologies such as live video conferencing, uploading of MRI or other scans to a cloud network, and remote monitoring of vital signs such as body temperature, blood pressure, and glucose levels, are all used by healthcare providers located at a distance and sometimes across the globe. All that is needed are a smartphone or computer device, and Internet access at both ends to connect patients with quality health care.

Figure 19.4
Healthcare Worker Using a Cellphone in Ghana

Source: Nana Kofi Acquah and Novartis Foundation

In emerging and developing nations around the globe, telemedicine is being used where doctors are scarce and resources are non-existent. Telemedicine is "the use of communication technologies to deliver healthcare, particularly in settings where access to medical services is insufficient" (GHWA & WHO, 2014). In China alone there are 900 million people living in rural areas struggling with scarce healthcare (Lin et al., 2014). With a projected 5.13 billion mobile phone users worldwide by 2017, innovations in telemedicine could be a solution in these underserved areas (eMarketer, Dec 2014).

Based on global use of mHealth, the uses of mobile phones in healthcare are still rudimentary and include telephone help lines (59%), emergency toll-free telephone services (55%), and mobile telemedicine (49%); however, the countries researched in this study were open to new ideas of eHealth applications (Kay, Santos, & Takane, 2011).

The major barriers to the progress of eHealth in developing countries are the lack of technological resources, the lack of knowledge and information, conflicting health system priorities, and the lack of policy and legal frameworks. The use of mHealth reported by member countries was low at just 12%; however, the implementation of standards of use are expected to propel eHealth forward in these developing countries (Kay et al., 2011). With a projected growth of mobile phone users worldwide and the adoption of new applications of use, mHealth can serve as foundational to economic growth in these developing countries by improving the health of its citizens.

Current Status

The health system in the United States is facing many challenges. The healthcare costs in the United States are the highest in the world with the average per capita cost reaching US $8,508 in 2013 (OECD, 2014), and aggregate healthcare spending in the United States is expected to reach US $4.8 trillion by 2021 (CMS, 2011). The *Patient Protection and Affordable Care Act* (PPACA), commonly known as "Obamacare," was signed into law on March 23, 2010 and took effect on January 1, 2014. This law prohibits insurers from denying coverage to individuals based on pre-existing conditions, and mandates that all individuals who are not already covered by a health plan secure an approved private insurance policy or pay a penalty.

PPACA is fueling eHealth by restricting health spending and creating a new business model for healthcare that is more value-based. Historically, the healthcare industry was not being rewarded for keeping patients healthy and avoiding expensive treatments; however, PPACA is forcing change to that model by propelling healthcare organizations into the adoption of new technologies to engage and empower patients.

The use of new technologies and apps that engage consumers is not enough. Full adoption of these new technologies as well as new applications of use will take time and will require a change in the consumer's behavior. These changes will require incentives for use and compatibility with current technologies, making is easy for users to adopt. As the proliferation of new technologies continues, it may be harder for consumers to navigate the many eHealth options available to them.

Security Issues in eHealth

Security concerns are the biggest inhibitor of adoption of new eHealth technologies. The key security issues in eHealth include security for patient confidentiality, security that enables authentication of electronic health records, and systems security that ensures secure transmission, processing, and storage of health data (Waegemann, 2003).

Table 19.1

Total Large PHI Breaches and Records Impacted, 2010-2013

Year	2010	2011	2012	2013
Total # of Incidents Reported	212	149	192	199
Total # of Patient Records Impacted	5,434,661	10,841,802	2,983,984	7,095,145

Source: Redspin (2014)

Electronic Health Records (EHR) were believed to be the solution for improved management of an individual's health, and as the tool that would give the patient control of their own health by providing access to his or her own medical history (Waegemann, 2002). EHRs have opened the doors for patient access, but have also opened the doors to security risks. The issues of protection of EHRs against intrusion, data corruption, fraud, and theft are of highest concern for health care professionals and the patients they serve. As illustrated in Table 19.1, since 2009 there have been more than 804 major breaches of protected health information affecting more than 29 million patient records (Redspin, 2014). The security of patient information will remain a priority for governmental agencies, healthcare, and the technology sectors.

Factors to Watch

We live in a highly personalized era where technologies have increased access to a wide range of goods and services. For example, you don't have to get in your car and drive to the store to get everything you need to survive in this world. You can now access millions of products online and have them delivered to your door, in some cases, the same day. Mobile phones have thousands of personalized accessories, case options, and more than a million apps to choose from. This culture of individualized access has created a generation of consumers that expect products and services that are customized to their individual needs, and this culture is quickly spilling over into the health care sector.

The convergence of digital technologies in mobile communication, wireless sensors, wearable wireless devices, and super networks of big data is creating an age of digital medicine that will be so highly individualized that diagnostics will not be based solely on symptoms, but rather on individual molecular and DNA structures. Dr. Eric Topol, leading eHealth expert says, "eHealth technology in the future will be used primarily as a prevention tool. The technology is being developed now and relies on a data network already available but not yet converged." (Topol, 2014b).

mHealth will continue to see rapid growth through the end of the decade as broadband services improve and smartphone adoption continues to grow. By 2017, mHealth is projected to be a mass market with a reach of more than 3.4 billion smartphone and tablet users with access to mobile applications that will expand their abilities to access health information. The number of people accessing eHealth apps will continue to grow with a projected 1.7 billion users downloading an eHealth app by 2017 (Jahns, 2013).

Wearables are an ever expanding area of eHealth and are expected to increase rapidly over the next few years. The number of wearable wireless health and fitness devices sold is expected to hit 170 million by 2017 (ABI, 2012). To put this in perspective, that would be an 800% increase in just five years. The technology will improve with the development of sensors with enhanced wireless connectivity. These technologies have the potential of improving quality of care and reducing health care costs by allowing patients to have greater control over their own health.

Watch for more widespread adoption of new eHealth technology by the healthcare sector driven by government regulations. The Federal Drug Administration (FDA) has recently issued guidelines for regulatory review of health apps intended to be used as an accessory to a regulated medical device, or transform a mobile platform into a regulated medical device (2013). Just over 100 eHealth apps have received FDA approval but that number is expected to grow as more devices are released that communicate wirelessly to mobile devices (McCray, 2014).

With the rapid diffusion of eHealth technologies, the revenue stream is expected to grow by 61% earning US $26 billion by 2018, much of which will come from mHealth and the hardware involved (Jahns, 2013). With major communications entities such as Google, Apple, Sony, and Samsung developing new communication devices for eHealth, those revenue projections may actually be low. The bulk of the revenue will come from eHealth hardware technologies with new applications driving the sales of the hardware and health services (Jahns, 2013).

eHealth Visionaries: Halle Tecco & Eric Topol, MD

Halle Tecco is a social entrepreneur passionate about creating a culture of innovation in eHealth. She is co-founder and CEO of Rock Health, which funds and supports early-stage eHealth startups. "We are currently working with diversely focused companies. Some of which are more enterprise focused, building tools for hospital administration, and other companies we are working with are creating digital therapies and digital treatments for chronic illnesses. The goal is to radically lower costs in the healthcare system by innovating products and services or improving on the delivery of more personalized and better care." (Personal interview, April 2014)

Halle doesn't see a proliferation of new technologies in eHealth, which has surprised her, "I love seeing how existing technologies can be used in eHealth. We haven't even gotten to the point that we are developing new technologies in healthcare yet. Literally, everything we are using are existing technologies—mobile, data mining, cloud computing are all examples of using existing technologies in newer applications. So what we need to focus on is creating great products and innovative companies that have the ability to shift a culture and shift the industry into the future." (Personal interview, April 2014)

When considering the challenges of eHealth, Tecco believes, "the healthcare system is a complex, siloed system with data confined to one space. If you were out of town and had a medical emergency, it would be very challenging for the doctor to get your health history. The challenge now is in the creation of a data platform that will incentivize the sharing of data within healthcare and improve patient access among all their healthcare providers."

Halle received a bachelor's degree from Case Western Reserve University and an MBA from Harvard Business School. She has been named one of Forbes "30 under 30," "12 Entrepreneurs Reinventing Healthcare" by CNN, "15 Women to Watch in Tech" by Inc. Magazine, and L'Oreal "Woman of Worth" (Rock Health, 2014).

Eric Topol, MD: Healthcare is in the process of exciting transformation in the rise of eHealth, and Eric Topol, MD is one of the key innovators leading the charge for the innovations and adoption of eHealth technologies. After more than a decade as the chief of cardiovascular medicine at the Cleveland Clinic, Eric Topol moved to La Jolla, California to become director of the Scripps Translational Science Institute. Topol also co-founded the West Wireless Health Institute which researches the use of innovative wireless technologies in the delivery of health care.

According to Topol, "There is nothing more important to the future of healthcare than eHealth. We are living in an era where our mobile wireless digital devices, including smartphones and tablets, will radically change lives and provide hyper-connectivity to healthcare data. This will give patients the ability to take control of their health."

Dr. Topol believes the biggest challenge is in giving patients access to their data, "getting doctors to let go and let patients have access to their own information is a big issue. If physicians will let go and adopt eHealth, then the access of patient's information can be innovated and healthcare can move forward much quicker."

In 2009, Topol was named one of the country's 12 "Rock Stars of Science" in GQ Magazine, he has published 1100 peer-reviewed articles and over 30 medical textbooks. In 2012, he was voted the most influential physician executive in the United States in a poll conducted by *Modern Healthcare*. His book The Creative Destruction of Medicine discusses the challenges and opportunities that new technologies bring to the field of healthcare.

Bibliography

ABI. (2012).Body Area Networks for Sports and Healthcare. ABI Research. Retrieved March 7, 2014 from http://www.abiresearch.com/research/1004149 .

Ahlberg, L. (2011). Smart skin: Electronics that stick and stretch like a temporary tattoo. Physical Sciences News Bureau: Illinois. Retrieved March 7, 2014 http://news.illinois.edu/news/11/0811skin_electronics_JohnRogers.html.

Aitken, M. (2014). Engaging patients through social media. IMS Institute for Healthcare Informatics. Retrieved March 9, 2014 from http://www.imshealth.com/cds/imshealth/Global/Content/Corporate/IMS%20Health%20Institute/Reports/Secure/IIHI_Social_Media_Report_2014.pdf.

Albrecht, U. V., von Jan, U., Kuebler, J., Zoeller, C., Lacher, M., Muensterer, O. J., & Hagemeier, L. (2014). Google Glass for Documentation of Medical Findings: Evaluation in Forensic Medicine. *Journal of medical Internet research,16*(2). Retrieved March 9, 2014 http://www.ncbi.nlm.nih.gov/pmc/articles/PMC3936278/.

American Diabetes Association. (2011). Statistics about Diabetes: Data from the 2011 National Diabetes Fact Sheet. Retrieved Retrieved March 9, 2014 from http://www.diabetes.org/diabetes-basics/statistics/.

Bayer. (2014). Didget Meter: Play with purpose. Bayer Diabetes. Retrieved March 7, 2014 from https://www.bayerdiabetes.ca/en/products/didget-meter.php.

Bellos, C. (2013, July 3-7). Clinical Validation of the CHRONIOUS Wearable System in Patients with Chronic Disease. Engineering in Medicine and Biology Society (EMBC), 2013 35th Annual International Conference of the IEEE Osaka (7084-7087). DOI:10.1109/EMBC.2013.6611190.

Business Wire. (2014). Fitbit Announces Leading Sales Across Digital Fitness Category for Full Body Activity Trackers (2014) Retrieved Retrieved March 14, 2014 from http://www.businesswire.com/news/home/20140107005987/en/Fitbit-Announces-Leading-Sales-Digital-Fitness-Category#.UzyOnKg7um4.

Cashen, M. S., Dykes, P., & Gerber, B. (2004). eHealth Technology and Internet resources: barriers for vulnerable populations. Journal of Cardiovascular Nursing, 19(3), 209-214.

Center for Medicare and Medicaid Services. (2011). National Health Expenditures Projections 2011-2021. CMS.gov. Retrieved March 19, 2014 from www.cms.gov/Research-Statistics-data-and-Systems/Statistics-Trends-and-Reports/NationalHealthExpendData/Downloads/Proj2011PDF.pdf.

Cityzen Sciences. (2014). Smart Sensing Industrial Project Smart Fabric. Smartsensing.fr/en Cityzen Sciences. Retrieved March 7, 2014 from http://www.cityzensciences.fr/en.

Cox, A. (2012). mHealth Users of Remote Health Monitoring to Reach 3 million by 2016: Smartphones Play Leading Role. Juniper Research Hampshire, UK Retrieved March 24, 2014 from http://www.juniperresearch.com/viewpressrelease.php?pr=285.

Comstock, J. (2014). Dexcom files for patent for smartphone-connected continuous glucose monitoring. Mobile Health News. Retrieved March 14, 2014 from http://mobihealthnews.com/28829/dexcom-files-for-patent-for-smartphone-connected-continuous-glucose-monitoring/.

Costello, S. (2013),How Many Apps are in the iPhone App Store. About.com/iPhone/iPod. Retrieved March 13, 2014. http://ipod.about.com/od/iphonesoftwareterms/qt/apps-in-app-store.htm .

DFC Intelligence. (2014). Global Video Game Industry to Reach $96B in 2018. February 12, 2014 REPORT.

Dexcom. (2014). Continuous Glucose Monitor Retrieved March 13, 2014 from http://www.dexcom.com/.

Dolan, B. (2014). Fitbit, Jawbone, Nike had 97 percent of fitness tracker retail sales in 2013. . Mobi Health News Retrieved April 1, 2014 from http://mobihealthnews.com/28825/fitbit-jawbone-nike-had-97-percent-of-fitness-tracker-retail-sales-in-2013/.

Donner, A., Goldstein, D., & Loughran, J., (2012). Health eGames Market Report: How Video Games, Social Media and Virtual Worlds will Revolutionize Health. Conecto. Health eGames Market Report..

eMarketer. (2014). Smartphone Users Worldwide Will Total 1.75 Billion in 2014. eMarketer.com Retrieved on March 16, 2014 from http://www.emarketer.com/Article/Smartphone-Users-Worldwide-Will-Total-175-Billion-2014/1010536#hRDqhsPlE3A7dIZA.99.

Fox, S., & Duggan, M. (2012). Mobile health 2012. Pew Research Center's Internet American Life Project.

Food and Drug Administration. (2013, October 22). Mobile Medical Applications. Retrieved March 16, 2014 from http://www.fda.gov/medicaldevices/productsandmedicalprocedures/connectedhealth/mobilemedicalapplications/default.htm.

Freedman, David H. (2014, February) The Startups Saving Health Care. Inc. Magazine Retrieved on March 29, 2014 from http://www.inc.com/magazine/201402/david-freedman/obamacare-health-technology-startups.html.

Gibbs, S. (2014). CES 2014: the best wearable smartwatches and fitness gadgets. The Guardian. Retrieved March 16, 2014 from http://www.theguardian.com/technology/2014/jan/07/ces-2014-wearable-smartwatches-fitness-gadgets.

Glauser, W. (2013). Doctors among early adopters of Google Glass. Canadian Medical Association Journal, cmaj-109. Retrieved March 16, 2014 from: http://www.cmaj.ca/content/early/2013/09/30/cmaj.109-4607.short.

Global Health Workforce Alliance & World Health Organization (2013) A Universal Truth: No Health Without a Workforce: Third Global Forum on Human Resources for Health Report. Retrieved March 16, 2014 from: http://www.who.int/workforcealliance/knowledge/resources/hrhreport2013/en/.

Hall, T. (2014). History of Medicine: All the Matters. New York, NY: McGraw-Hill.

Iwata, S. (2014). Letter from the President to Shareholders and Investors. Nintendo. Retrieved on March 24, 2014 from http://www.nintendo.co.jp/ir/en/management/message.html.

Jahns, R.G. (2013). 500m people will be using healthcare mobile applications in 2015. research2guidance Global Mobile Health Trends and Figures Market Report 2013-2017: The Commercialization of mHealth Applications (Vol.3) Retrieved March 16, 2014 from https://research2guidance.com/500m-people-will-be-using-healthcare-mobile-applications-in-2015/.

Kay, M., Santos, J., & Takane, M. (2011). mHealth: New horizons for health through mobile technologies. *World Health Organization.* http://whqlibdoc.who.int/publications/2011/9789241564250_eng.pdf?ua=1.

Kim, D. H., Lu, N., Ma, R., Kim, Y. S., Kim, R. H., Wang, S., & Rogers, J. A. (2011). Epidermal Electronics. Science, 333(6044), 838-843.

Klonoff, D. C. (2014). New Wearable Computers Move Ahead Google Glass and Smart Wigs. *Journal of Diabetes Science and Technology, 8*(1), 3-5. http://dst.sagepub.com/content/8/1/3.short.

Kreps, G. L., Bonaguro, E. W., & Query Jr, J. L. (2003). The history and development of the field of health communication. HISTORY, 10, 12..

Lin, C. W., Abdul, S. S., Clinciu, D. L., Scholl, J., Jin, X., Lu, H., ... & Li, Y. C. (2013). Empowering village doctors and enhancing rural healthcare using cloud computing in a rural area of mainland China. Computer methods and programs in biomedicine. Retrieved March 16, 2014 from: http://www.sciencedirect.com/science/article/pii/S0169260713003453.

McCray, R.B. (2014). Wireless-Life Sciences Alliance (2014) FDA Guidelines on Approving mHealth Apps Feb. 20, 2014 Retrieved March 19, 2014 from http://wirelesslifesciences.org/2014/02/fda-guidelines-on-approving-mhealth-apps/.

NPD. (2014a). The NPD Group/Weekly Tracking Service Digital Fitness Devices, November 17, 2013—December 21, 2013 www.npd.com.

NPD. (2014b). Retail Sales Of Sporting Goods' Run/Fitness Products In The US Increased 6 Percent To $10.5 Billion In 2013. NPD Group. Retrieved March 19, 2014 from https://www.npd.com/wps/portal/npd/us/news/press-releases/retail-sales-of-sporting-goods-run-fitness-products-in-the-u-s-increased-6-percent-to-10-point-5-billion-dollars-in-2013/.

Nintendo. (2014) What is Wii U? Retrieved March 25, 2014 from http://www.nintendo.com/wiiu/what-is-wiiu/.

Nintendo. (2013). Nintendo Top Selling Software Sales Units: Wii Fit. Nintendo. 2013-03-31. Retrieved March 25, 2014 from http://www.nintendo.co.jp/ir/en/sales/software/wii.html.

OECD. (2014). Health Policies and Data OECD Health Statistics 2013. Retrieved March 16, 2014 from http://www.oecd.org/health/health-systems/oecdhealthdata.htm.

Protio Research. (2013). Mobile Applications Futures 2013-2017. Retrieved March 16, 2014 from http://www.portioresearch.com/en/major-reports/current-portfolio/mobile-applications-futures-2013-2017.aspx.

Robert Wood Johnson Foundation. (2013, July 8). Hospitals, Physicians Make Major Strides in Electronic Health Record Adoption Retrieved March 29, 2014 from http://www.rwjf.org/en/about-rwjf/newsroom/newsroom-content/2013/07/hospitals--physicians-make-major-strides-in-electronic-health-re.html.

Rock Health (2014) Halle Tecco Bio. Rockhealth.com Retrieved March 27 from rockhealth.com/about/team.

Redspin. (2014). Breach Report 2013: Protected Health Information (PHI) Carpinteria, CA: Retrieved March 16, 2014 from http://www.redspin.com/docs/Redspin-2013-Breach-Report-Protected-Health-Information-PHI.pdf.

Smart Monitor. (2014). Smart Watch Monitoring Device. Retrieved March 14, 2014 from http://www.smart-monitor.com/.

Steinhubl S, R., Muse E, D,, & Topol E, J. (2013). Can Mobile Health Technologies Transform Health Care? JAMA. 2013;310(22):2395-2396. Retrieved March 24, 2014 from http://jama.jamanetwork.com/article.aspx?articleID=1762473.

Tecco, H. (2014a). Rock Health Funding: A Year in Review 2013. Rock Health. Retrieved March 16, 2014 from http://rockhealth.com/resources/rock-reports/.

Tecco, H. (2014b, March 31). Telephone Interview..

Topol, E. (2013). The Creative Destruction of Medicine: How the Digital Revolution will Create Better Healthcare. 2013.

Topol, E. (2014, March 24).Telephone Interview..

Trost, S. G., Sundal, D., Foster, G. D., Lent, M. R., & Vojta, D. (2014). Effects of a Pediatric Weight Management Program With and Without Active Video Games: A Randomized Trial. JAMA pediatrics.

Watson, T. (2013). National Business Group on Health. Reshaping Health Care: Best Performers Leading the Way. 18th Annual Employer Survey on Purchasing Value in Health Care.

Waegemann, C. P., Status Report (2002). Electronic Health Records, Medical Records Institute, Retrieve March 16, 2014 from www.medrecinst.com/.

Waegemann, C. P. (2003, September) .Confidentiality and Security for e-Health. Paper presented at Mobile Health Conference, Minneapolis, MN.

Weintraub, A. (2012). Gamification Hits Healthcare as Startups Vie for Cash and Partners. Retrieved March 16, 2014 from http://www.xconomy.com/new-york/2012/06/21/gamification-hits-healthcare-as-startups-vie-for-cash-and-partners/.

Woods, M., & Woods, M. B. (2000). Ancient Medicine: From Sorcery to Surgery. Twenty-First Century Books.

IV

Networking Technologies

20

Broadband & Home Networks

John J. Lombardi, Ph.D.*

Why study broadband and home networks?

- Broadband technologies are becoming intrinsic to our communication, entertainment, and work needs.
- Broadband technologies are fast becoming the backbone to our national infrastructure. They are increasingly used to send, receive, and track the distribution of our nation's energy needs.
- Broadband technologies are estimated to create hundreds of thousands of new jobs, shape our educational system, and administer high quality healthcare.

Introduction

Tim Berners-Lee, the man credited with developing the World Wide Web, has said "Anyone who has lost track of time when using a computer knows the propensity to dream, the urge to make dreams come true, and the tendency to miss lunch" (FAQ, n.d.). The World Wide Web is just a bit more than two decades old. However, it is deeply engrained in the daily lives of millions of people worldwide. But the World Wide Web is just part of the growing Internet experience. The Internet is used in virtually all aspects of our lives. In addition to surfing the Web, the Internet allows users to share information with one another.

The Internet can be used to send and receive photos, music, videos, phone calls, or any other type of data. At the beginning of the current decade there was more than one zettabyte of digital information in the world...a number that is expected to double every two years (ITU, 2012a). With benefits to healthcare, energy consumption, and an improved global economy the importance of high speed Internet access around the globe will continue to grow.

The term "broadband" is generally used to describe high speed Internet. What constitutes "high speed," however, varies a bit from country-to-country. In the U.S., the FCC considers broadband that which has speeds greater than or equal to 4 Mbps download and 1 Mbps upload. The Organisation for Economic Co-Operation and Development (OECD), an organization based in France that collects and distributes international economic and social data, considers broadband any connection with speeds of at least 256 Kb/s. Meanwhile the International Telecommunication Union considers broadband to be connection speeds of between 1.5 and 2.0 Mb/s (ITU, 2003).

While the FCC defines broadband as connections equal to or greater than 4 Mbps upload and 1 Mbps download, actual speeds in the U.S. are generally much faster, and they continue to rise. One source ranks the U.S. 31st internationally with an average download speed of 12.75 Mb/s. South Korea is at the top of the international list at 32.58 Mb/s download speeds (Household download index, 2012).

Broadband penetration rates continue to grow. The OECD estimates that the penetration rate has

* Professor of Mass Communication Frostburg State University (Frostburg, Maryland)

increased to nearly 70% within the 34 country OECD group. Six countries (Australia, Denmark, Finland, Korea, Japan, and Sweden) have eclipsed the 100% penetration mark. With the exponential growth of tablets and smartphones, it is becoming more and more common for people to have multiple access points to broadband services (OECD Broadband, 2014).

The increasing connection speeds associated with broadband technology allow for users to engage in such bandwidth intensive activities such as "voice-over-Internet-protocol" (VoIP) including video phone usage, "Internet protocol television" (IPTV) including increasingly complex video services such as Verison's FiOS service, and interactive gaming. Additionally, the "always on" approach to broadband allows for consumers to easily create wireless home networks.

A wireless home network can allow for multiple computers or other devices to connect to the Internet at one time. Such configurations can allow for wireless data sharing between numerous Internet protocol (IP) devices within the home. Such setups can allow for information to easily flow from and between devices such as desktop computers, laptop computers, tablets, smartphones, and audio/video devices such as stereo systems, televisions, and DVD players.

As an example, with a home wireless broadband network you could view videos on your television that are stored on your computer or listen to music that is stored on your computer through your stereo system. Additionally, video services such as Netflix allow subscribers to access certain content instantly. In this case subscribers could add movies to their "instant queue" and then access them directly through their television. The key device in most home networks is a residential gateway, or router. Routers are devices that link all IP devices to one another and to the home broadband connection.

This chapter briefly reviews the development of broadband and home networks, discusses the types and uses of these technologies, and examines the current status and future developments of these exciting technologies.

Background

Broadband networks can use a number of different technologies to deliver service. The most common broadband technologies include digital subscriber line (DSL), cable modem, satellite, fiber cable networks, and wireless technologies. Thanks in part to the *Telecommunication Act of 1996*, broadband providers include telephone companies, cable operators, public utilities, and private corporations.

DSL

Digital subscriber line (DSL) is a technology that supplies broadband Internet access over regular telephone lines with service being provided by various local carriers nationwide. There are several types of DSL available, but asymmetrical DSL (ADSL) is the most widely used for broadband Internet access. "Asymmetrical" refers to the fact that download speeds are faster than upload speeds. This is a common feature in most broadband Internet network technologies because the assumption is that people download more frequently than upload, and they download larger amounts of data.

With DSL, the customer has a modem that connects to a phone jack. Data moves over the telephone network to the central office. At the central office, the telephone line is connected to a DSL access multiplexer (DSLAM). The DSLAM aggregates all of the data coming in from multiple lines and connects them to a high-bandwidth Internet connection.

A DSL connection from the home (or office) to the central office is not shared. As such, individual connection speeds are not affected by other users. However, ADSL is a distance sensitive technology. This means that the farther your home is from the central office, the slower your connection speed will be. Also, this technology only works within 18,000 feet (about 3 ½ miles) of the central office (though "bridge taps" may be used to extend this range a bit).

ASDL typically offers download speeds up to 1.5 Mb/s and upload speeds from 64 Kb/s to 640 Kb/s. Some areas have more advanced DSL services called ADSL2, ADSL2+, and more recently ADSL 2++ (sometimes referred to ADSL 3 or ADSL 4). These services offer higher bandwidth, up to 12 Mb/s with ADSL2, 24 Mb/s with ADSL2+, and 45 Mb/s with ADSL 2++. Prices vary from a low of $29.95 per month for 3 Mb/s download to $65 per month for up to 45 Mb/s with AT&T's DSL service (AT&T, n.d.).

FTTN

Fiber-to-the-node is a hybrid form of DSL often times referred to as VDSL (very high bit-rate DSL). This service, used for services such as Verizon's FiOS and AT&T's U-Verse, offers speeds up to 50 Mb/s downstream and 20 Mb/s upstream. This system employs a fiber optic cable that runs from the central office to a node in individual neighborhoods. The neighborhood node is a junction box that contains a VDSL gateway that converts the digital signal on the fiber optic network to a signal that is carried on ordinary copper wires to the residence.

Verizon's FiOS service offers speeds from 15 Mb/s downstream and 5 Mb/s upstream for $49.99 a month to 500 Mb/s downstream and 100 Mb/s upstream for $299.99 per month (Verizon, n.d.).

FTTH

Fiber-to-the-home employs fiber optic networks all the way to the home. Fiber optic cables have the advantage of being extremely fast (speeds up to 1 Gb/s) and are the backbone of both cable and telecommunications networks. Extending these networks to the home is still somewhat rare due to cost constraints. However, costs are coming down, and at least one Internet company made a metropolitan area a high speed Internet guinea pig.

In February 2012 Google announced that it would begin its test of high speed FTTH by wiring the twin cities of Kansas City, Kansas and Kansas City, Missouri with fiber optic cabling and networks cable of generating speeds up to 1 Gb/s (Kansas City is Fiber-Ready, 2012). As of early 2014 Google is planning to expand this service to 34 other cities across the country (Exploring new cities, 2014).

Cable Modem

Cable television providers also offer Internet service. In their systems, a customer's Internet service can come into the home on the same cable that provides cable television service and for some, telephone service.

With the upgrade to hybrid fiber/coaxial cable networks, cable television operators began offering broadband Internet access. But how can the same cable that supplies your cable television signals also have enough bandwidth to also supply high speed Internet access? They can do this because it is possible to fit the download data (the data going from the Internet to the home) into the 6 MHz bandwidth space of a single television channel. The upload speed (the data going from the computer back to the Internet) requires only about 2 MHz of bandwidth space.

In the case of Internet through a cable service, the signal travels to the cable headend via the cable modem termination system (CMTS). The CMST acts like the DSLAM of a DSL service. From the cable headend, the signal travels to a cable node in a given neighborhood. A coaxial cable then runs from the neighborhood node to the home.

Cable modems use a standard called data over cable service interface specifications (DOCSIS). First generation DOCSIS 1.0 , which was used with first-generation hybrid fiber/coax networks, was capable of providing bandwidth between 320 Kb/s and 10 Mb/s. DOCSIS 2.0 raised that bandwidth to up to 30 Mb/s (DOCSIS, n.d.). DOCSIS 3.0 provides bandwidth well in excess of 100 Mb/s. In fact, some modem chipsets can bond up to eight downstream channels thus creating the possibility of delivering up to 320 Mb/s (Docsis 3.0, 2009).

Figure 20.1

Fixed (Wired) Broadband Subscriptions, by Technology, June 2013

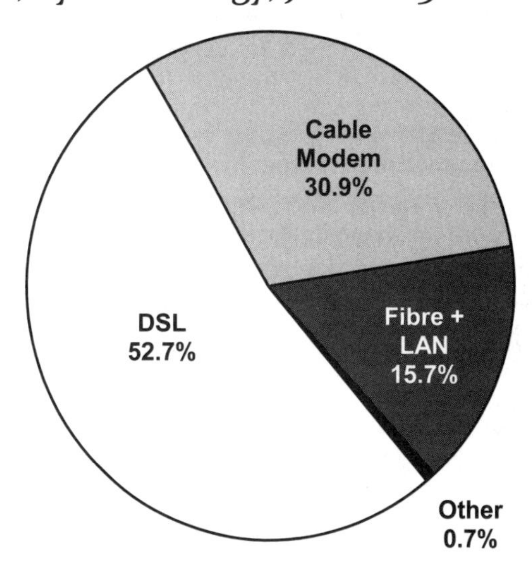

Source : OECD (2013)

A DOCSIS 3.1 modem is currently in the works that would allow for download speeds approaching 10 Gb/s. The offerings from ISPs are not expected to catch up to that speed in the near future, however (Baumgartner, 2013). According to the Organisation for Economic Co-operation and Development (OECD), approximately 17% of U.S. Internet households receive their service from cable providers (see Figure 20.1).

More and more cable Internet providers are moving toward FTTN systems (e.g. Verison's FiOS and Comcast's XFINITY). As such it is becoming increasingly difficult to accurately compare costs and speeds of traditional cable modem Internet access. However, Internet services provided by one of the nation's largest ISPs, Comcast, range in cost from about $50 per month for 6 Mb/s download speeds to up to $90 per month for 105 Mb/s service. Prices for cable broadband service are usually lower when bundled with other services (XFINITY, n.d.).

Although cable Internet provides fast speeds and, arguably, reasonable rates, this service is not without problems. Unlike DSL, cable Internet users share bandwidth. This means that the useable speed of individual subscribers varies depending upon the number of simultaneous users in their neighborhood.

Satellite

For those people who live out of DSL's reach and in rural areas without cable, satellite broadband Internet access is an option. With this service, a modem is connected to a small satellite dish which then communicates with the service providers' satellite. That satellite, in turn, directs the data to a provider center that has a high-capacity connection to the Internet.

Satellite Internet service cannot deliver the bandwidth of cable or DSL, but speeds are a great improvement over dial-up. For example, HughesNet offers home service with 5 Mb/s download and 1 Mb/s upload for $39.99 a month. Its highest speed service for the residential market, Power Max, offers 15 Mb/s download and 2 Mb/s upload for $129.99 a month (HughesNet, n.d.).

The two most popular satellite services in the U.S. are HughesNet and Wild Blue. HughesNet uses three DBS satellites on the high-power Ku band, while Wild Blue uses the Ka-band and 11 gateways located throughout the U.S. (Wild Blue, n.d).

Wireless

There are two different types of wireless broadband networks: mobile and fixed. Mobile broadband networks are offered by wireless telephony companies and employ 3G and 4G networks (discussed in more detail in Chapter 21). Second generation (2G) mobile broadband networks generally use the Enhanced Data GSM Environment (EDGE) protocol (some refer to this as 2.75G because it is better than traditional 2G, but not quite at the level of true 3G networks). Third generation (3G) networks generally use Evolution, Data Optimized (EVDO), or High-Speed Uplink Packet Access (HSUPA). Fourth generation (4G) networks use Long Term Evolution (LTE).

In January 2012 the International Telecommunication Union agreed on specifications for "IMT-Advanced" mobile wireless technologies (this includes what is commonly referred to as LTE-Advanced in the U.S.). It is this technological standard that is employed in 4G mobile broadband networks and is touted as being "at least 100 times faster than today's 3G smartphones" (ITU, 2012b, para. 5).

Fixed broadband wireless networks use either Wi-Fi or WiMAX. Wi-Fi uses a group of standards in the IEEE 802.11 group to provide short-range, wireless Internet access to a range of devices such as laptops, cellphones, and tablets. Wi-Fi "hotspots" can be found in many public and private locations. Some businesses and municipalities provide Wi-Fi access for free. Other places charge a fee.

WiMAX, which stands for worldwide interoperability for microwave access, is also known as IEEE 802.16. There are two versions: a fixed point-to-multipoint version and a mobile version. Unlike Wi-Fi which has a range of 100 to 300 feet, WiMAX can provide wireless access up to 30 miles for fixed stations and three to ten miles for mobile stations (What is wimax, n.d.). As of early 2012, Clear, now part of Sprint, provides WiMAX service in selected locations throughout the United States, but service availability is still modest.

BPL

Broadband over power line (BPL) was, at one time, thought to be the wave of the future. Given that power lines went into every home it was easy to understand how convenient it would be to use these cables to send broadband data into homes. The modem would actually be plugged into an electrical outlet in the subscriber's home as a means of obtaining the service. However, several factors have caused this technology to lose its appeal. BPL is quite susceptible to interference from radio frequencies, and other broadband services (listed above) provide a faster and more reliable connection. Today this technology, at least as it applies to bring Internet into the home, is pretty much obsolete. The use of this technology within the home, however, is still ongoing (Courtney, 2013).

Home Networks

Computer networking was, at one time, only found in large organizations such as businesses, government offices, and schools. The complexity and cost of such networking facilities was beyond the scope of most home computer owners. At one time, a computer network required the use of an Ethernet network and expensive wiring called "Cat" (category) 5. Additionally, a server, hub, and router were needed. And all of this required someone in the household to have computer networking expertise as network maintenance was regularly needed.

Several factors changed the environment to allow home networks to take off: broadband Internet access, multiple computer households, and new, networked consumer devices and services. Because of these advances, a router (costing as little as $30) can be quickly installed. This router essentially splits the incoming Internet signal and sends it (either through a wired or wireless connection) to other equipment in the house. Computers, cellphones, televisions, DVD players, stereo receivers, and other devices can be included within the home network.

With a home network users can, among other things, send video files from their computer to their television; they can send audio files from their computer to their stereo receiver; they can send a print job from their cellphone to their printer; or with some additional equipment, they could use their cellphone to control home lighting and other electrical devices within the home.

There are two broad types of home networks:

- Wired networks (including ethernet, phone lines, and power lines)
- Wireless networks (including Wi-Fi, Bluetooth, and Zigbee)

When discussing each type of home network, it is important to consider the transmission rate, or speed, of the network. Regular file sharing and low-bandwidth applications such as home control may require a speed of 1 Mb/s or less. The MPEG-2 digital video and audio from DBS services requires a speed of 3 Mb/s, DVD-quality video requires between 3 Mb/s and 8 Mb/s, and compressed high-definition television (HDTV) requires around 20 Mb/s. Not all ISPs provide these higher speeds and not all home networking technology supports the transmission of these higher speeds.

Wired Networks

Traditional networks use Ethernet, which has a data transmission rate of 10 Mb/s to 100 Mb/s. There is also Gigabit Ethernet, used mostly in business, that has transmission speeds up to 1 Gb/s. Ethernet is the kind of networking commonly found in offices and universities. As discussed earlier, traditional Ethernet has not been popular for home networking because it is expensive to install and maintain and difficult to use. To direct the data, the network must have a server, hub, and router. Each device on the network must be connected, and many computers and devices require add-on devices to enable them to work with Ethernet. Thus, despite the speed of this kind of network, its expense and complicated nature make it somewhat unpopular in the home networking market, except among those who build and maintain these networks at the office.

Many new housing developments come with "structured wiring" that includes wiring for home networks, home theatre systems, and other digital data networking services such as utility management and security. One of the popular features of structured wiring is home automation including the ability to unlock doors or adjust the temperature or lights. New homes represent a small fraction of the potential market for home networking services and equipment, so manufacturers have turned their attention to solutions for existing homes. These solutions almost always are based upon "no new wires" networking solutions that use existing phone lines or power lines, or they are wireless.

Phone lines are ideal for home networking. This technology uses the existing random tree wiring typically found in homes and runs over regular telephone wire—there is no need for Cat 5 wiring. The technology uses frequency division multiplexing (FDM) to allow data to travel through the phone line without interfering with regular telephone calls or DSL service. There is no interference because each service is assigned a different frequency.

The Home Phone Line Networking Alliance (HomePNA) has presented several standards for phone line networking. HomePNA 1.0 boasted data transmission rates up to 1 Mb/s. It was replaced by HomePNA (HPNA) 2.0, which boasts data transmission rates up to 10 Mb/s and is backward-compatible with HPNA 1.0. HomePNA 3.1 provides data rates up to 320 Mb/s and operates over phone wires and coaxial cables, which makes it a solution to deliver video and data services (320 Mbps, n.d.).

The home's power lines can also be used to distribute a signal around the house. In this scenario devices would be plugged in to outlets in various parts of the home to send and/or receive a signal. The main advantage to this approach is that it requires no new wires. Most homes have phone jacks in numerous rooms, but all rooms have electric lines. The primary downside to this approach is the speed. Data rates generally do not exceed 350 Mb/s. Homes with older wiring can experience slower rates if it works at all.

Wireless

The most popular type of home network is wireless. Currently, there are several types of wireless home networking technologies: Wi-Fi (otherwise known as IEEE 802.11a, 802.11b 802.11g, 802.11n, and 802.11ac), Bluetooth, and wireless mesh technologies such as ZigBee. Mesh technologies are those that do not require a central control unit.

Wi-Fi, Bluetooth, and ZigBee are based on the same premise: low-frequency radio signals from the instrumentation, science, and medical (ISM) bands of spectrum are used to transmit and receive data. The ISM bands, around 2.4 GHz, not licensed by the FCC, are used mostly for microwave ovens and cordless telephones (except for 802.11a and 802.11ac, which operate at 5 GHz).

Wireless networks utilize a transceiver (combination transmitter and receiver) that is connected to a wired network or gateway (generally a router) at a fixed location. Much like cellular telephones, wireless networks use microcells to extend the connectivity range by overlapping to allow the user to roam without losing the connection (Wi-Fi Alliance, n.d.).

Wi-Fi is the most common type of wireless networking. It uses a series of similarly labeled transmission protocols (802.11a, 802.11b, 802.11g, 802.11n, and 802.11ac). Wi-Fi was originally the consumer-friendly label attached to IEEE 802.11b, the specification for wireless Ethernet. 802.11b was created in July 1999. It can transfer data up to 11 Mb/s and is supported by the Wi-Fi Alliance. A couple years later 802.11a was introduced, providing bandwidth up to 54 Mb/s. This was soon followed by the release of 802.11g, which combines the best of 802.11a and 802.11b, providing bandwidth up to 54 Mb/s. The 802.11n standard was released in 2007 and amended in 2009 and provides bandwidth over 100 Mb/s (Mitchell, n.d.). The 802.11ac standard, released in 2012 allows for speeds up to 1.3 Gb/s. However, because this standard will operate only in the 5GHz frequencies, the transmission range of this Wi-Fi standard could be smaller than that of 802.11n Wi-Fi (Vaughan-Nichols, 2012).

There are two other emerging standards as well. Currently, the 802.11ad standard operates in the 60GHz frequencies with throughput speeds up to 7 Gb/s (Poole, n.d.a). Another standard is also in the developmental stages. The 802.11af or "White-Fi" standard would utilize low power systems working within "white space" (the unused frequency spectrum space between television signals). There are two primary benefits to this approach. Using frequencies in this portion of the spectrum would allow for greater coverage areas. Additionally, this standard could accommodate greater bandwidth. While this standard is still being developed, it does look promising (Poole, n.d.b).

Because wireless networks use so much of their available bandwidth for coordination among the devices on the network, it is difficult to compare the rated speeds of these networks with the rated speeds of wired networks. For example, 802.11b is rated at 11 Mb/s, but the actual throughput (the amount of data that can be effectively transmitted) is only about 6 Mb/s. Similarly, 802.11g's rated speed of 54 Mb/s yields a data throughput of only about 25 Mb/s (802.11 wireless, 2004). Tests of 802.11n have confirmed speeds from 100 Mb/s to 140 Mb/s (Haskin, 2007). Actual speed of the 802.11ac protocol are expected to top out at about 800 Mb/s (Marshall, 2012).

Security is an issue with any network. Wi-Fi uses two types of encryption: WEP (Wired Equivalent Privacy) and WPA (Wi-Fi Protected Access). WEP has security flaws and is easily hacked. WPA fixes those flaws in WEP and uses a 128-bit encryption. There are two versions: WPA-Personal that uses a password and WPA-Enterprise that uses a server to verify network users (Wi-Fi Alliance, n.d.). WPA2 is an upgrade to WPA and is now required of all Wi-Fi Alliance certified products (WPA2, n.d.).

While Wi-Fi can transmit data up to 140 Mb/s for up to 150 feet (depending upon which protocol), Bluetooth was developed for short-range communication at a data rate of up to 3 Mb/s and is geared primarily toward voice and data applications. Bluetooth technologies are good for transmitting data up to 10 meters. Bluetooth technology is built into devices such as laptop computers, music players (including car stereo systems), and cellphones.

Bluetooth-enhanced devices can communicate with each other and create an ad hoc network. The technology works with and enhances other networking technologies. Bluetooth 4.0 is the current standard and is slowly making its way into more and more consumer products. Bluetooth 4.0 is backwards compatible so there should be no communication issues between new and old devices. The main advantage of Bluetooth 4.0 is that it requires less power to run thus making it useable in more and more (and smaller and smaller) devices (Lee, 2011).

ZigBee, also known as IEEE 802.15.4, is classified, along with Bluetooth, as a technology for wireless personal area networks (WPANs). Like Bluetooth 4.0 ZigBee's transmission standard uses little power. It uses the 2.4 GHz radio frequency to deliver data in numerous home and commercial devices (ZigBee, n.d.).

Usually, a home network will involve not just one of the technologies discussed above, but several. It is not unusual for a home network to be configured for HPNA, Wi-Fi, and even traditional Ethernet. Table 20.1 compares each of the home networking technologies discussed in this section.

Table 20.1

Comparison of Home Networking Technologies

Protocol	How it Works	Standard(s)	Specifications
Ethernet	Uses Cat 5, 5e, 6, 6a, or 7 wiring with a server and hub to direct traffic	IEEE 802.3xx IEEE 802.3.1	10 Mb/s to 10 Gb/s
HomePNA	Uses existing phone lines and OFDM	HPNA 1.0 HPNA 2.0 HPNA 3.0 HPNA 3.1	1.0, up to 1 Mb/s 2.0, 10 Mb/s 3.0, 128 Mb/s 3.1, 320 Mb/s
IEEE 802.11a Wi-Fi	Wireless. Uses electro-magnetic radio signals to transmit between access point and users.	IEEE 802.11a 5 GHz	Up to 54 Mb/s
IEEE 802.11b Wi-Fi	Wireless. Uses electro-magnetic radio signals to transmit between access point and users.	IEEE 802.11b 2.4 GHz	Up to 11 Mb/s
IEEE 802.11g Wi-Fi	Wireless. Uses electro-magnetic radio signals to transmit between access point and users.	IEEE 802.11g 2.4 GHz	Up to 54 Mb/s
IEEE 802.11n Wi-Fi	Wireless. Uses electro-magnetic radio signals to transmit between access point and users.	IEEE 802.11n 2.4 GHz	Up to 140 Mb/s
IEEE 802.11ac Wi-Fi	Wireless. Uses electro-magnetic radio signals to transmit between access point and users.	IEEE 802.11ac 5 GHz	Up to 1.3 Gb/s
IEEE 802.11ad Wi-Fi	Wireless. Uses electro-magnetic radio signals to transmit between access point and users.	IEEE 802.11ad 60 GHz	Up to 7 Gb/s
Bluetooth	Wireless.	v. 1.0 (2.4 GHz) v. 2.0 + (EDR) v. 3.0 (802.11) v. 4.0 (802.11)	v. 1.0 (1 Mb/s) v. 2.0 (3 Mb/s) v. 3. 0 (24 Mb/s) v. 4.0 (24 Mb/s + lower power)
Powerline	Uses existing power lines in home.	HomePlug v1.0 HomePlug AV HPCC HomePlug AV2	v. 1 (Up to 14 Mb/s) AV (Up to 200 Mb/s) AV2 (Up to 500 Mb/s)
ZigBee	Wireless. Uses Electro-magnet radio signals to transmit between access point and users	IEEE 802.15.4	250 Kb/s
Z-wave	Uses 908 MHz 2-way RF	Proprietary	100 Kbps

Source: J. Lombardi and J. Meadows

Residential Gateways

The residential gateway, also known as the broadband router, is what makes the home network infinitely more useful. This is the device that allows users on a home network to share access to their broadband connection. As broadband connections become more common, the one "pipe" coming into the home will most likely carry numerous services such as the Internet, phone, and entertainment. A residential gateway seamlessly connects the home network to a broadband network so all network devices in the home can be used at the same time.

Working Together—The Home Network and Residential Gateway

A home network controlled by a residential gateway or central router allows multiple users to access a broadband connection at the same time. Household members do not have to compete for access to the Internet, printer, television content, music files, or movies. The home network allows for shared access of all controllable devices.

Technological innovations have made it possible to access computer devices through a home network in the same way as you would access the Internet. Televisions and Blu-ray DVD players regularly come configured with hardware that allows for accessing streamed audio and video content without having to funnel it through a computer. Cellphone and tablet technology is more regularly being used to access home networks remotely. With this technology it is now possible to set your DVR to record a show or to turn lights on and off without being in the home. The residential gateway or router also allows multiple computers to access the Internet at the same time. This is accomplished by creating a "virtual" IP address for each computer. The residential gateway routes different signals to appropriate devices in the home.

Home networks and residential gateways are key to what industry pundits are calling the "smart home." Although having our washing machine tell us when our clothes are done may not be a top priority for many of us, utility management, security, and enhanced telephone services are just a few of the useful applications for this technology. Before these applications can be implemented, however, two developments are necessary. First, appropriate devices for each application (appliance controls, security cameras, telephones, etc.) have to be configured to connect to one or more of the different home networking topologies (wireless, HPNA, or power line). Next, software, including user interfaces, control modules, etc., needs to be created and installed. It is easy to conceive of being able to go to a web page for your home to adjust the air conditioner, turn on the lights, or monitor the security system, but these types of services will not be widely available until consumers have proven that they are willing to pay for them.

Broadband technology can be used to improve home networks to allow for the control of home appliances, heating/cooling systems, sprinkler systems, and more. This allows for homeowners to continually and expeditiously monitor resource consumption. According to the FCC's National Broadband Plan (discussed in more depth later in this chapter), consumers who can easily monitor their own consumption are more likely to modify their usage thus eliminating or at least reducing waste.

The National Broadband Plan

On March 16, 2010, the Federal Communications Commission unveiled its National Broadband Plan. The primary purpose of the nearly 376 page plan is to "create a high-performance America—a more productive, creative, efficient America in which affordable broadband is available everywhere, and everyone has the means and skills to use valuable broadband applications" (FCC, 2010, p. 9). The plan was drafted after the U.S. Congress, in early 2009, directed the FCC to develop a plan to ensure every American has access to broadband capabilities (FCC, 2010)

The assumption, according to the report, is that the U.S. government can influence the broadband landscape in four general ways. First, the government can create policies that ensure robust competition among broadband players. The thought is that this competition will maximize consumer welfare, innovation, and investment. Second, the government can ensure efficient allocation and management of broadband assets. Third, the government can reform current service procedures that will support the launching of broadband in more affluent areas, and ensure that low-income Americans have physical and financial access to broadband technology. Fourth, the government can reform laws and policies to maximize broadband usage in areas traditionally overseen by the government such as public education, healthcare, and other government operations (FCC, 2010a).

In the National Broadband Plan, the FCC outlined six primary goals. They are:

Goal 1: At least 100 million U.S. homes should have affordable access to actual download speeds of at least 100 megabits per second and actual upload speeds of at least 50 megabits per second.

Goal 2: The United States should lead the world in mobile innovation, with the fastest and most extensive wireless networks of any nation.

Goal 3: Every American should have affordable access to robust broadband service, and the means and skills to subscribe if they so choose.

Goal 4: Every community should have affordable access to at least 1 Gb/s broadband service to anchor institutions such as schools, hospitals, and government buildings.

Goal 5: To ensure the safety of Americans, every first responder should have access to a nationwide public safety wireless network.

Goal 6: To ensure that America leads in the clean energy economy, every American should be able to use broadband to track and manage their real-time energy consumption (FCC 2010a).

There are approximately 200 specific recommendations within the plan directed toward President Obama's administration, the U.S. Congress, or the commission itself. Nonetheless, each of these recommendations must be dealt with separately (Gross, 2010).

Many consider the 2013-2014 Congress to be one of the most inactive in history. It should, therefore, be no surprise that there has been little movement in addressing the primary goals of the National Broadband Plan. A 2013 Congressional Research Service report suggests only minimal progress has been made. The good news, however, is that the overall initiative is still very much within the legislative discourse. The report suggests that the U.S. is making some progress in achieving the target of connecting 100 million American homes with high speed (100 Mb/s download speed) broadband access. Only marginal progress has been noted regarding the Plan's goal of having every community having affordable access to at least 1 Gb/s broadband service. FirstNet and Smart Grid are two initiatives currently underway that should help improve and expand the nationwide public safety network and enable better tracking and managing of energy consumption (Kruger, 2013).

One area where significant growth has emerged is in mobile innovation and reach. According to Organisation for Economic Co-operation and Development (OECD) data the U.S. has seen an approximately 35% increase in mobile penetration rates since 2010.

Net Neutrality

The issue of "net neutrality" continues to circulate. Currently it is possible for Internet Service Providers to block or prioritize access to web content. The Obama administration and the FCC, however, want to prevent this from happening. The proponents of net neutrality believe this is a free speech issue. They suggest that ISPs who block or prioritize access to certain Web content can easily direct users to certain sites and away from others or increase the prices for access to certain content. Supporters of the current system believe that new net neutrality regulations would serve only to minimize investment (Bradley, 2009). On April 6, 2010 the court ruled that the FCC has only limited power over Web traffic under current law. As such, the FCC cannot tell ISPs to provide equal access to all Web content (Wyatt, 2010).

Despite this setback, the FCC remains committed to a free and open Internet and will take another crack at the net neutrality issue. In early 2014 regulators reaffirmed this commitment when they unveiled a new plan. The new plan is an attempt to accomplish virtually the same thing as the plan the U.S. Court of Appeals disliked, but with some technical differences the Commission hopes will pass muster. Regulators are concerned that large broadband players such as Verizon and Comcast could have an unfair advantage. The fear is that allowing ISPs to give preferential treatment to some content companies could stifle innovation (Wyatt, 2014).

The issue of "peering" agreements is likely to draw continued attention to the issue of net neutrality. The main fear, that ISPs will deliberately alter pass through rates of various content providers, is being realized. Netflix, one of the largest online video content providers in the U.S., is increasingly finding itself battling some of the nation's largest ISPs. Because of the increased popularity of Netflix and similar content providers, the amount of data being passed through is quickly and exponentially increasing. Internet providers, such as Comcast, that also provide video content, dislike the idea of allocating so much of their bandwidth to competing program providers. This has led to behind the scenes negotiations taking place in order to reach what are being made called "peering" agreements. In exchange for some level of compensation, an ISP will increase the pass through speed for certain content providers (Gustin, 2014). This is the type of deal Netflix made with Comcast in spring 2014.

Recent Developments

In order for the FCC to make significant inroads regarding the *National Broadband Plan* significant funds will need to be raised and more spectrum space will need to be allocated for wireless networking (or, more likely, reallocated).

In his 2014 State of the Union address President Obama reaffirmed his commitment to the *National Broadband Plan* when he called for the connection of 99% of that nation's schools to high-speed broadband service. This announcement met with quick support from the Fiber to the Home Council. The FTTH Council president, Heather Burnett Gold said "As we noted in our petition recommending this action to the Commission: these experiments will enable local creativity to identify the best options for the future, spurring innovation and job creation, AND empowering and connecting communities in areas of the country often left behind" (Brunner, 2014, para. 6).

Additionally, the FCC is reforming the universal service and intercarrier compensation systems. The "universal service fund" is basically a surcharge placed on all phone and Internet services. Intercarrier compensation is basically a fee that one carrier pays another to originate, transport, or terminate various telephony related signals. Jointly these fees generate about $4.5 billion annually. These fees will now go into a new "Connect America Fund" which is designed to expand high-speed Internet and voice service to approximately 7 million Americans in rural areas in a six-year period. Additionally, about 500,000 jobs and $50 billion in economic growth is expected (FCC, 2011a). The implementation of this is ongoing.

In terms of accessing additional spectrum space, there are two ways this can be done. The first way is to tap into the unused space that is located between allocated frequencies. This is referred to as "white space." In December 2011 the FCC announced that the Office of Engineering and Technology (OET) approved the use of a "white spaces database system." This is thought to be the first step toward using this available spectrum space. Expanding wireless services would be a primary use for this space (FCC, 2011b). In 2013 Google launched its "white space database." Through the Internet giant, people can quickly search for unused white space (Fitchard, 2013).

The second way, reclaiming space currently allocated to broadcasters, is a bit more challenging. Since the television transition from analog to digital transmissions in June 2009, there has been some unused spectrum space. Because digital television signals take up less spectrum space than analog signals, television broadcasters are generally using only part of the allocated spectrum space. However, very few broadcasters appear willing to give up their currently unused spectrum space, even if they would receive some financial compensation. Nonetheless, in February, 2012, Congress passed a bill that gives the FCC the ability to reclaim and auction spectrum space. Additionally, the legislation creates a second digital television (DTV) transition that will allow for current signals to be "repacked" (Eggerton, 2012).

Current Status

The statistics on broadband penetration vary widely. The OECD keeps track of worldwide broadband penetration. According to the International Telecommunications Union (ITU), broadband subscribers in the U.S. have gradually increased over the last few years. As Table 20.2 illustrates, an estimated 28.4% of Americans had access to broadband service by mid-2012. This is more than a 400% increase since 2002. Despite this increase, the U.S.' world standing for broadband access is considered low. The OECD has the U.S. ranked 16th in the world in broadband penetration (see Table 20.3).

Table 20.2
U.S. Broadband Penetration History

2002	2003	2004	2005	2006	2007	2008	2009	2010	2011	2012
6.9	9.5	12.6	17.2	20.0	23.1	24.7	25.3	25.6	27.5	28.4

Source: ITU (2014)

Table 20.3

Fixed (Wired) Broadband Subscribers per 100 Inhabitants, by Technology, June 2013

	Rank	DSL	Cable	Fiber/LAN	Other	Total	Total Subs
Switzerland	1	27.8	12.8	2.9	0.3	43.8	3,475,000
Netherlands	2	19.1	18.1	2.7	0.0	40.0	6,701,000
Denmark	3	21.9	11.4	7.3	0.0	39.7	2,218,925
Korea	4	4.0	9.8	23.3	0.0	37.1	18,529,845
France	5	34.0	2.4	0.6	0.0	37.0	24,210,000
Norway	6	16.4	11.5	8.7	0.0	36.6	1,836,872
Iceland	7	27.8	0.0	7.4	0.0	35.1	112,658
U.K.	8	25.5	6.8	2.5	0.0	34.9	22,069,673
Germany	9	28.4	5.7	0.3	0.1	34.5	28,289,051
Belgium	10	17.0	16.9	0.0	0.0	34.0	3,758,266
Canada	11	14.1	18.1	0.7	0.0	32.8	11,457,845
Luxembourg	12	27.4	3.3	1.7	0.1	32.7	173,533
Sweden	13	14.5	6.1	11.6	0.1	32.3	3,077,000
Finland	14	19.5	5.5	0.8	4.6	30.5	1,649,100
New Zealand	15	27.7	1.5	0.3	0.0	29.5	1,315,901
United States	16	9.9	16.8	2.3	0.3	29.3	91,342,000
Japan	17	4.0	4.7	19.1	0.0	27.8	35,494,373
Australia	18	21.0	4.1	0.5	0.0	25.6	5,836,000
Austria	18	17.2	8.0	0.4	0.0	25.6	2,168,666
Spain	20	19.7	4.6	1.0	0.0	25.3	11,681,800
OECD	Average	14.1	8.3	4.2	0.2	26.7	

Source: OECD (2013)

The United States currently ranks 16th in the world for broadband penetration with 28.4 subscribers per 100 inhabitants (up from just 6.9 in 2002). The rankings are presented in Table 20.3. The United States has the largest number of broadband subscribers with over 91 million. Fiber connections were most numerous in Korea (23.3) and Japan (19.1). The U.S. is above the OECD average penetration rates for overall broadband usage and for the usage of cable modems. However, the U.S. is below the OECD average for DSL and fiber penetration.

Approximately 321 million people worldwide subscribe to some type of broadband service (up from approximately 271 million in 2009). Worldwide DSL subscribers are the most abundant, with nearly 169 million subscribers. Approximately 99 million people have cable Internet and over 50 million have fiber (see Figure 20.2). As mentioned above, Japan and Korea lead the way in terms of fiber usage. Figure 20.2 shows in Japan, approximately 70% of all broadband subscribers utilize fiber networks. In Korea that number is about 63%. In the U.S., however, the number is 8%. The OECD global average for fiber usage is 16%

Factors to Watch

Home networking offers countless global opportunities. In Korea, for instance, it is expected that the majority of homes will be connected to advanced home networking systems by 2015. This advanced system goes beyond what has been described above. The idea here is to network all home appliances as well as traditional electronic devices. Additionally, the advanced home networking system should support telemedicine and teletherapy processes (Taegyu, 2010). The development of home networking applications will continue to escalate if the overall market share continues to expand. The international home networking device market totaled nearly $5 billion in the second half of 2012 with North America capturing 45% of the global market (Infonetics Research, 2013).

Figure 20.2
Percentage of Fiber Connections in Total Broadband Subscriptions, June 2013

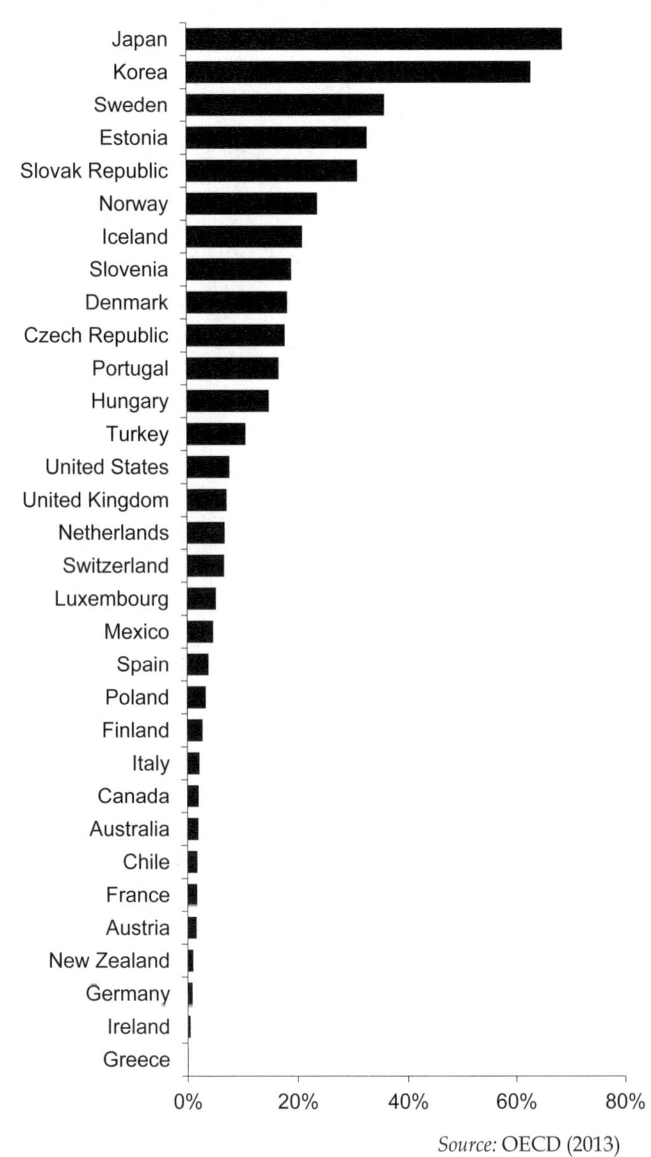

Source: OECD (2013)

Mobile broadband speeds will continue to increase. Currently, 4G LTE networks are becoming more and more prevalent. In theory, 4G can reach speeds of 100 Mb/s down and 50 Mb/s up. More realistically, however, consumers can expect speeds of 5 to 12 Mb/s downstream and 2 to 5 Mb/s upstream (Smith, 2012). Regardless, this is a significant increase over 3G speeds (generally less than 1 Mb/s down) and there is already talk of 5G technology, though it's still thought to be several years away. The primary limiting factor for advancing 5G technology is the immense electrical requirements.

Aside from continued market growth and increased speeds, there continue to be several topics that must be continually monitored. They include:

Spectrum Auctions

This issue is a long way from being fully resolved, but there has been some movement. While the FCC now has the authority to conduct auctions of unused spectrum space, they may have a difficult time finding anyone willing to give up such space (at least within the television spectrum). At this point it seems unlikely that any large market (or network owned) television stations will be willing to return any of their unused spectrum space. The resistance from broadcasters is that, while they may not be using the space now, they could very well use it in the future. Because digital transmissions take up considerably less spectrum space than analog transmissions broadcasters can send out multiple channels of content using their spectrum space. Given that the analog-to-digital transition was only completed in the middle of 2009, there has been relatively little time for broadcasters to fully utilize their spectrum space.

Where the FCC may find some willing participants is with owners of small stations. In Los Angeles, for instance, there are numerous independent stations that cater to various ethnic groups. It's possible that such stations could join forces using one station's spectrum allocation for multiple channels of programming, thus freeing up spectrum originally assigned to another station (Flint, 2012).

As of early 2014, the FCC was gearing up for its first spectrum auction since 2008. The 2014 auction will include two high-frequency bands, but other auctions of more sought after (generally lower) frequencies should take place later in the year. That spectrum will generally be in the 600 MHz band, which is currently being used by television stations. The thought is that wireless carriers will make a big push at these auctions in order to use the lower frequency spectrum to expand wireless coverage (Gryta & Nagesh, 2014).

Usage-Based Pricing

Tied to the net neutrality issue, at least indirectly, is the issue of "usage-based pricing." Usage-based pricing, where users pay based on how much data they send/receive is common with cellphone companies. However, it may become a reality with ISPs as well. Currently in the U.S. about 80% of broadband service is provided by cable companies. The increased demand for high bandwidth throughput for video services such as Netflix, along with the content competition such services provide, some cable companies are experiencing revenue drops. If net neutrality becomes a reality and ISPs are prohibited from tweaking the throughput speeds of some content providers, they may be even more inclined to explore usage-based pricing (Engebretson, 2013).

Broadband & Home Networks Visionary: Tim Berners-Lee

In 1989 **Tim Berners-Lee** invented the World Wide Web while working at CERN (the European Particle Physics laboratory in Switzerland). While at CERN, in 1980, he wrote a program called "Enquire" which he called a "memory substitute." It was basically a computer accessible notebook that allowed him to store and access notes from various locations. This was his basis for the World Wide Web.

Berners-Lee utilized two other technological advances to construct the theoretic and physical backbone what would become the World Wide Web. One technology was hypertext. This was a means of creating documents to be published in a nonlinear format. Today "hypertext" links allow Web surfers to instantly jump from document to another. The second technology was the Internet. While often used synonymously, the terms "Internet" and "World Wide Web" are not the same. The Internet is the manner in which computers from around the world are interconnected. The World Wide Web is a way or presenting, accessing, and storing content to be accessed on the Internet. It is this system that was developed by Berners-Lee.

At CERN (and certainly within other organizations around the world) people would create documents to share that they organized and formatted, but compatibility was a problem. (Even today it's often times difficult to share files among people who have different versions of the same software). But the use of a specific computer language (specifically hypertext markup language or HTML) allowed for near universal sharing of all types of documents (print, photo, audio, video, etc.).

Sir Tim Berners-Lee (he was knighted in 2004) was born in 1955 in London, England, Berners-Lee is a 1976 graduate of Oxford University. He is currently the director of the World Wide Web Consortium (W3C).

Bibliography

320 Mbps home networking specification released (n.d.). Retrieved,from HomePna website: http://www.homepna.org/products/specifications/.

AT&T. (n.d.). Retrieved from http://www.att.com/u-verse/shop/index.jsp?wtSlotClick=1-006ASL-0-1&shopFilterId=500001&ref_from=shop&addressId=&zip=22722#fbid=5CdOI9UJfIb.

Baumgartner, J. (2013, October 24). First DOCSIS 3.1 modems will have 4-5 gbps potential. *Multichannel News*. Retrieved from http://www.multichannel.com/distribution/first-docsis-31-modems-will-have-4-5-gbps-potential/146255.

Bradley, T. (2009, September 22). Battle lines drawn in FCC net neutrality fight. *PC World*. Retrieved from http://www.pcworld.com/businesscenter/article/172391/battle_lines_drawn_in_fcc_net_neutrality_fight.html.

Brunner, C. (2014, January 24). FTTH council hails fcc order to boost deployment of fiber broadband networks in rural areas. *FTTH Council Americas*. Retrieved from http://www.ftthcouncil.org/p/bl/et/blogid=3&blogaid=254.

Courtney, M. (2013, October 15). Whatever happened to broadband over power line? *Engineering and Technology Magazine*. Retrieved from http://eandt.theiet.org/magazine/2013/10/broadband-over-power-line.cfm.

DOCSIS (n.d.): Data over cable service interface specifications. (n.d.). Retrieved from Javvin website: http://www.javvin.com/protocolDOCSIS.html.

Docsis 3.0. (2009, March 13). Retrieved from Light Reading website: http://www.lightreading.com/document.asp?doc_id=173525.

Eggerton, J. (2012, February 17). Congress approves spectrum incentive auction. *Broadcasting and Cable*. Retrieved from http://www.broadcastingcable.com/article/480721-Congress_Approves_Spectrum_Incentive_Auctions.php.

Engebretson, J. (2013, June 7). Moffett: Cablecos should impose usage-based pricing; but can they? *Telecompetitor*. Retrieved from http://www.telecompetitor.com/moffett-cablecos-should-impose-usage-based-pricing-but-can-they/.

Environmental sustainability (n.d.). *Broadband for America.* Retrieved from http://www.broadbandforamerica.com/benefits/environmental.

Exploring new cities for google fiber (2014, February 19) Retrieved from http://googleblog.blogspot.com/2014/02/exploring-new-cities-for-google-fiber.html.

FAQ (n.d.) w3.org. Retrieved from http://www.w3.org/People/Berners-Lee/FAQ.html.

Federal Communications Commission. (2008, June 12). Report and order and further notice of proposed rulemaking (WC Docket No. 07-38). Washington, DC: Federal Register.

Federal Communications Commission. (2010). The national broadband plan. Retrieved from http://download.broadband.gov/plan/national-broadband-plan.pdf.

Federal Communications Commission (2011a, November 18). FCC releases "connect America fund" order to help expand broadband, create jobs, benefit consumers. [Press Release]. Retrieved from http://www.fcc.gov/document/press-release-fcc-releases-connect-america-fund-order.

Federal Communications Commission (2011b, December 22). FCC chairman Genachowski announces approval of first television white space database and device. [Press Release]. Retrieved from http://www.fcc.gov/document/chairman-announces-approval-white-spaces-database-spectrum-bridge

Fehrenbacher, K. (2009, August 26). How high-speed broadband can fight climate change. *Benton Foundation.* Retrieved from http://www.benton.org/node/27386.

Fitchard, K. (2013, November 14). *White spaces anyone? Google opens its spectrum database to developers.* Retrieved from http://gigaom.com/2013/11/14/white-spaces-anyone-google-opens-its-spectrum-database-to-developers/.

Flint, J. (2012, February 17). FCC can auction spectrum, but will broadcasters sell? LA Times. Retrieved from http://latimesblogs.latimes.com/entertainmentnewsbuzz/2012/02/broadcast-spectrum.html.

Gross, G. (2010, March 16). FCC officially releases national broadband plan. *PC World.* Retrieved from http://www.pcworld.com/businesscenter/article/191666/fcc_officially_releases_national_broadband_plan.html.

Gryta, T. & Nagesh, G. (2014, January 21). FCC to hold major auction of wireless airwaves. *The Wall Street Journal.* Retrieved from http://online.wsj.com/news/articles/SB10001424052702304027204579335181451959904.

Gustin, S. (2014, February 19). Here's why your netflix is slowing down. *Technology and Media.* Retrieved from http://business.time.com/2014/02/19/netflix-verizon-peering/.

Haskin, D. (2007). FAQ: 802.11n wireless networking. *MacWorld.* Retrieved from http://www.macworld.com/article/57940/2007/05/80211nfaq.html.

Household download index (February 24, 2012) Retrieved from Net Index website: http://www.netindex.com/download/allcountries/.

HughesNet, (n.d.). Retrieved from http://www.hughesnet.com/?page=Plans-Pricing.

Infonetics Research (2013, May 20). MoCA heating up home networking device market; 802.11ac/n wifi broadband routers on the rise. [Press release]. Retrieved from http://www.infonetics.com/pr/2013/2H12-Home-Networking-Devices-Market-Highlights.asp.

International Telecommunication Union (2012a). *Building our networked future based on broadband.* Retrieved from http://www.itu.int/en/broadband/Pages/overview.aspx.

International Telecommunication Union (2012b, January 18). IMT-Advanced standards announced for next-generation mobile technology. [Press Release]. Retrieved from http://www.itu.int/net/pressoffice/press_releases/2012/02.aspx.

International Telecommunications Union. (2003). *Birth of broadband.* Retrieved from http://www.itu.int/osg/spu/publications/birthofbroadband/faq.html.

Kansas City is Fiber-Ready. (2012, February 6). Retrieved from http://googlefiberblog.blogspot.com/2012/02/weve-measured-utility-poles-weve.html.

Kruger, L. (2013). The national broadband plan goals: Where do we stand? [PDF file]. Available from http://www.fas.org/sgp/crs/misc/.

Lee, N. (2011, October 5). Bluetooth 4.0: What is it, and does it matter? Retrieved from http://reviews.cnet.com/8301-19512_7-20116316-233/bluetooth-4.0-what-is-it-and-does-it-matter/.

Marshall, G. (2012, February 1). 802.11ac: what you need to know. Retrieved from http://www.techradar.com/news/networking/wi-fi/802-11ac-what-you-need-to-know-1059194.

Mitchell, B. (n.d.). Wireless standards- 802.11b 802.11a 802.11g and 802.11n: The 802.11 family explained. About.com. Retrieved from http://compnetworking.about.com/cs/wireless80211/a/aa80211standard.htm.

OECD Broadband (2014, January 9). Retrieved from the Organisation for Economic Co-operation and Development website: http://www.oecd.org/sti/broadband/broadband-statistics-update.htm.

Poole, I. (n.d.a). IEEE 802.11ad microwave wi-fi/wigig tutorial. *Radio-Electronics.com.* Retrieved from http://www.radio-electronics.com/info/wireless/wi-fi/ieee-802-11ad-microwave.php.
Poole, I. (n.d.b). IEEE 802.11af white-fi. *Radio-Electronics.com.* Retrieved from http://www.radio-electronics.com/info/wireless/wi-fi/ieee-802-11af-white-fi-tv-space.php .
Smith, J. (2012, January 1). Ridiculous 4G LTE speeds hit Indianapolis for super bowl 46. *Gotta Be Mobile.* Retrieved from http://www.gottabemobile.com/2012/01/19/ridiculous-4g-lte-speeds-hit-indianapolis-for-super-bowl-46/.
Tae-gyu, K. (2010, February 7). Seamless home networking to debut in 2011. *Korea Times.* Retrieved from http://www.koreatimes.co.kr/www/news/biz/2010/02/123_60437.html.
Vaughan-Nichols, S.J. (2012, January 9). 802.11ac: Gigabit wi-fi devices will be shipping in 2012. ZDNet. Retrieved from http://www.zdnet.com/blog/networking/80211ac-gigabit-wi-fi-devices-will-be-shipping-in-2012/1867.
Verizon. (n.d.). *FiOS Internet.* Retrieved from http://www.verizon.com/home/fios-fastest-internet/.
What is wimax (n.d.). Retrieved from http://www.wimax.com/general/what-is-wimax.
Wi-Fi Alliance. (n.d.). *Wi-Fi overview.* Retrieved from http://www.wi-fi.org/OpenSection/ why_Wi-Fi.asp?TID=2.
Wild Blue. (n.d.). *About Wild Blue.* Retrieved from http://www.wildblue.com/aboutwildblue/index.jsp.
WPA2. (n.d.). Retrieved from http://www.wi-fi.org/knowledge_center/wpa2/.
Wyatt, E. (2010, April 6). U.S. court of curbs F.C.C. authority on web traffic. *The New York Times.* Retrieved from http://www.nytimes.com/2010/04/07/technology/07net.html.
Wyatt, E. (2014, February 19). FCC seeks a new path on 'net neutrality' rules. *The New York Times.* Retrieved from http://www.nytimes.com/2014/02/20/business/fcc-to-propose-new-rules-on-open-internet.html?_r=0.
XFINITY (n.d.). Retrieved from https://www.comcast.com/internet-service.html.
ZigBee. (n.d.). *ZigBee Alliance.* Retrieved from http://www.zigbee.org/About/AboutTechnology/ZigBeeTechnology.aspx.

21

Telephony

William R. Davie, Ph.D.*

Why Study Telephony?

- Telephones have become so accepted that in 1985 the US government created the Lifeline program to supply service to those in need.
- The mobile phone era has produced a global shift in telecommunications technology and interpersonal communications.
- The level of mobile and smart phone adoption around the world commands social, cultural, and economic attention.
- Mobile telephony's use of spectrum affects the dissemination of other digital media.

Introduction

Theodore Twombly is a star-crossed lover in writer/director Spike Jonze's 2013 feature film, *Her*. Unfortunately, Twombly's love interest is his smartphone's operating system named Samantha, who finds other operating systems more attractive than his human capacity. Twombly's hapless fate might exaggerate how people grow attached to their voice-operated smartphones, but the constant scanning and touching of its glass surface, the intense reliance on its apps, and the anxiety of leaving home without it are quite real. Smartphones beckon users day and night with entertainment, information, and conversation—the human sort. We text. We tweet. We email. We scan headlines. We go shopping. We pay bills. We play games. We create music libraries. We share photos and status updates, and we even on occasion make a phone call—all with the same handheld device.

Not everyone in the world owns a mobile phone, but based on the diffusion model discussed in Chapter 3 the developed world has reached enough users to enter the final phase of adoption. In the United States, the number of subscribers (according to smartphone and "feature" (cellular) phone billing records) exceeds the population size (CTIA, 2014).

Background

The telephone, whether mobile or tethered, is an essential link in the contemporary world of telecommunications. The evolution of this 19th century invention rooted in Alexander Graham Bell's vision owes much to his associates, his rivals, and certain elements of technological innovation, corporate hubris, government control, and cultural heritage. It seems apt that a Scottish inventor, who migrated to Canada, then moved to the USA in order to make Boston his home, became the telephone's patron saint for inventing the "talking telegraph," a machine that brought friends and families together over distances. Bell's dramatic race to the U.S. patent office to beat his rival inventor Elisha Gray in 1876 is evidence of his enduring legacy (AT&T, 2010a).

The company, American Telephone and Telegraph, was born on March 3, 1885 (AT&T, 2010b). Theodore Vail actually served twice as its president—during its formation until 1888, and again from 1907-1919. It was his desire to build AT&T into a centralized monopoly, which he accomplished by buying out smaller phone companies, and creating a transcontinental phone line in 1915 (John, 1999). By the time Vail retired in 1919, AT&T was reputed to own every telephone pole, telephone switch, and telephone instrument in the country. Its status as a protected

* Professor, University of Louisiana, Lafayette. Acknowledgement to Dr. Steven J. Dick for editorial assistance.

monopoly was secured under the Federal Communications Commission in 1934 (Thierer, 1994). AT&T eventually secured end-to-end ownership of the entire network including, in most cases, the actual telephone inside the home.

As technology advanced, traditional telephony was called plain old telephone service (POTS). POTS circulates human voices via copper wires that connect to an exchange center where multiple calls are rapidly switched to trunk and branch lines to reach a certain destination. This switching takes place over the Public-Switched Telephone Network (PSTN) (Livengood, Lin & Vaishnav, 2006).

Telephony evolved through its conversion to computer technology and digital processing. Early on, the switching of telephone numbers was done through conversion to computer memory, known as Storage Program Control (SPC). This conversion made switching easier by means of centralized and distributed control systems where digital microprocessors routed vast numbers of phone calls through a block or series of exchanges (Viswanathan, 1992). In the late twentieth century, telephone engineers modified the system of digital transmission to offer data and video services along with the voice telephone circuitry common to PSTN. New fiber optic glass strands replaced copper wires, and computer modems translated telephone calls into binary signals.

Birth of Mobile Phones

Wired telephones were too limiting for some people. Thomas Carter of Dallas designed his two-way radio attachment so that Texas ranchers and oil field workers could talk over distances using AT&T phones and lines. The homemade Carterfone sold about 3,500 units between 1955-66, but AT&T demanded that the FCC put a stop to it. The court decision that followed handed a win to the Carterfone, and the case precedent established the right to allow "any lawful device" to connect with phone company equipment (Lasar, 2008). This early mobile phone service, dubbed land mobile telephone, was a precursor for the cellular revolution to follow.

The mobile phone owes much to the genius of an electrical engineer working on car phones for Motorola's research and development lab. Martin Cooper viewed the "communicator" used on the TV series *Star Trek* (Time, 2007), and adopted that vision to create a completely portable phone independent of the automobile. In 1973, he demonstrated its use by calling rival engineer, Joel S. Engel, head of research at Bell Labs. This early mobile phone was clearly bulkier than Captain Kirk's communicator and was dubbed "the brick" since it weighed almost two pounds (Economist, 2009). Perfecting the mobile phone was a major challenge, but a transmitting network for relaying calls on the move was an even greater feat.

Early land-mobile telephone used a group of frequencies broadcast over a broad area, but lack of capacity limited its usefulness and kept prices high. For more conventional mobile phones, AT&T needed a network of carefully spaced transmission centers, and Bell Telephone Labs began working on that issue. Joseph Engel and Richard Frenkiel designed a cellular map that afforded growth in mobility links (Lemels, 2000). The map caught the attention of the FCC's early mobile communications engineers, but it would take another generation before cellular phones and transmission networks were widely available to most Americans. Telecom engineers were working instead toward the next generation (Oehmk, 2000). In place of broad coverage antennas cellular telephone became a service of small transmission reception areas that allowed the reuse of frequencies over a large area.

Smartphone Evolution

Cellular transition to smartphone technology began in 1993 with the introduction of the IBM Simon, and continued to evolve over the next two decades with additional features based on software applications. BellSouth customers who purchased a Simon used a touch screen to open an address book, a calculator, and a sketchpad. It was not a market success, however, only a digital harbinger.

In 1996 Nokia moved forward in this race with its 9000 model that merged mobile phone features with those of Hewlett-Packard's Personal Digital Assistant (PDA). Motorola's engineers further advanced the smart-phone competition in 2003 when they introduced the MPx200. This joint venture with Microsoft had a Windows-based operating system with a full package of AT&T wireless services such as email, instant messaging, along with other apps.

The smartphone actually is a "stack" of four technical layers (Grimmelmann, 2011). At the first level there is the app itself, which might be a social media site, a video game for amusement, a calendar for appointments, an online newspaper for current events, a music store for favorite tunes, or any square icon's program easily pressed into service. At the second tier, mobile apps rely on an **operating system** such as Android, Apple's iOS, or Windows, which are the leading ones. The third tier is the **smartphone** itself, which could be a Galaxy by Samsung, an iPhone by Apple, or a Nokia Lumia. Finally, the fourth tier is the **cellular network** that connects mobile devices to the rest of the world through networks identified by their acronyms like CDMA, GSM, EDGE, EVDO, and LTE, and are numbered according to generational technology (1G, 2G, 3G, 4G).

BlackBerry to iPhone

The Blackberry grew out of Research in Motion's (RIM) line of two-way pagers. Its initial success was credited to its compact features that included wireless email, mobile faxing, and a tiny keyboard (Connors, 2012). On city streets, groups of business professionals known as "Blackberry Jam" were fixated on their handsets nicknamed "Crackberry" to indicate their addictive quality. Competition grew fierce over the years and by 2014, Blackberry's fortunes had fallen fast with quarterly losses approaching a billion dollars (Austen, 2013). Despite the Canadian firm's efforts to regain viability with new models like the Z10, the company faced a daunting uphill fight to regain its market share and profitability.

Apple's iPhone, went on sale in 2007 at the spectacular price of $499. Rather than focus on sticker shock, Apple capitalized on its customers' loyalty and affection for the iPod that made iPhone's price tag seem more palatable (Vogelstein, 2008). Its download capacity and the apps encased in its multi-touch screen that users could press, pinch, tap, or flick without hardware buttons or styluses convinced them it would be worth it. Plus iPhone's reliance on a revolutionary operating system (iOS) with the responsive Web browser sold customers on the idea of placing this new rectangle compact of aluminum and ceramic glass in their pocket or purse and dispose of whatever cell phone they had been using before.

Recent Developments

When measured by the growth of model handset sales, operating systems, and broadband networks the global reach of smartphones achieved another pinnacle in 2013. By the end of the year, a billion smartphones had reached consumers, reflecting 54% of worldwide sales (Zekaria, 2014), which made 2013 the first years in which smartphone sales exceeded the sale of feature phones. U.S. data showed the smartphone majority grew to be about 65% of American mobile subscribers, and facilitating the market penetration of smart-phones was the growth of high-speed (4G) networks and Wi-Fi access. See Table 21.1.

Table 21.1

Cellphone and Smartphone Owners in 2014

Type of phone owned:	Cellphone	Smartphone
All Adults	90%	58%
Sex		
Men	93%	61%
Women	88%	57%
Race/Ethnicity*		
White	90%	53%
African-American	90%	59%
Hispanic	92%	61%
Age group		
18-29	98%	83%
30-49	97%	74%
50-64	88%	49%
65+	74%	19%
Education Level		
High School Grad or less	87%	44%
Some College	93%	67%
College+	93%	71%
Household Income		
Less than $30,000/yr	84%	47%
$30,000-49,999	90%	53%
$50,000-74,999	99%	61%
$75,000+	98%	81%
Community Type		
Urban	88%	64%
Suburban	92%	60%
Rural	88%	43%

Source: Pew Research Center

In terms of operating systems, Google's Android grew to 81% of the world-wide market share followed by Apple's iOS at 12.9%. Microsoft's operating system moved up to third place (3.6%) following the news of its merger with Nokia. This step placed it ahead of BlackBerry, which fell to about 2%; Bada and Symbian respectively shared only 0.3% and 0.2% of the market (Gabriel, 2013). Android's supremacy remained virtually unchallenged since its operating systems surpassed Apple's iOS and RIM's BlackBerry several years earlier.

Current Status

Ninety-eight percent of the American population has some sort of telephone service. By the end of 2012, the FCC reported that landline phones covered 96 million subscribers compared to 305 million mobile subscriptions and 42 million VoIP customers, for a total of 443 million local phone service connections (FCC, 2013). Some landline phone customers maintain they need to keep their cord phone in place for emergency 911 and call stability. They argue their landline calls reach the whole family at home, so that even voice mail is more accessible. Clearer reception; freedom from re-charging cell batteries; and ease of access to 911 service are primary reasons landline customers stubbornly hold to their tethered phones (Donahue, 2010). On the other hand, mobile subscriptions are advantageous to American adults living at or near poverty. In fact, the likelihood that a cellular only decision was made for economic reasons is roughly between 33% and 40% (Blumberg & Luke, 2013).

A national survey in 2013 illustrates the U.S. trend toward mobile only users with more than a third of American households reporting they opted to cut the landline phone cord during the previous year. See Figure 21.1. About 38% of all American households were managing to survive on cellular phone service only (Blumberg & Luke, 2013, para. 1), while U.S. households maintaining landline service had to rely on a wired infrastructure that was aging and decaying in some areas. Rather than replace aging systems, telephone executives insisted it was time to move away from POTS. However, the carrier of last resort principle requires that the telephone company serve even hard-to-reach rural areas with phone service at prevailing rates. AT&T petitioned the FCC for permission to abandon its traditional switched-circuit phone systems and complete the transition to all-Internet and wireless. AT&T predicted that could be accomplished for all but one percent (1%) of its customers by 2014.

Cellular Penetration

Mobile subscriptions worldwide will exceed the global population in 2014 and reach 7.3 billion handsets, according to the International Telecommunications Union (Pramis,, 2013). Mobile subscriptions reached 4.7 billion worldwide in 2009, and then jumped to 5.4 billion in 2010, before accelerating to 6 billion in 2011. The global numbers took a huge leap forward to 7.1 billion in 2013. More than half of the mobile subscribers in the world are found in the Asia Pacific region, and two of the most populous countries, India and China, added 300 million mobile subscriptions in 2010. Today's combined total is 1,218 million Chinese subscribers and 870.6 million in India. Thus, the top three cellular network providers in the world are China Mobile, China Unicom, and Bharti Airtel India (mobiThinking, 2013).

Figure 21.1

Percentage of Wireless-Only U.S Households

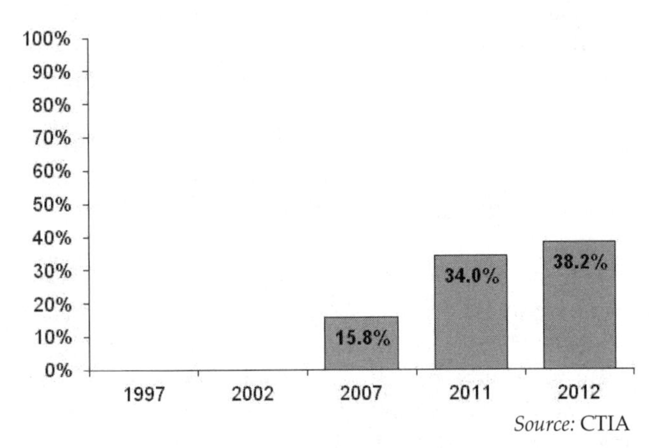

Source: CTIA

In the United States, the adoption curve for personal mobile phones ran a quick course. The early adoption rate was a mere blip at 13% of the U.S. population in 1995, but by 2013 wireless broadband subscriptions topped 89% with many Americans reporting multiple mobile phones per household (CTIA, 2011). See Figure 21.2. In sheer numbers, 33 million Americans had a mobile phone in 1995 but by 2013 subscriptions easily topped 300 million. See Figure 21.3. Pew Research data confirmed that 91% of all adults in the United States had a cell phone (Rainie, 2013). Another way to view the 20-year mobile phone adoption curve is to compare it with landline phones. It took about 90 years for landlines to reach 100 million households and become integral to American life (CTIA, 2011). In just 15 years, 47% of older teenagers felt that their mobile phone was a necessity while fewer Americans felt their landline phone was still essential (Taylor & Wang, 2010).

In 2013, smartphone penetration continued to rise in the United States with estimates pegged at 56% of the population (Smith, A., 2013). Pew Center's data showed 91% of all Americans owned some type of mobile phone. The adoption curve trends toward age and income variables. The smartphone rate for 25-34 year olds reached 81% of the population. "Younger adults—regardless of income level—are very likely to be smartphone owners. Conversely, for older adults smartphone ownership was somewhat of an "elite" phenomenon tied to education and income levels. Smartphones at the upper end of the income distribution level were common, but less so among lower income Americans (Smith, A., 2013, para. 7).

Figure 21.2

Wireless Penetration in U.S. Households

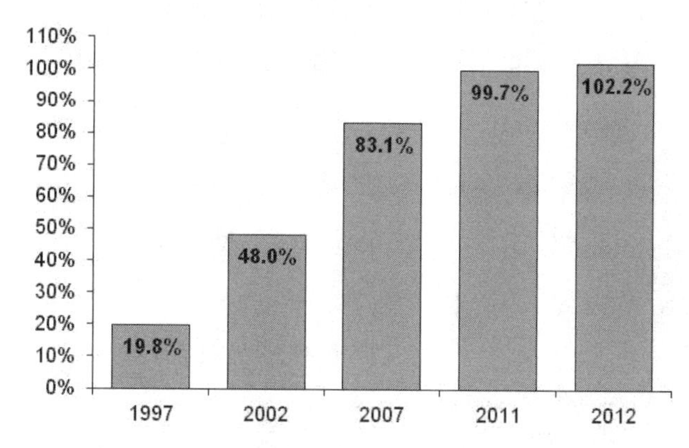

Equals number of active units divided by the total U.S. and territorial population (Puerto Rico, Guam, and the U.S. Virgin Islands

Source: CTIA

Figure 21.3

U.S. Wireless Subscriber Connections

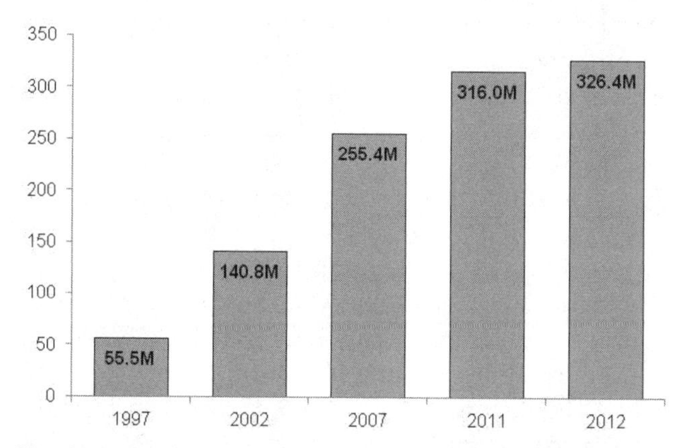

Number of active devices, including smartphones, feature phones, tablets, etc. Since users may have more than one wireless device, it is not equal to individual subscribers

Source: CTIA

The platform battle in the United States remains primarily between Android and iPhone with figures showing Android leading iPhone by 52-to-41% (Nielsen, 2012). From a global perspective 80% of smartphones shipped in 2013 were Android-based, but the mobile web traffic did not seem to add up to the same level of Android penetration (Sterling, 2014, para. 4). American smartphone users rely principally on the Android models (28%) first, and then iPhone (25%), followed by BlackBerry (4%) and Windows (1%) units, according to the Pew Center data (Rainie, 2013).

Nokia took aim at the smartphone market following its announced merger with Microsoft in 2013, which touted the Windows Phone 8 and its uses of hubs to integrate features and apps such as Facebook and Twitter. The Nokia Lumia 920 handset relied on the Windows-based operating system, and was larger than Samsung's Galaxy models that were competing with iPhone 5s and 5c. Critics lauded its spectacular touch screen, larger capacity battery, but criticized its weight and camera, which prompted Nokia to introduce the Lumia 925 at a lighter weight and with new camera technology. BlackBerry lost ground in the smartphone race. Despite a new mobile operating system and its first fully touch screen model, the BlackBerry Z10 lagged behind its competition (Scott, J. 2013).

The worldwide smartphone competition in 2014 was seen by some as a contest between Apple's iPhone 5s and Samsung's Galaxy's S5 (Huffington Post, 2014), which focused on screen sizes, operating systems, and overall design plus special features. Samsung's 13-megapixel lens played into the hands of Galaxy *selfie* takers, and took center stage at the Academy Awards and White House sports presentations. Apple's iPhone 5s and 5c boasted a larger aperture and bolder pixels with "True Tone" dual flashes also with the *selfie* crowd in mind. One sour bite of the Apple came from China following a long-awaited distribution agreement between iPhone and China Mobile. Sluggish sales followed iPhone 5s and iPhone 5c's Chinese debut in 2013. Prior smartphone campaigns from Samsung, Lenovo, Coolpad and Huawei had sewn up a significant share of China's huge mobile market (Pfanner, 2014).

4G Networks

The days are over when cellular networks advertised 3G systems for mobile customers, but the fact is that 3G connections are often used as backups for overloaded 4G systems. Third-generation formats include Verizon and Sprint's EVDO (Evolution Data

Optimized) and HSPA (High Speed Packet Access), the format used by AT&T and T-Mobile. Even though Verizon and Sprint opted for EVDO mobile broadband, HSPA proponents preferred its network efficiency and capacity for handling peak data loads (Goldstein, 2013).

What constitutes a 4G technology depends on the standards set by the International Telecommunications Union (ITU), which lists peak speed requirements of 100 megabits per second for high-mobility communication and 1 gigabit per second for relatively stationary users (ITU, 2008). In the United States, Verizon Wireless promoted its LTE network that was launched in 2010 as the leader in 4G technology using 700 MHz bandwidth. AT&T started its rollout of LTE almost a year after Verizon did but boasted multiple networks to reach its customers. The aim of Sprint is to give customers "better signal strength, faster data speeds, expanded coverage, and better in-building performance" as part of the LTE experience (Sprint, 2012). T-Mobile has a smaller footprint than either Verizon or AT&T, but has impressive 4G speeds (Fitchard, 2014). It should be noted that the 4G networks in the United States do not meet the standards set by the ITU mentioned above.

Factors to Watch

More people are fascinated by their smartphone than ever before—regularly texting friends, downloading apps, and sharing photo—but the demand for bandwidth is of greatest concern. See Figure 21.4 for examples of applications. Smartphone manufacturers and software programmers—those who design new models and make new apps—show tidy profits. But it is actually the cellular networks that support billions of mobile users, and those networks are paying more to build and maintain 4G networks than ever before. Unless the business model changes, cellular providers predict a data crunch by 2015 (mobiThinking, 2012). The need for spectrum is that pressing.

Based on a White House initiative, the FCC identified broadcasters as "inefficient" spectrum users and began to coax them to give up their channels in exchange for a share of the proceeds from the auction of frequency bandwidth. The Obama administration first announced plans to secure at least 500 MHz of additional spectrum for mobile- and fixed-wireless broadband use in 2010, but the mechanism for achieving his goal met with resistance (Obama, 2010). In 2012, Congress approved a 600 MHz spectrum auction, and recommended 2014 as the year for holding it under the auspices of the Federal Communications Commission. A new FCC chair took office in 2013, and that auction was postponed until 2015 in order to work out the details (Reardon, 2013).

Figure 21.4

Percentage of Cellphone Owners Who Use Their Cell Phone to...

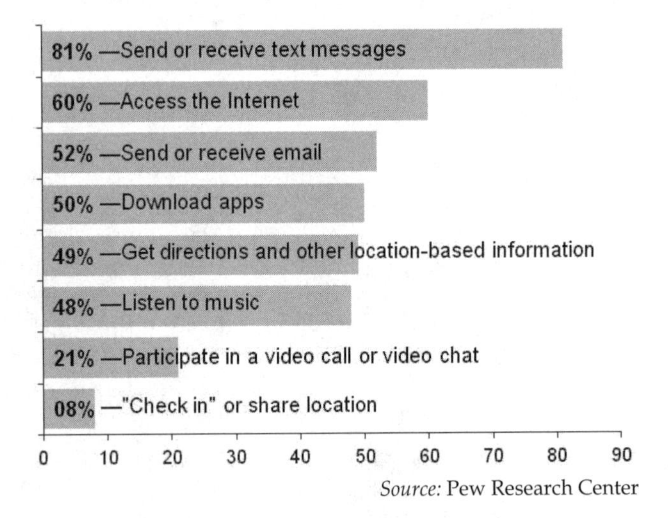

Source: Pew Research Center

Voice Over Internet Protocol (VoIP)

A new generation of telephone systems have taken advantage of increased data rates combined with a relatively old Internet service. VoIP calls either come directly from a computer, a VoIP phone, or a phone adaptor that makes the conversion from analog to digital Internet transmission. What these services have in common is the capacity to convert the human voice into a digital signal for Internet relay, thus eliminating the cost of PSTN transmission. Credit for pioneering VoIP is given to two electrical engineers, Alon Cohen and Lior Haramaty, co-founders of VocalTec Inc. (Marcial, 2000). Voice-over Internet Protocol was recommended by the Federal Communications Commission as a "cheap and easy way to make international calls" through such popular services as Skype (DeLaTorre, 2010). A growing group of cable cutters have abandoned traditional cable television and telephone services for IPTV and VoIP. While the trend started in POTS, higher mobile bandwidth has caused some mobile users to migrate to VoIP.

Peer-to-peer Teleconferencing

Sky-peer-to-peer or Skype is the brainchild of Swedish inventors Niklas Zennstrom and Janus Friis Skype in 2003. Skype started as an early voice over Internet protocol (VoIP) service Its registered users could take and receive calls via computer using a microphone and webcam to reach distant locales in the United States, the United Kingdom, and select countries in Europe and Asia. Skype aggressively marketed its VoIP applications for cellular phones, as well as its instant messaging service (Thomann, 2006). While low cost phone calls remains an important part of its business. Skype's video capability represents a new type of peer-to-peer teleconferencing service. Some new services are built as a video component to social networks (Google Hangouts, Line), while others are directed more toward business applications (Go-to-Meeting).

Privacy & Safety

Key factors to watch for in the future involve mobile phone privacy and personal safety. Mobile phone tracking does not require the active use of the handset to locate a person. Locating an individual is easily done by checking the signal strength of the mobile unit at the closest antenna as it sends out a signal to local base stations several times a minute, but some agencies go even further to get GPS data that is more precise. This form of surveillance allows law enforcement authorities to follow criminal suspects or innocent bystanders without notifying them once they become the targets of police surveillance. The practice is so common that companies such as Sprint provide law enforcement agencies manuals showing them what customer cell phone data they have and how much it will cost the police officers to obtain that data (ACLU, 2012).

Distracted driving is a leading cause of death according to the National Transportation Safety Board, which estimates than 3,000 fatalities in the United States each year can be attributed to either texting, mobile phone talking, or related distractions. The CDC reported that daily in the United States, an average of nine people are killed and more than a thousand injured in crashes involving a distracted driver. Forty-two states now ban text messaging for all drivers, and as of 2013, twelve states now prohibit all drivers from using handheld mobile phones, although hands-free mobile phones are still permitted (GHSA, 2013).

Telephony Visionary: Sir Jony Ive

OPEN BOX **Sir Jony Ive**: Apple Design's British Guru. When the world woke up on Jan. 9, 2007 to Steve Jobs' announcement that Apple was reinventing the phone, few realized how fast and how far that reinvention would go in changing their lives. Not even its chief designer was fully aware of what was in store for his latest creation in terms of its customers, competitors, and his legacy. This designer was to be knighted into British royalty for his genius that is part of people's lives around the world. Sir Jonathan Paul Ive was honored in 2012 as the Knight Commander of the Order of the British Empire for his service in design and enterprise largely thanks to the iPhone and other Apple creations.

Steve Jobs once referred to Sir Jonathan (Jony) Ive as his spiritual partner at Apple, and for good reason. The son of a British silversmith educated at Newcastle Polytechnic would take design credit for iconic Apple products such as the iMac, MacBook Pro, the MacBook Air, the iPod, the iPad, and of course, the iPhone. While still a student in England, Ive discovered Apple's MacIntosh computer, which was a personal revelation. Rather than feeling like a digital idiot, Ive found the user-friendly features of the Mac perfectly in keeping with the principles that he had learned from Deiter Rams of Braun Inc. They included functional utility and innovation, simplicity and aesthetics, and user friendliness. If a technical device cannot be understood and embraced by its least sophisticated customers, the how can it win the market?

By 1992, Ive was working for Apple for a consultant and would soon be putting his pencil and paper to work on the forerunner of the iPad, the MessagePad 110. Within a few years, he was senior vice president of industrial design for Apple. Moving to northern California, Ive began to enjoy a great deal of operational freedom at Apple. He largely worked in secret - keeping his design creations confidential until they are ready for public debut, often as the centerpiece of one of Apple's vaunted marketing campaigns. It is not surprising that Ive's name is set to appear in 2014 on the cover of a biography that refer to him as "the genius behind Apple's greatest products" because he plans to continue designing Apple products that customers will find attractive to hold, simple to use, and built to last.

Bibliography

ACLU (2012, April 2). "ACLU Releases Cell Phone Tracking Documents From Some 200 Police Departments Nationwide." Retrieved from https://www.aclu.org/national-security/aclu-releases-cell-phone-tracking-documents-some-200-police-departments-nationwide.

AT&T. (2010a). Inventing the Telephone. *AT&T Corporate History.* Retrieved from http://www.corp.att.com/history/inventing.html.

AT&T. (2010b). Milestones in AT&T History. *AT&T Corporate History.* Retrieved from http://www.corp.att.com/history/milestones.html.

Austen, I. (2013, Sept. 27). BlackBerry's Future in Doubt, Keyboard Lovers Bemoan Their Own. *The New York Times.* Retrieved from http://222.nytimes.com/2013/09/28/technology/blackberry-loses-nearly-1-billion-in-quarter.html?action=click&module=Search®ion=searchResults%230&ve.

Blumberg, S.J., & Luke, J.V. (2013, Dec. 20). Wireless Substitution: Early Release of Estimates from the National Center for Health Statistics. *National Center for Health Statistics.* Retrieved from *http://www.cdc.gov/nchs/data/nhis/earlyrelease/wireless201312.pdf*

Connors, W. (2012, March 29). Can CEO revive blackberry? *Wall Street Journal Marketplace,* B1, B4.

CTIA. (2014). Wireless Quick Facts. CTIA—The Wireless Association. Retrieved from http://www.ctia.org/your-wireless-life/how-wireless-works/wireless-quick-facts.

CTIA. (2011). *Wireless Quick Facts.* CTIA—The Wireless Association. Retrieved from http://www.ctia.org/media/industry_info/index.cfm/AID/10323.

DeLaTorre, M. (2010). Can I make calls over the Internet from Wi-Fi hotspots? *Reboot.FCC.Gov. The Official Blog of the Federal Communications Commission.* Retrieved from http://reboot.fcc.gov/blog?entryId=524120.

Donahue, W. (2010, July 22). Landlines vs. cell phones - Is it time to cut the cord? *Tribune Newspapers.* Retrieved from http://articles.chicagotribune.com/2010-07-22/business/sc-cons-0722-save-landline-vs-cell-20100722_1_landline-cell-phones-outages.

Economist. (2009). Brain Scan—Father of the cell phone. *The Economist* website. Retrieved from http://www.economist.com/node/13725793?story_id=13725793.

FCC. (2013, November). Local Telephone Competition: Status as of December 31, 2012. *Wireline Competition Bureau Statistical Reports* Retrieved from http://hraunfoss.fcc.gov/edocs_public/attachmatch/DOC-324413A1.pdf.

Fitchard, K. (2014, Jan. 30). The State of LTE in the U.S.: How the Carriers' 4G Networks Stack Up. *Gigaom.* Retrieved from http://gigaom.com/2014/01/30/4g-vs-4g-comparing-lte-networks-in-the-us/.

Gabriel, C. (2013, Nov. 15). Windows Phone makes small gains on iOS Haptics. *Tactile Feedback.* Retrieved from http://www.rethink-wireless.com/print.asp?article_id=25123.

GHSA. (2013). Distracted driving what the research shows. *GHSA Governor's Highway Safety Association.* Retrieved from http://www.ghsa.org/html/stateinfo/laws/cellphone_laws.html; One text or call could wreck it all. Retrieved from http://www.distraction.gov/.

Goldstein, P. (2013, June 14). Sprint Changes Terms of Service to Give WiMax Customers No More Than Flexibility to Switch to LTE. *Fierce Wireless.* Retrieved from http://www.fiercewireless.com/story/sprint-changes-terms-service-give-wimax-customers-more-flexibility-switch-l/2013-06-14.

Grimmelmann, J. (2011). Owning the stack: The legal war to control the smartphone platform. *Ars technical.* Retrieved from http://arstechnica.com/tech-policy/news/2011/09/owning-the-stack-the-legal-war-for-control-of-the-smartphone-platform.ars.

Huffington Post. (2014, March 24). Samsung Galaxy S5 Vs. iPhone 5S: Which Should You Get? *Huffpost Tech United Kingdom.* Retrieved from http://www.huffingtonpost.co.uk/2014/03/24/iphone-5s-vs-galaxy-s5_n_5021226.html

ITU. (2008, March 27). International Telecommunications Union to Radiocommunication Bureau (Direct Fax N°. +41 22 730 57 85) Circular Letter 5/LCCE/27 March 2008 Administrations of Member States of the ITU and Radiocommunication Sector Members participating in the work of Working Parties 5A, 5B, 5C and 5D of Radiocommunication Study Group 5, IEEE L802.16-08/008.

John, R. (1999). Theodore N. Vail and the civic origins of universal service. *Business and Economic History,* 28:2, 71-81.

Lasar, M. (2008). *Any lawful device: 40 years after the Carterfone decision. Ars technical.* Retrieved from http://arstechnica.com/tech-policy/news/2008/06/carterfone-40-years.ars.

Lemels. (2000). LEMELS N-MIT *Inventor of the Week Archive – Cellular Technology.* Retrieved from http://web.mit.edu/invent/iow/freneng.html.

Livengood, D., Lin, J. & Vaishnav, C. (2006, May 16). Public Switched Telephone Networks: An Analysis of Emerging Networks. Engineering Systems Division, Massachusetts Institute of Technology. Retrieved from http://ocw.mit.edu/courses/engineering-systems-division/esd-342-advanced-system-architecture-spring-2006/projects/report_pstn.pdf.

Marcial, G. (2000, March 20). For a two-for-one deal, Dial VocalTec: This net telephony pioneer is that rare bird: A tech stock that's also a value play. *BusinessWeek*. Retrieved from http://www.businessweek.com/bwdaily/dnflash/mar2000/nf00321g.htm.

mobiThinking. (2012). Global mobile statistics 2012: all quality mobile marketing research, mobile Web stats, subscribers, ad revenue, usage, trends… Retrieved from http://mobithinking.com/mobile-marketing-tools/latest-mobile-stats.

Nielsen. (2012). More US Consumers choosing smartphones as Apple closes the gap on Android. Retrieved from http://blog.nielsen.com/nielsenwire/consumer/more-us-consumers-choosing-smartphones-as-apple-closes-the-gap-on-android/.

Obama, B. (2010, 28 June). Presidential Memorandum: Unleashing the wireless broadband revolution. Memorandum for the heads of executive departments and agencies. Retrieved from http://www.whitehouse.gov/the-press-office/presidential-memorandum-unleashing-wireless-broadband-.

Oehmk, T. (2000). Cell phones ruin the Opera? Meet the culprit. *New York Times - Technology*. Retrieved from http://www.nytimes.com/2000/01/06/technology/cell-phones-ruin-the-opera-meet-the-culprit.html.

Pfanner, E. (2014, Jan. 17). Apple's China Mobile iPhone debut sees mediocre reactions. *The New York Times*. Retrieved from http://www.newsobserver.com/2014/01/17/3542336/apples-china-mobile-iphone-debut.html#storylink=cpy.

Pramis, J. (2013, Feb. 28). Number of Mobile Phones to Exceed World Population by 2014. *Digital Trends*. Retrieved from http://www.digitaltrends.com/mobile/mobile-phone-world-population-2014/#!BXAXj.

Rainie, L. (2013, June 6). Cell phone ownership hits 91% of adults. *Pew Research Center FacTank News in the Numbers*. Retrieved at http://www.pewresearch.org/fact-tank/2013/06/06/cell-phone-ownership-hits-91-of-adults/.

Reardon, M. (2013, Dec. 6). FCC delays broadcast spectrum auction until 2015. *C/NET*. Retrieved from http://www.cnet.com/news/fcc-delays-broadcast-spectrum-auction-until-2015/.

Scott, J. (2013, Jan. 30). BlackBerry drops RIM brand at BB10 launch. *Computer Weekly.com*. Retrieved from http://www.computerweekly.com/news/2240177242/BlackBerry-drops-RIM-brand-at-BB10-launch).

Smith, A. (2013, June 5). Smartphone Ownership 2013. *Pew Research Center's Internet & American Life Project*. Retrieved from http://www.pewinternet.org/2013/06/05/smartphone-ownership-2013/.

Sprint. (2012, Feb. 8). Baltimore and Kansas City Sprint customers to benefit from 4G LTE and 3G enhancements in 2012 Sprint adds to the list of cities to benefit from new and improved network technology by mid-year. *Sprint. News Releases*. Retrieved from http://newsroom.sprint.com/article_display.cfm?article_id=2180 .

Sterling, G. (2014, Jan. 31). The Big Smartphone Market Share Data Roundup. *Marketing Land*. Retrieved from http://marketingland.com/big-roundup-smartphone-smart-share-numbers-72502.

Taylor, P. & Wang, W. (2010). The fading glory of the television and telephone. *Pew Research Center*. Retrieved from http://pewresearch.org/pubs/1702/luxury-necessity-television-landline-cell-phone.

Thierer, A.D. (1994). *Unnatural Monopoly: Critical Moments in the Development of the Bell System Monopoly*. The Cato Journal 14:2.

Thomann, A. (2006). *Skype – A Baltic Success Story. Corporate Communications*. Retrieved from https://infocus.creditsuisse.com/app/article/index.cfm?fuseaction=OpenArticle&aoid=163167&coid=7805&lang=EN.

Time. (2007). Best Inventions of 2007. Best Inventors – Martin Cooper - 1926. *Time Specials*. Retrieved from http://www.time.com/time/specials/2007/article/0,28804,1677329_1677708_1677825,00.html.

Viswanathan, T. (1992). *Telecommunication Switching Systems and Networks*. New Delhi: Prentice Hall of India. Retrieved from http://www.certified-easy.com/aa.php?isbn=ISBN:8120307135&name=Telecommunication_Switching_Systems_and_Networks.

Vogelstein, F. (2008). The untold story: how the iPhone blew up the wireless industry. *Wired Magazine 16:02 – Gadgets – Wireless*. Retrieved at http://www.wired.com/gadgets/wireless/magazine/16-02/ff_iphone?currentPage=1.

Zekaria, S. (2014, Jan. 7). Gartner Says Android Device Sales to Top 1 Billion in 2014. WSJ.D Tech. Retrieved from http://blogs.wsj.com/digits/2014/01/07/gartner-says-android-devices-to-pass-1-billion-in-2014/.

22

The Internet

Stephanie Bor, Ph.D.*

Why study the Internet?

- The Internet is a global phenomenon that is used by nearly 2.5 billion of people living on six different continents.
- The Internet is a constantly evolving technology that stimulates a seemingly infinite number of uses that continue to proliferate each day.

Introduction

There is no denying the significance of the Internet on human culture, as it has virtually infiltrated almost every aspect of society. The pervasiveness of this technology is illustrated by recent statistics, which reveal that nearly 2.5 billion people use the Internet throughout the globe (Miniwatts Marketing Group, 2012). Since its humble beginnings as a military project during the late 1960s, the Internet has emerged a crucial part of everyday life as people have come to rely on this technology for work, education, relationships, and entertainment.

So what exactly is this technology that has taken over our lives? According to Martin Irvine of Georgetown University, the Internet can best be understood in three components. It is "a worldwide computing system using a common means of linking hardware and transmitting digital information, a community of people using a common communication technology, and a globally distributed system of information" (DeFleur and Dennis, 2002, p. 219). It is important to note, however, that the Internet does not act alone in providing us with seemingly endless information-seeking and communication opportunities. An integral part of this technology is the World Wide Web. While the Internet is a network of computers, the World Wide Web allows users to access that network in a user-friendly way. It provides an audio-visual format and a graphical interface that is easier to use than remembering lines of computer code, allowing people the ability to browse, search, and share information among vast networks.

The impact of the Internet on its users' lives is widespread and diverse, as it influences the ways in which people understand salient issues in their lives, such as their health, government, and communities. It has disrupted traditional social conventions by changing the style of scope of communication performed by people in their interactions with friends and family, as well as with strangers. Further, it is becoming more apparent that the network structure of the Internet enhances individuals' personal autonomy by allowing people to function more effectively on their own; it is no longer necessary for people to rely on physical institutions such as banks and post offices to perform daily tasks such as paying bills and sending messages (Rainie & Wellman, 2012). The scene at your local coffee shop in the middle of the day exemplifies this point, as patrons are seen hunched behind their laptop computers using public wireless Internet to perform job functions that were once accomplished in traditional office settings.

This chapter examines Internet technology by beginning with a review of its origins and rise to popularity. Next, recent developments in online marketing, social interaction, and politics will be discussed in relation to their impact on the current state of the Internet. We will conclude by briefly highlighting several issues related to the Internet that are anticipated to receive attention in future debate and research.

* Assistant Professor, Reynolds School of Journalism, University of Nevada, Reno (Reno, Nevada)

Background

Though it is now accessible to virtually anyone who has a compatible device, the Internet began as a military project. During the Cold War, the United States government wanted to maintain a communication system that would still function if the country was attacked by missiles, and existing radio transmitters and telephone poles were disabled. The solution was to transmit information in small bits so that it could travel faster and be sent again more easily if its path were disrupted. This concept is known as packet-switching.

Many sources consider the birth of the Internet to have occurred in 1968 when the Advanced Research Projects Agency Network (ARPANET) was founded. Several universities, including UCLA and Stanford, were collaborating on military projects and needed a fast, easy way to send and receive information about those projects. Thus ARPANET became the first collection of networked computers to transfer information to and from remote locations using packet switching.

ARPANET users discovered that, in addition to sending information to each other for collaboration and research, they were also using the computer network for personal communication, so individual electronic mail (email), accounts were established. Email accounts allow users to have a personally identifiable user name, followed by the @ sign, followed by the name of the host computer system.

USENET was developed in 1976 to serve as a way for students at The University of North Carolina and Duke University to communicate through computer networks. It served as an electronic bulletin board that allowed users on the network to post thoughts on different topics through email. USENET then expanded to include other computers that were not allowed to use ARPANET.

In 1986, ARPANET was replaced by NSFNET (sponsored by the National Science Foundation) which featured upgraded high-speed, fiber-optic technology. This upgrade allowed for more bandwidth and faster network connections because the network was connected to supercomputers throughout the country. This technology is what we now refer to as the modern-day Internet. The general public could now access the Internet through Internet service providers (ISPs) such as America Online, CompuServe, and Prodigy. Every computer and server on the Internet was assigned a unique IP (Internet Protocol) address that consisted of a series of numbers (for example, 290.152.74.113).

The theory of Diffusion of Innovation points out that people look for low levels of complexity in an innovation to determine whether or not they want to adopt it; in other words, they want to know how easy the innovation is to use. That qualification presented a problem for the early versions of the Internet—much of it was still being run on "text-based" commands. Even though the public could now access the Internet, they needed a more user-friendly way to receive the information it contained, and send information to others, that didn't involve learning text based commands.

In 1989, Tim Berners-Lee created a graphical interface for accessing the Internet and named his innovation the "World Wide Web." One of the key features of the World Wide Web was the concept of hyperlinks and common-language web addresses known as uniform resource locators (URLs). This innovation allows a user to simply click on a certain word or picture and automatically retrieve the information that is tied to that link. The hyperlink sends a request to a special server known as a "domain name server," the server locates the IP address of the information, and sends that back to the original computer, which then sends a request for information to that IP address. The user's computer is then able to display text, video, images, and audio that has been requested.

Today we know this as simple "point-and-click" access to information, but in 1989 it was revolutionary. Users were no longer forced to memorize codes or commands to get from one place to the next on the Internet—they could simply point to the content they wanted and access it.

It is worthwhile at this point to explain the IP address and domain name system in more detail. The domain name (e.g., google.com) is how we navigate the World Wide Web, but on the back end (which we don't see), the IP address—numbers—are the actual addresses. The Internet Corporation for Assigned Names and Numbers, or ICANN, is responsible for assigning domain names and numbers to specific websites and servers. With 1.6 billion users on the Internet, that can be quite a task (ICANN, 2010). To try and keep things simple, ICANN maintains two different sets of "top level domain" names:

generic TLD names (gTLD) such as .edu, .com, and .org, and country codes (ccTLD) such as .br for Brazil, .ca for Canada, and .ru for Russia.

IP addresses used to consist of a set of four numbers (e.g., 209.152.74.113), in a system known as IPv4. With 256 values for each number, more than four billion addresses could be designated. The problem is that these addresses have been allocated, requiring a new address system. IPv6 is the designation for these new addresses, offering 340,282,366,920,938,000,000,000,000,000,000,000,000 separate addresses (Parr, 2011). Without getting too detailed, it is doubtful that these addresses will be used up any time soon.

So what made the Internet so popular in the first place? During the late 1990s and the early 2000s, the Internet became one of the most rapidly adopted mass consumer technologies in history (see Figure 22.1). By comparison, radio took thirty-eight years to attract 50 million Americans, while the Internet took only four years to attract a comparable size audience (Rainie and Wellman, 2012). Advancements in hardware and software exist as primary factors that stimulated the widespread adoption of the Internet. Additionally, enthusiasm displayed by the U.S. federal government, which imposed minimal legal regulations on this technology, also contributed to early penetration of this technology.

Figure 22.1

Percent of U.S. Adults Who Use the Internet 1995-2014

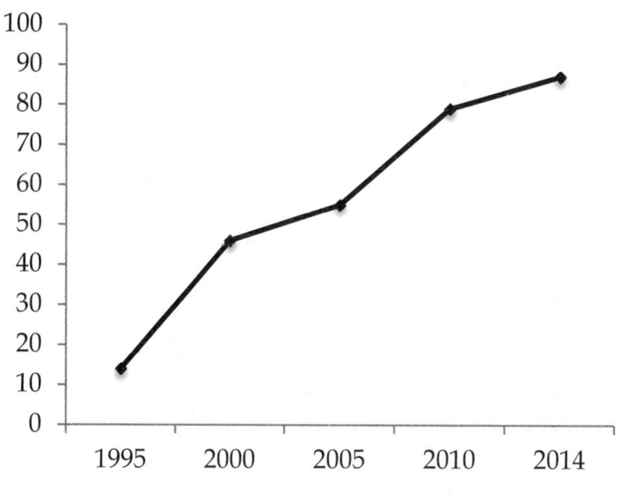

Source: Pew Research, 2014

Internet growth in the early 2000s can also be attributed to consumers' attraction to certain Web features and applications. For example, online gaming, radio, instant messaging, health-focused websites, and pornography were all highly instrumental in enticing new Internet users (Rainie & Wellman, 2012). Retail shopping was another activity that attracted new users to the Internet, as businesses quickly capitalized on markets of consumers that preferred buying products online. It is worth emphasizing that the impact of the Internet on business and commerce has been significant. A new concept—e-commerce—was created to describe any transaction completed over a computer-mediated network that involves the transfer of ownership or rights to use goods or services. For example, if you purchase a song from iTunes you are engaging in e-commerce.

E-commerce is not to be confused with e-business, which is a different term that encompasses procedures for business that are conducted over a computer-mediated network, such as ordering new materials to aid in the production of goods, as well as marketing to customers and processing their orders (Mesenbourg, 1999). E-business and e-commerce continue to constitute prominent activities that engage Internet users, as both have emerged as established fields in business school curriculum and the workforce.

Recent Developments

Computer-Mediated Relationships

The Internet continues to be increasingly important for social interactions. Especially with the rise in social networking sites, people use the Internet to cultivate new relationships and to maintain existing bonds with friends and family. Computer-mediated interpersonal communications constitute a major area field of research, which ultimately reflects conflicting evidence as to whether the Internet enhances or is harmful to relationships.

Research concludes that several Internet activities such as game playing and social media can improve online and face-to-face peer relationships (Lai and Gwung, 2013). It is also suggested that online communication can have a positive influence on adolescents' sense of identity and the quality of their friendships (Davis, 2013). In regard to the parent-child social dynamic, there is some evidence that certain activities such as watching videos online can enhance relationships (Lai and Gwung, 2013). However, the Internet has also been identified as a source of tension between parents and children, as growing concerns over safety have challenged parents to negotiate rules for monitoring and controlling their children's Internet use.

In terms of romantic relationships, research shows that the Internet can have varying levels of impact for couples in committed relationships (Lenhart & Duggan, 2014). Populations that are most influenced by the Internet included young adults ages 18 to 29, smartphone owners, and social network site users. Among these populations, the percentage of couples that said that the Internet positively impacts their relationship has declined over time, and the percentage of people negatively impacted has increased. While the Internet has undoubtedly provided more diversified channels for couples to communicate and cultivate intimacy, it also exists as a source of frustration and distraction, especially for younger adults and those in relatively new relationships.

Perhaps the most significant evidence of the influence of the Internet on romantic relationships can be recognized in the $2 billion online dating industry that consists of more than 1,400 websites devoted to helping adults meet a partner. Results from a study conducted in 2013 conclude that one in 10 Americans have used an online dating site or mobile dating application, and an even higher percentage know at least one other person who has (Smith & Duggan, 2013). Among the most popular sites are Match, eHarmony, and Plenty of Fish; however, hundreds of other niche online dating sites have emerged that target specific religions, ethnicities, and other narrow demographic categories. According to journalist Dan Slater (2013), online dating has had a profound change on society by modifying our perceptions of commitment, as well as the potential for romantic chemistry to be determined by mathematical algorithms. Research reveals that attitudes towards online dating have become more positive over time, suggesting their increased prominence in the future of Internet use.

Social Networking Sites and Politics

Social networking sites have become an essential tool for political influence. Websites such as Facebook and Twitter have provided the technological infrastructure for people to organize and activate massive political movements that have influenced significant events, such as elections and government upheavals.

In the context of electoral politics, social networking sites have become a standard tool in politicians' campaign toolboxes. They provide political candidates with an inexpensive means to spread messages and generate voter support in the months leading up to Election Day. Since their emergence during the 2008 United States election, the use of social networking sites has expanded significantly as candidates in the 2012 election maintained accounts on an array of sites such as Facebook, Twitter, Tumbler, and Instagram. Politicians used these platforms to collect information about voters such as their email addresses, geographic locations, and personal interests, which could subsequently be used to customize messages to be sent to individual voters (Bor, 2013). Evidence of political candidates' success in generating attention on social networking sites was confirmed by this research, which showed that nearly 40% of American adults engaged in civic or political activities using social networking sites during the 2012 election (Rainie, Smith, Schlozman, Brady, & Verba, 2012). These activities included posting thoughts about civic and political issues, encouraging others to act on issues and vote, or belonging to political or social groups working to advance a cause.

Beyond democratic elections, social networking sites have provided coordinating tools for political movements throughout the world. In commenting on the powerful role of social networking sites in civil society, Internet critic Clay Shirky (2010) explains, "As the communications landscape gets denser, more complex, and more participatory, the networked population is gaining greater access to information, more opportunities to engage in public speech, and an enhanced ability to undertake collective action."

Social networking sites such as Twitter and Facebook have proven to be an especially effective tool for citizens during political protests and revolutions because they can provide an efficient means for conveying warnings and updates about dramatically shifting ground events such as violent conflict and home evacuations. For more on social networking see Chapter 23.

Advancements in Online Marketing

With the ongoing changes in the online experience, the premises for Internet marketers are in constant flux. The sheer amount of traditional advertising messages on the Web has led consumers to develop an immunity against ads, causing the conversion rates for traditional advertisements to decline. Additionally, the growing focus on social connections that has been stimulated by the popularity of social networking sites has made it evident that marketing now requires a true two-way dialogue with consumers. It is not enough to write press releases or even to simply post advertisements to social networking sites. Marketers now have to identify conversations about their companies, products and competitors - and then actively engage in them. This "social listening" is part of the increased efforts to monitor and analyze the outcome of online marketing campaigns, which allows marketers to gain more information about potential customers and target their efforts to specific user groups.

A key factor in engaging with potential customers is content marketing. While the concept has been around for years, creating brand-specific content that actually has a value for potential customers has become even more valuable in times of online social media. A recent study revealed that content marketing produces three times more leads than regular marketing efforts, even outperforming paid search results on search engines (Eloqua & Kapost, 2012).

Current Status

Recent findings from the Pew Research Internet and American Life Project illustrate the remarkable growth in Internet adoption since the turn of the century (Fox & Rainie, 2014). According to their 2014 report, 87% of American adults use the Internet, and 71% of these users go online daily. While there was an increase in Internet use observed across all demographic groups, it is interesting to point out distinctions among certain user populations. For example, when comparing different age groups it is evident that younger adults are considerably more likely to use the Internet. The percentage of people between ages 18 to 29 who use the Internet reaches near-saturation at 97%; 93% of people ages 30-49 used the Internet; 88% of people ages 50-64 used the Internet; and only 57% of people over 65 years of age used the Internet.

Education level also appears to be a factor in predicting Internet usage (Fox & Rainie, 2014). More than 90% of people with at least some college education use the Internet, while only 76% of high school graduates or less use the Internet. When it comes to economic income, nearly all households (99%) with a reported income of $75,000 or more are Internet users. This percentage steadily declines as household income decreases, but still 77% of households that make less than $30,000 per year reporting using the Internet.

Other demographic categories such as gender, race, and ethnicity reflect minimal variation between groups. However, it is interesting to note that a gap still remains when comparing different ethnic groups' broadband connections at home. A survey completed in September 2013 revealed that while 74% of white, non-Hispanic Internet users have high-speed Internet at their home, only 62% of black, non-Hispanic and 56% of Hispanic adults use high-speed Internet at home (Broadband Technology Fact Sheet, 2013).

In addition to tracking user penetration statistics, it seems equally important to understand what people are actually *doing* online. According to *The Digital Future Report* (2013) that conducts an annual survey of Internet trends and issues, Internet users go online to engage in four main activities which include: 1) communication services (i.e. checking email, instant messaging, posting on message boards), 2) fact-finding, information sources, and education (i.e. looking up a definition, distance learning), 3) posting information and uploads (i.e. posting photos, uploading music videos), and 4) information gathering (seeking news, looking for health information). Additionally, over the past decade there has been a significant increase in the percentage of Internet users making online purchases. In 2013, 78% of Internet users bought something online, with clothes and travel being the most popular items purchased (see Table 22.1).

Although Internet users generally agree that this technology has positive implications for individuals and society as a whole, a study of non-users reveals that 15% of American adults still choose not to use the Internet or email (Zickuhr, 2013). Among this percentage of non-users, irrelevance and difficulty in using the technology were reported as the top two reasons for Internet avoidance.

Table 22.1
10 Most Popular Online Purchases in 2013

Item(s) Purchased Online	% of Internet users who have purchased item online
Travel	66
Clothes	66
Books	63
Gifts	60
Electronics	51
Videos/DVDs	42
Computers/peripherals	40
Software/games	37
CDs	35
Products for hobbies	34

Source: The Digital Future Report, USC Annenberg School Center for the Digital Future

Factors to Watch

Privacy and Personal Data

With the increasing sociability and personalization of the Internet, protecting privacy online has become an important topic. The vast majority of Americans agrees that safe practices on the Internet are central not only to their own, but also the nation's, safety (National Cyber Security Alliance, 2012). And as more and more routine tasks (e.g. banking, social security administration, healthcare, bill payments) move online, the importance of safely handling personal data in an online environment will only increase in years to come.

The rise of social media has made personal information (such as photos, birth dates, addresses, and phone numbers) available to third parties. Oftentimes, this information is given away willingly by the individuals who control the information, or is collected by third parties without expressed consent. And while publicly posting vacation photos on Facebook might not seem like a serious privacy and security threat, the consequences can be severe. To exemplify, researchers have been able to successfully predict individuals' social security numbers using publicly available data (such as Facebook profiles) and other over-the-counter software (Acquisti & Gross, 2009).

More than 85% of Internet users have taken steps to reduce the amount of data they make available online by setting stricter privacy settings in social networking sites, changing their browsing behaviour, or installing specific security software (Rainie, Kiesler, Kang, Madden, Duggan, Brown & Dabbish, 2013). Still, 59% of Internet users don't think it is possible to be completely anonymous online.

The Rise in Mobile Connectivity

The widespread adoption of smartphones has dramatically changed the way the Internet is used. Since 2009, the proportion of cell phone owners who use their phone to go online has doubled. And as of 2013, almost two-thirds (63%) of all cell phone owners use their phone to go online (Duggan & Smith, 2013). This shift from the stationary use of the Internet, which has been the standard for most of the Internet's history, is having a great impact on the way content is presented and consumed. Due to smaller screen sizes and different usage patterns (shorter, but re-occurring usage), the question becomes whether the information presented to mobile Internet users should replicate the regular Internet content, or if it should be an extension? The trend currently points towards a converged model, in which both worlds are closely related.

This convergence becomes especially important considering another trend in mobile Internet use: "leapfrogging." Leapfrogging refers to the process of obtaining Internet access by mobile devices only, and not through the more traditional way of a personal computer (Napoli & Obar, 2013). This is an especially salient trend among minorities and economically disadvantaged populations that often do not own a personal computer, and only use cell phones to go online. For 21% of the total cell owner population, going online has become the main use of their mobile phone (Duggan & Smith, 2013).

To conclude, the Internet is clearly a constantly evolving technology that will continue to be used by humans in new and creative ways. A survey revealed that the importance of the Internet for its users continues to increase over time, as more than half of Internet users in 2013 claimed that the Internet would be, at a minimum, "very hard" to give up

(Fox & Rainie, 2013). While its capacity to make life easier remains debatable, the Internet unarguably makes information and communication more accessible. It will be important to continuing monitoring the unanticipated outcomes of Internet use, and to analyze the influence of these behaviors on society.

The Internet Visionary: Tim Berners-Lee

Tim Berners-Lee is credited as being the inventor of the World Wide Web. As explained in this chapter, the World Wide Web is the code system that makes it simpler for people to navigate the mass network of linked computers that make up the composition of the Internet. As a British computer scientist working at the European Particle Physics Laboratory CERN in 1989, Berners-Lee wrote a paper proposing an "information management system" that would eventually become the conceptual and architectural structure for the Web. A year later in 1990. he released the code for this system free for the world.

Since its initial inception, Berners-Lee has continued to play an active role in the technical development and expansion of the Web. He is the founder of two leading Internet authorities—the World Wide Consortium (WC3) and The Web Foundation. Initially launched at Massachusetts Institute of Technology (MIT) in 1994, WC3 works with the world's leading academic institutions and software developers to determine standards for all Web infrastructure. In 2009 Berners-Lee founded the World Wide Web Foundation, which strives to oversee the spread and ethical application of the Web, and to ensure that this technology is being used throughout the world to empower humanity mediate positive change.

Berners-Lee remains an enthusiastic supporter of Internet freedom and the importance of recognizing human rights on the Web. In light of recent controversy regarding government censorship and surveillance, Berners-Lee launched the "Web We Want" initiative. This project aims to create a universal "Internet Users Bill of Rights" that would establish clear rights to Internet access and provide defined protections for personal user information. Looking to the future, Berners-Lee recognizes that society's increased reliance on the Web also raises the potential for abuse of this technology. Ultimately, he envisions that a more mature Web will continue to emerge that will ultimately empower users by giving them the tools to further enhance the personalization of their Web experiences.

Bibliography

Acquisti, A., & Gross, R. (2009). Predicting social security numbers from public data. *Proceedings of the National Academy of Sciences, 106* (27), 10975-10980.

Bor, S. (2013). Using social network sites to improve communication between political campaigns and citizens in the 2012 election. *American Behavioral Scientist,* doi:10.1177/0002764213490698.

Broadband Technologies Face Sheet. (2013). *Pew Internet and American Life Project* (Report). Pew Research Center, Washington, D.C. Retrieved from www.pewinternet.org/fact-sheets/broadband-technology-fact-sheet.

Cannarella, J., & Spechler, J. A. (2014). Epidemiological modeling of online social network dynamics. Unpublished manuscript. Department of Mechanical and Aerospace Engineering, Princeton University. Retrieved from http://arxiv.org/pdf/1401.4208v1.pdf.

Davis, K. (2013). Young people's digital lives: The impact of interpersonal relationships and digital media use on adolescents' sense of identity. *Computers in Human Behavior, 29,* 2281-2293.

DeFleur, M. L. & Dennis, E. E. (2002) *Understanding mass communication: A liberal arts perspective.* Boston: Houghton-Mifflin.

Duggan, M., & Smith, A. (2013). Cell Internet use 2013. *Pew Internet and American Life Project* (Report). Pew Research Center, Washington, D.C. Retrieved from http://www.pewinternet.org/2013/09/16/cell-internet-use-2013/.

Eloqua, & Kapost. (2012). *Content Marketing ROI: Why content marketing can become your most productive channel.* [eBook] Retrieved from http://marketeer.kapost.com/.

Fox, S. & Rainie, L. (2014). The web at 25 in the U.S. *Pew Internet and American Life Project* (Report). Pew Research Center, Washington, D.C. Retrieved from http://www.pewinternet.org/2014/02/27/the-web-at-25-in-the-u-s.

ICANN Internet Corporation for Assignment Names and Numbers (2010). Accessed from http://www.icann.org/.

Lai, C., & Gwung, H. (2013). The effect of gender and Internet usage on physical and cyber interpersonal relationships. *Computers & Education, 69,* 303-309.

Lenhart, A. & Duggan, M. (2014). Couples, the Internet, and social media. *Pew Internet and American Life Project* (Report). Pew Research Center, Washington, D.C. Retrieved from www.perinternet.org/2014/02/11/couples-the-internet-and-social-media.

Mesenbourg, T. L. (1999). *Measuring electronic business: Definitions, underlying concepts, and measurement plans.* Retrieved from http://www.census.gov/epdc/www/ebusiness.hum.

Miniwatts Marketing Group. (2012). Internet usage statistics. Retrieved from www.internetworldstats.com/stats.htm.
Napoli, P. & Obar, J. (2013). Mobile Leapfrogging and Digital Divide Policy: Assessing the Limitations of Mobile Internet Access. Fordham University Schools of Business Research Paper No. 2263800. Available at SSRN: http://ssrn.com/abstract=2263800.
National Cyber Security Alliance, 2012. (2012). NCSA / McAfee Online Safety Survey. Retrieved from http://staysafeonline.org/stay-safe-online/resources/ [Accessed October 12 2013].
Parr, B. (2011). IPv4 and IPv6: A short guide. Retrieved from http://mashable.com/2011/02/03/ipv4-ipv6-guide/.
Rainie, L., Kiesler, S., Kang, R., Madden, M., Duggan, M., Brown, S., & Dabbish, L. (2013). Anonymity, Privacy, and Security Online. *Pew Internet and American Life Project* (Report). Pew Research Center, Washington, D.C. Retrieved from http://www.pewinternet.org/2013/09/05/anonymity-privacy-and-security-online/.
Rainie, L., Smith, A., Schlozman, K. L., Brady, H. & Verba, S. (2012). Social media and political engagement. *Pew Internet and American Life Project* (Report). Pew Research Center, Washington, D.C. Retrieved from http://pewinternet.org/Reports/2012/Political-Engagement.aspx.
Rainie, L. & Wellman, B. (2012). *Networked: The new social operating system.* Cambridge, MA: MIT Press.
Slater, D. (2013). *Love in the Time of Algorithms: What Technology Does to Meeting and Mating.* New York: Penguin Group.
Shirky, C. (2010, December 20). The political power of social media. *Foreign Affairs.* Retrieved from http://www.foreignaffairs.com/articles/shirky/the-political-power-of-social-media.
Smith, A. & Duggan, M. (2013). Online dating & relationships. *Pew Internet and American Life Project* (Report). Pew Research Center, Washington, D.C. Retrieved from http://www.pewinternet.org/2013/10/21/online-dating-relationships.
The Digital Future Report. (2013). The 2013 digital future report. *USC Annenberg School Center for the Digital Future* (Report). Center for the Digital Future. Los Angeles, CA. Retrieved from www.digitalcenter.org/wp-content/uploads/2013/06/2013-Report.pdf.
Zickuhr, K. (2013). Who's not online and why. *Pew Internet and American Life Project* (Report). Pew Research Center, Washington, D.C. Retrieved from http://www.pewinternet.org/2013/09/25/whos-not-online-and-why.

23

Social Networking

Rachel Sauerbier, M.A.*

Why Study Social Networking?

- Social networking sites account for almost 27% of all time spent online—over 10% more than entertainment sites, the second most popular kind of site.
- Four of the top ten apps downloaded for smart phones last year were social media apps.
- Over 70% of adults online in the United States have at least one kind of social media profile.
- Near saturation of social media has occurred in the 18-29 age demographic, with 90% of online users in that age range having one or more social media profiles.

Introduction

Your phone vibrates and the screen lights up with a new notification. Your best friend on vacation in Spain has just posted a new Vine video to her Twitter account. Her seven second video shows a mob of people running from bulls in Pamplona. Switching over to your Facebook account, you notice this same friend has posted a few new photos of the bull run to her Instagram account, with the hashtags: #beef #itswhatsfordinner. From its humble beginnings with SixDegrees.com, to Friendster, to Facebook, Twitter, LinkedIn, Vine, Snapchat, and everything in between, the role of social networking websites has come to permeate almost every aspect of the online experience. Even with so much exposure, there is still some confusion as to what constitutes a social networking site (SNS), and which of the literally millions of Web pages on the Internet can be considered SNSs.

What is an SNS? According to boyd and Ellison (2008), there are three criteria that a website must meet to be considered an SNS. A website must allow users to "(1) construct a public or semi-public profile within a bounded system, (2) articulate a list of other users with whom they share a connection, and (3) view and traverse their list of connections and those made by others within the system" (boyd & Ellison, 2008, p. 211). These guidelines may seem to restrict what can be considered an SNS, however there are still literally hundreds of vastly diverse websites that are functioning as such.

As Facebook penetration starts to level off and Twitter continues to grow, there are numerous other SNSs that have burst onto the scene. One trend that is continuing to grow amongst successful new SNSs is the move from static text updates to dynamic multimedia updates that rely heavily on mobile telephone technology, with sites like Vine (video), Instagram (picture), Pinterest, and Snapchat (video and picture) usurping the giants like Facebook and Twitter as favorites amongst the Millennial generation. Just a year after going live, Vine boasted over 40 million users (Crook, 2014) and a year after Facebook acquired Instagram, it had more than 150 million active users every month (Hof, 2013). As of mid-2014, Pinterest had over 70 million users (Smith, 2014) while the surprising newcomer to this quartet, Snapchat, boasted 100 million unique users (Costill, 2014). As social media continues to grow and evolve to offer users a mobile, multimedia experience, the number of choices available to Internet users is as varied as the directions SNSs have gone. To understand this social media explosion, we must first understand how it began.

* Doctoral Candidate, Edward R. Murrow School of Journalism, Washington State University (Pullman, Washington)

Background

SNSs have taken on many forms during their evolution. Social networking on the Internet can trace its roots back to listservs such as CompuServe, BBS, and AOL, where people would converge to share computer files and ideas (Nickson, 2009). CompuServe was started in 1969 by Jeff Wilkins, who wanted to help streamline his father-in-law's insurance business (Banks, 2007).

During the 1960s, computers were still prohibitively expensive; so many small, private businesses could not afford a computer of their own. During that time, it was common practice to "timeshare" computers with other companies (Banks, 2007). Timesharing, in this sense, meant that there was one central computer that allowed several different companies to share access in order to remotely use it for general computing purposes. Wilkins saw the potential in this market, and with the help of two college friends, talked the board of directors at his father-in-law's insurance company into buying a computer for timesharing purposes. With this first computer, Wilkins and his two partners, Alexander Trevor and John Goltz, started up CompuServe Networks, Inc. By taking the basic concept of timesharing already in place and improving upon it, Wilkins, Trevor, and Goltz created the first centralized site for computer networking and sharing. In 1977, as home computers started to become popular, Wilkins started designing an application that would connect those home computers to the centralized CompuServe computer. The home computer owner could use the central computer for access, for storage and—most importantly—for "person-to-person communications—both public and private" (Banks, 2007).

Another two decades would go by before the first identifiable SNS would appear on the Internet. Throughout the 1980s and early 1990s, there were several different bulletin board systems (BBSs) and sites including America Online (AOL) that provided convergence points for people to meet and share online. In 1996, the first "identifiable" SNS was created—SixDegrees.com (boyd & Ellison, 2008) SixDegrees was originally based upon the concept that no two people are separated by more than six degrees of separation. The concept of the website was fairly simple—sign up, provide some personal background, and supply the e-mail addresses of ten friends, family, or colleagues. Each person had his or her own profile, could search for friends, and for the friends of friends (Caslon Analytics, 2006). It was completely free and relatively easy to use. SixDegrees shut down in 2001 after the dot com bubble popped. What was left in its wake, however, was the beginning of SNSs as they are known today. There have been literally hundreds of different SNSs that have sprung from the footprints of SixDegrees. In the decade following the demise of SixDegrees, SNSs such as Friendster, MySpace, LinkedIn, Facebook, and Twitter have become Internet zeitgeists.

Friendster was created in 2002 by a former Netscape engineer, Jonathan Abrams (Milian, 2009). The website was designed for people to create profiles that included personal information—everything from gender to birth date to favorite foods—and the ability to connect with friends that they might not otherwise be able to connect to easily. The original design of Friendster was fresh and innovative, and personal privacy was an important consideration. In order to add someone as a Friendster contact, the friend requester needed to know either the last name or the e-mail address of the requested. It was Abrams' original intention to have a website that hosted pages for close friends and family to be able to connect, not as a virtual popularity contest to see who could get the most "Friendsters" (Milian, 2009).

Shortly after the debut of Friendster, a new SNS hit the Internet, MySpace. From its inception by Tom Anderson and Chris DeWolfe in 2003, MySpace was markedly different from Friendster. While Friendster focused on making and maintaining connections with people who already knew each other, MySpace was busy turning the online social networking phenomenon into a multimedia experience. It was the first SNS to allow members to customize their profiles using HyperText Markup Language (HTML). So, instead of having "cookie-cutter" profiles like Friendster offered, MySpace users could completely adapt their profiles to their own tastes, right down to the font of the page and music playing in the background. As Nickson (2009) stated, "it looked and felt hipper than the major competitor Friendster right from the start, and it conducted a campaign of sorts in the early days to show alienated Friendster users just what they were missing" (par. 15). This competition signaled trouble for Friendster, which was slow to adapt to this new form of social networking. A stroke of good fortune for MySpace also came in the form of rumors being spread that Friendster was going to start charging fees for its services. In 2005, with 22 million users, MySpace was sold to News Corp. for $580 million (BusinessWeek, 2005); News Corp. later sold it for only $35 million. MySpace, now with approximately 36 million users, has since been toppled as the number one SNS by Facebook, which has over 1.31 billion users worldwide (Smith, 2013; "Facebook Statistics," 2014).

From its humble roots as a way for Harvard students to stay connected to one another, Facebook has come a long way. Facebook was created in 2004 by Mark Zuckerberg with the help of Dustin Moskovitz, Chris Hughes and Eduardo Saverin (Facebook, 2012). Originally, Facebook was only open to Harvard students, however, by the end of the year, it had expanded to Yale University, Columbia University, and Stanford University—the latter's hometown providing the new headquarters for the company in Palo Alto, CA. In 2005, the company started providing social networking services to anyone who had a valid e-mail address ending in .edu. By 2006, Facebook was offering its website to anyone over the age of 13 who had a valid e-mail address (Facebook, 2014). What made Facebook unique, at the time, was that it was the first SNS to offer the "news feed" on a user's home page.

In all other SNSs before Facebook, in order to see what friends were doing, the user would have to click to that friend's page. Facebook, instead, put a live feed of all changes users were posting—everything from relationship changes, to job changes, to updates of their status. In essence, Facebook made microblogging popular. This was a huge shift from MySpace, which had placed a tremendous amount of emphasis on traditional blogging, where people could type as much as they wanted to. Interestingly, up until July 2011, Facebook users were limited to 420 character status updates. Since November 2011, however, Facebook users have a staggering 63,206 characters to say what's on their mind (Protalinski, 2011). The fact still remains, however, that Facebook's original limit on characters used in status updates changed the way people were using websites for social networking and paved the way for sites such as Twitter, with its limitation of 140-character "tweets" (Dsouza, 2010).

Twitter was formed in 2006 by three employees of podcasting company Odeo, Inc.: Jack Dorsey, Evan Williams, and Biz Stone (Beaumont, 2008). It was created out of a desire to be able to stay in touch with friends easier than allowed by Facebook, MySpace, and LinkedIn. Taking the concept of the 160 character limit text messaging imposed on users, Twitter shortened the message length down to 140 (to allow the extra 20 characters to be used for a user name) (Twitter, 2014). Twitter attributes a huge amount of its initial success to its usage at festivals like Austin, Texas' South by Southwest (SXSW). Twitter had already gone live a year before, but it was at the 2007 SXSW that the SNS exploded (Mayfield, 2007). Another helpful, if not totally unexpected, promotion of Twitter came from the unusually large number of celebrities who adopted the use of Twitter fairly early—so much so that websites including followfamous.com emerged to track which celebrities use Twitter, even going so far as to have the tagline of their website as "find-follow-spy" (FollowFamous, 2009).

Not all SNSs have been used for social, personal, dating, or celebrity-stalking purposes. LinkedIn was created in 2003 by Reid Hoffman, Allen Blue, Jean-Luc Vaillant, Eric Ly, and Konstantin Guericke (LinkedIn, 2014). LinkedIn returned the concept of SNSs to its old CompuServe roots. According to Stross (2012), LinkedIn is unique because "among online networking sites, LinkedIn stands out as the specialized one—it's for professional connections only" (par. 1). So instead of helping the user find a long lost friend from high school, LinkedIn helps build professional connections, which in turn could lead to better job opportunities and more productivity. As of November of 2013, LinkedIn had over 259 million members worldwide (Marrouat, 2013) and is leading the way in the unique section of SNSs that deals strictly with business relationships.

The final giant in the world of SNSs in the western world is Google+ (aka, Google Plus). Created by the Internet giant, Google, Google+ is trying to capitalize on creating a more "realistic experience" of social interaction for its users. According to Google+ (n.d.), "Google+ makes connecting on the web more like connecting in the real world" (par. 1). Their unique approach to making the social experience online more closely match real world interactions comes from their use of social "circles," or as Google+ (n.d.) put it: "Circles make it easy to put your friends from Saturday night in one circle, your parents in another, and your boss in a circle by himself, just like real life" (par. 2). The concept of circles is to keep Google+ users from having the embarrassing experience of sharing inappropriate information about what they did Saturday night with their grandparents, alluding to the issues that have come up with "over share" mishaps that were occurring with a relatively high frequency on Facebook and MySpace (Zukerman, 2011).

In response to Google+'s circles concept, Facebook has created the option of putting different friends into different groups. Opened to the public in September 2011, Google+ had more than 300 million active users in the "stream" as of the end of October 2013 (Isaac, 2013). As it seems to be the norm every time Google releases a new set of user statistics, that number has been called into question. According to Issac (2013), "the 'stream' is more broadly defined than one would think. It also means clicking on the little red bell or share icons you see across *all* of Google's properties." (par. 4). It should be pointed out, however, that websites such as Facebook also use broad metrics when reporting their user base, just not quite as broad and creatively as Google does when reporting their users "in the stream." (Issac, 2013). One thing is for certain, Google+ has not gone gently into the good night as many reported it would after its sluggish start a few years ago.

Recent Developments

There have been two huge trends seen in relation to SNSs since 2012: multimedia social media and social networking "on the go" via mobile applications. Multimedia SNSs have taken the Internet by storm with two of the social networking giants, Facebook and Twitter, each acquiring smaller startup multimedia SNSs. Facebook purchased Instagram in 2012 from Kevin Systrom and Mike Krieger for $1 billion in cash and stock (Langer, 2013). Instagram was originally developed by the aforementioned duo in 2010 as a photo-taking application where users could apply various filters to the photographs and then upload them to the Internet. In 2011, Instagram added the ability to hashtag the photographs uploaded as a way to find both users and photographs ("Introducing hashtags," n.d.). By the time Facebook acquired the filtered photo app giant, Instagram had over 30 million users (Upbin, 2012). While many in the tech industry saw Facebook's extravagant price paid for Instagram as a bad business decision, the move has proved to be lucrative as Instagram quickly grew to more than 100 million users and has become one of the most popular social media sites on the planet.

Not to be outdone, Twitter bought the video-sharing start up Vine for $30 million in 2012 (Crunch Base, 2014). Vine was created by Dominik Hoffman, Rus Yusupov, and Colin Kroll the year before and operated as a private, invite-only application (Dave, 2013). After purchasing Vine, Twitter released the video app on Apple's app store in January 2013. By June 2013 it already had 13 million iOS users when the company released the app for Android on Google Play (Crook, 2014).

Snapchat, originally released under the name Picaboo, was created by two Stanford University students, Evan Spiegel and Robert Murphy, in 2011 (Colao, 2012). The premise of Snapchat is simple: users send friends "snaps," photographs and videos that last anywhere from one to ten seconds, and when the time expires the photos or videos disappear. In addition to the fleeting nature of the snap, if the recipient of a snap screen captures it, Snapchat will let the sender know that the person they sent it to saved it. To say that Snapchat has become popular would be an understatement. By October 2012, one billion snaps had been sent, and the app averaged more than 20 million snaps a day (Gannes, 2012). In November 2013, Snapchat turned down a $3 billion offer in cash from Facebook, as some speculate that the company could be worth more than $4 billion before 2015 (Rusli & MacMillan, 2013).

While it is obvious that the three aforementioned SNSs are centered on a more multimedia experience for the user, their popularity also signals the trend that social media is becoming increasingly mobile. All three of the apps, Instagram, Vine, and Snapchat, were designed for smart phones. Beyond the increasing popularity of SNSs designed specifically for smart phones and other mobile technologies, there is an increase in other forms of mobile apps linking up with social media. For example, there is a growing trend of fitness apps including Fitocracy, Map My Run, and Runtastic allowing users to sign-in with their Facebook, Twitter, Pinterest, Google+ accounts, to name a few and allow the SNSs to update their profiles with their fitness progress (Davidson, 2012).

From check-ins on Foursquare to posting a new radio station from Spotify, social media has become most effective when it is on the go. This trend toward fast updating on the go is signaled not only by the creation of apps tailored for mobile technology, but also by the way we disseminate and consume media. It is no coincidence that Vine videos, Instagram photos, and Snapchat snaps last less than ten seconds. These updates are multimedia forms of the Tweet, where users are limited to 140 characters or less. The days of the long post on Facebook and terrible poetry written by 16 year olds on MySpace are gone. Successful social media is the media that can be written and consumed on the go.

Though social media is moving toward a quicker, more multimedia experience, there is still plenty going on for the "veteran" SNSs including Facebook and Twitter. Facebook's stock went public with its initial public offering (IPO) on the NASDAQ stock exchange on May 18, 2012, however due to technical glitches, the company's stock was not available for public trade until 11:30 AM. Even then there was considerable confusion amongst buyers and traders as to whether or not the company's shares were in fact being sold (Strasburg & Bunge, 2012). The first day jitters and glitches that plagued the social networking giant were merely a portent of things to come. In the month following the company's initial market filing, it faced more than 40 lawsuits because of allegations that it gave preferred buyers a reduced market forecast for the stock, in addition to charges that investors lost money due to the technical glitches that the NASDAQ suffered on the first day of trading (Tsukayama, 2012). Within that first month of trading, the Facebook stock price dropped 20%, falling from its initial offering of $38 a share to around $29 a share (Tsukayama, 2012). The company and its stock would eventually rebound nicely from the controversy surrounding its IPO and as of March 2014, traded at approximately $69 a share.

Learning from his own debacle with Facebook's IPO, Mark Zuckerberg showed some social media brotherly love and gave Twitter CEO Dick Costolo some sage advice about taking the other social networking giant public. Rather than resist taking the company public as Zuckerberg did with Facebook, Zuckerberg urged that large social networking companies like Twitter take their companies public sooner rather than later to avoid confusion and complications they went through. Twitter and its board did just that and had its IPO on November 7, 2013 with an opening price of $45.10 on the New York Stock Exchange (Shefrin, 2013). Twitter's IPO suffered none of the same problems that Facebook's IPO did and came out technical glitch and lawsuit free. As of March 2014, Twitter's stock sells for $54.28 a share.

Current Status

It should come as no surprise that the use of SNSs is still on the rise. According to research conducted by the Pew Research Internet Project, the top five social networks saw an increase in adults who have added profiles to their sites from 2012 to 2013 (Duggan & Smith, 2013). More than 70% of all adults online had a Facebook profile in 2013, a 4% increase from the year before. In addition, SNSs are still the number one type of site that users spend their time on when they are online. Social media accounts for almost 27% of all time spent online, followed distantly by entertainments sites, which account for 15% of time online (Experian Marketing Services, 2013). Contrary to the popular belief of many of this author's New Communication Technology students, porn sites are not the most frequented sites online, making up a mere 3.33% of time spent on the Internet.

Figure 23.1

Percent of Adults Following Social Media Sites 2012 vs. 2013

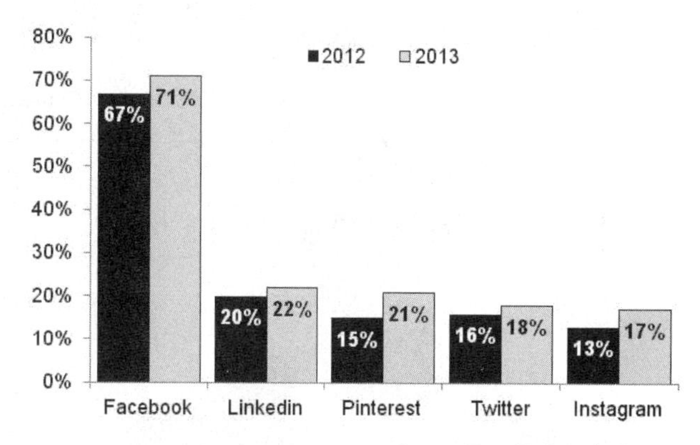

Source: Pew Research Center

As mentioned previously, one of the largest shifts seen in social media has been to mobile access. In research conducted by Nielsen in 2013, the average user spends nine hours and sixteen minutes amonth on SNSs accessed from their smart phones (Chmielewski, 2013). In comparison, smart phone users spend approximately one hour and eleven minutes a month streaming video from their smart phone (Chmielewski, 2013). These findings correlate to data released last year by Facebook, where the SNS found that over 78% of its users in the United States access the site via their smart phone (Constine, 2013). In addition, in 2013, four of the top ten mobile apps downloaded for smart phones were social media apps (Graham, 2013).

Figure 23.2

Social Networking Site Use by Age Group, 2005-2013

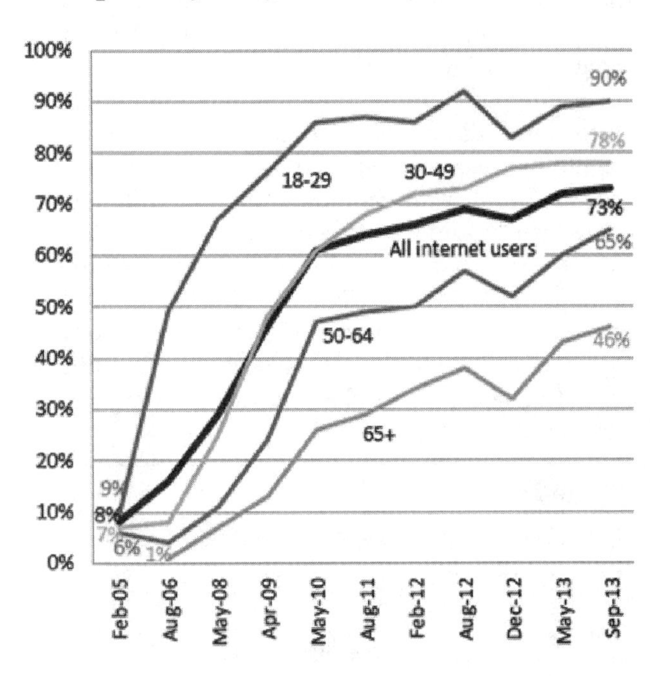

Source: Pew Research Center

Figure 23.3

Percent of Time Spent Online by Category

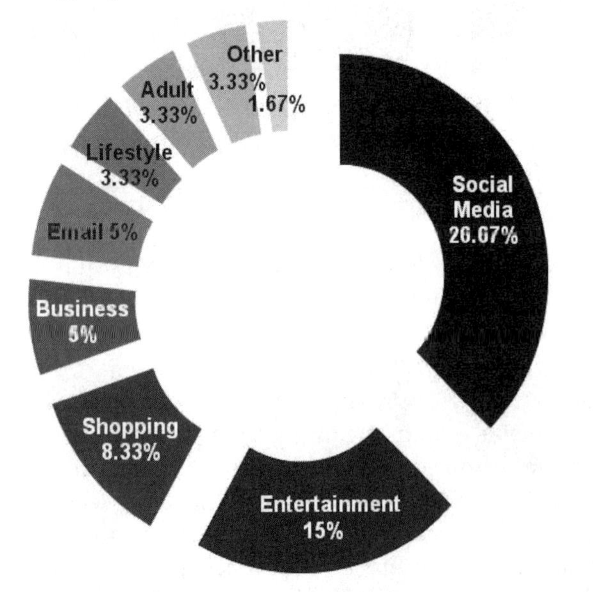

Source: Pew Research Center

Even though social media use is on the rise by most metrics, mainstream SNSs overall could be suffering from the phenomenon of the long tail, as more people are utilizing niche social networks such as Vine, Snapchat, and Dogster/Catster, the latter of which is for the dog or cat lover seeking people of the same orientation. In addition, the increase in the number of users of the older generations on sites like Facebook have lead younger generation users to flee from the site, not willing to share the virtual air with their parents, grandparents, and gym teachers. According to research conducted by the Pew Center for Internet and American life, from 2012 to 2013, there was a 35% increase in users aged 65 and over on Facebook (Risen, 2014). In contrast, during that same time period, there was a 2% decrease in Facebook users between the ages of 18 to 29 (Risen, 2014).

Factors to Watch

Expect to see more integration of pictures and video across all social network platforms. As has been evidenced by the increase in popularity of SNSs that cater to visual media, other SNSs like Facebook and Twitter will start to integrate multimedia messages into their overall design and layout. In addition to a more general increase of pictures and video, look for more multimedia advertisements on the social media giants. This is because now that Facebook and Twitter are publicly traded companies, they are constantly going to be looking for the most effective ways to monetize their sites. Facebook and Twitter have advertisements in the form of videos popping up, and a dramatic increase in their use is expected in the near future.

Also, expect to see a much more pronounced shift to mobile social media in the next couple years. As has been explored at length in this chapter (and the rest of this book), mobile computing *is* the way of the future and will continue to be so for the next several years. The trend toward mobile devices of all shapes and sizes—from the smart phone to the phablet (the unfortunate love child of the smart phone and tablet) to the tablet—allows people to access all forms of Internet media on the go. Because of this, existing SNSs will continue to gear their content and layout toward mobile technology, and more niche SNSs will pop up that cater toward the social media on-the-go experience.

Finally, good, bad, or indifferent, be prepared for an increase in the selfie. Selfies are the product of a person taking a picture of oneself with the camera on his or her phone (or with his or her tablet if he or she wants to look completely ridiculous) which is then often posted onto social media profiles. If the 2014 Academy Award ceremony is any indication, the popularity of the selfie should not be underestimated. Within one hour of being uploaded to Twitter, Ellen Degeneres' mega-selfie with a laundry list of who's-who in Hollywood completely crashed Twitter and became the most re-tweeted picture on the social media site's history (Bulik, 2014). According to Stinson (2014), there were 79 million photographs on Instagram alone labeled #selfie a short time later. This is not including the literally millions of other photographs on Instagram that have some variation of that hashtag, including #selfies and #selfiefordays (Stinson, 2014). Selfies, for many within the virtual realm, are seen as the newest form of self-expression and self-understanding as people who post them are often trying to not only present themselves to the rest of the world in a positive light, but also to get a better understanding of how they look to others in the world around them (Schulten, 2013). The selfie has permeated all aspects of pop culture—from video games like Grand Theft Auto V to television shows like *Homeland* (Schulten, 2013). As we all integrate more of the virtual realm into our everyday interactions, the selfie may become the natural stand-in for our actual faces in the real world.

Social Networking Visionary: Andrew Weinreich

Be honest. How many of you have heard of **Andrew Weinreich**? When asked to name a visionary in the world of social networking, many people will be quick to name someone like Mark Zuckerberg or Biz Stone. Some of the more savvy may name someone like Eduardo Saverin or Johnathan Abrams. However, much of what we owe to the modern iteration of the social networking site can be attributed to Andrew Weinreich. In 1997, Weinreich was a financial analyst who created the first modern social networking site, SixDegrees.com (Plymale, 2012). He based this new website on Hungarian author Frigyes Karinthy's theory that everyone in this world is linked to everyone else in this world by six degrees or less of separation. The website allowed users to list family, friends, and acquaintances and allowed those people to join, thereby creating the connections online. It was this first articulation of the social networking site that not only paved the way for most other social media sites that followed, but also lead to boyd and Ellison's (2008) definition of what is and is not a social networking site online. SixDegrees only lasted until 2001, but by that point it had made an indelible impact on the virtual realm. Weinreich would move on to other ventures, including websites like I Stand For... in 2003, which allowed for political fundraising to be transferred to online transactions, and MeetMoi, a mobile, location-based dating website. MeetMoi has had modest success, but like SixDegrees, its biggest success has been in paving the way for other mobile dating sites like Grindr and Tinder. Andrew Weinreich may not be the household name that Mark Zuckerberg is, but without his visionary work, the realm of social networking could be quite different than we know it today.

References

Banks, M.A. (2007, January 1). The Internet, ARPANet, and consumer online. *All Business.* Retrieved from http://www.allbusiness.com/media-telecommunications/Internet-www/10555321-1.html.

Beaumont, C. (2008, November 25). The team behind Twitter: Jack Dorsey, Biz Stone and Evan Williams. *Telegraph.* Retrieved from http://www.telegraph.co.uk/technology/3520024/The-team-behind-Twitter-Jack-Dorsey-Biz-Stone-and-Evan-Williams.html.

boyd, d.m. & Ellison, N.B. (2008). Social networking sites: Definition, history, and scholarship. *Journal of Computer-Mediated Communication, 13,* 210-230.

Bulik, B. S. (2014, March 5). Ellen Degeneres' Samsung selfie ups social-marketing game. *Advertising Age.* Retrieved from http://adage.com/article/media/ellen-degeneres-samsung-selfie-ups-social-marketing-game/291989/.

BusinessWeek. (2005, July 29). MySpace: WhoseSpace?. *BusinessWeek.* Retrieved from http://www.businessweek.com/technology/content/jul2005/tc20050729_0719_tc057.htm.

Caslon Analytics. (2006). Caslon Analytics social networking services. *Caslon Analytics.* Retrieved from http://www.caslon.com.au/socialspacesprofile2.htm.

Chmielewski, D. C. (2013, June 9). Nielsen study: Social networking dominates smartphone, tablet use. *The Los Angeles Times*. Retrieved from http://articles.latimes.com/2013/jun/09/entertainment/la-et-ct-nielsen-study-social-networking-smartphone-tablet-20130609

Colao, J. J. (2012, November 27). Snapchat: The biggest no-revenue mobile app since Instagram. *Forbes*. Retrieved from http://www.forbes.com/sites/jjcolao/2012/11/27/snapchat-the-biggest-no-revenue-mobile-app-since-instagram/.

Constine, J. (2013, August 13). Facebook reveals 78% of US users are mobile as it starts sharing user counts by country. *TechCrunch*. Retrieved from http://techcrunch.com/2013/08/13/facebook-mobile-user-count/.

Costill, A. (2014, February 28). 25 things you didn't know about Snapchat. *Search Engine Journal*. Retrieved from http://www.searchenginejournal.com/25-things-didnt-know-snapchat/91169/.

Crook, J. (2014, January 23). One year in, Vine's battle has just begun. *Tech Crunch*. Retrieved from http://techcrunch.com/2014/01/23/one-year-in-vines-battle-has-just-begun/.

Crunch Base. (2014, January 29). Vine. *Crunch Base*. Retrieved from http://www.crunchbase.com/company/vine.

Dave, P. (2013, June 20). Video app Vine's popularity is spreading, six seconds at a time. *The Los Angeles Times*. Retrieved from http://articles.latimes.com/2013/jun/20/business/la-fi-vine-20130620.

Davidson, S. (2012, March 29). Fitness apps: Fitocracy social media site goes mobile with new—free—iPhone app. *PT 365*. Retrieved from http://blogs.militarytimes.com/pt365/2012/03/29/fitness-apps-fitocracy-social-media-site-goes-mobile-with-new-free-iphone-app/.

Dsouza, K. (2010, March 14). Facebook status update has 420 character limit too. *Techie Buzz*. Retrieved from http://techie-buzz.com/social-networking/facebook-has-a-420-character-status-update-limit-too.html.

Duggan, M., & Smith, A. (2014). Social media update 2013. Pew Research Center. Retrieved from http://pewinternet.org/Reports/2013/Social-Media-Update.aspx.

Experian Marketing Services. (2013, April 16). Experian marketing services reveals 27 percent of time spent online is on social networking. *Experian*. Retrieved from http://press.experian.com/United-States/Press-Release/experian-marketing-services-reveals-27-percent-of-time-spent-online-is-on-social-networking.aspx.

Facebook. (2014). Newsroom timeline. Retrieved from http://newsroom.fb.com/content/default.aspx?NewsAreaId=20.

Facebook. (2012). Newsroom fact sheet. Facebook. Retrieved from http://newsroom.fb.com/content/default.aspx?NewsAreaId=22.

Facebook Statistics. (2014, January 1). Facebook statistics. *Statistic Brain*. Retrieved from http://www.statisticbrain.com/facebook-statistics/.

FollowFamous. (2009). Find famous celebrities on Twitter. *Follow Famous*. Retrieved from http://www.followfamous.com/.

Gannes, L. (2012, October 29). Fast-growing photo-messaging app Snapchat launches on Android. *All Things D*. Retrieved from http://allthingsd.com/20121029/fast-growing-photo-messaging-app-snapchat-launches-on-android/.

Google+. (n.d.). Overview. *Google+*. Retrieved from http://www.google.com/+/learnmore/.

Graham, J. (2013, December 17). Apple reveals top app downloads of 2013: Candy wins. *USA Today*. Retrieved from http://www.usatoday.com/story/tech/2013/12/17/top-apple-downloads-of-the-year/4042057/.

Hof, R. (2013, September 8). So much for Facebook ruining Instagram—It just hit 150 million monthly active users. *Forbes*. Retrieved from http://www.forbes.com/sites/roberthof/2013/09/08/so-much-for-facebook-ruining-instagram-it-just-hit-150-million-monthly-active-users/.

Introducing hashtags on Instagram (n.d.). Instagram. Retrieved from http://blog.instagram.com/post/8755963247/introducing-hashtags-on-instagram.

Issac, M. (2013, October 31). About those Google+ numbers… All Things D. Retrieved from http://allthingsd.com/20131031/about-those-google-user-numbers/.

Langer, A. (2013, June 15). Six things you didn't know about the Vine app. *Yahoo! Finance*. Retrieved from http://finance.yahoo.com/news/six-things-didnt-know-vine-192222105.html.

LinkedIn. (2014). Company history. *LinkedIn*. Retrieved from http://press.linkedin.com/about.

Marrouat, C. (2013, November 3). Facebook and LinkedIn release new traffic numbers. *Social Media Slant*. Retrieved from http://socialmediaslant.com/facebook-linkedin-stats-q3-2013/.

Mayfield, R. (2007, March 10). Twitter tips the tuna. *Ross Mayfield's Weblog*. Retrieved from http://ross.typepad.com/blog/2007/03/twitter_tips_th.html.

Milian, M. (2009, July 22). Friendster founder on social networking: I invented this stuff (updated). *The Los Angeles Times*. Retrieved from http://latimesblogs.latimes.com/technology/2009/07/friendster-jonathan-abrams.html.

Nickson, C. (2009, January 21). The history of social networking. *Digital Trends*. Retrieved from http://www.digitaltrends.com/features/the-history-of-social-networking/.

Plymale, S. (2012, May 26). A forefather of social media: Andrew Weinreich and SixDegrees.com. *Eastern Michigan University Public Relations Student Society of America*. Retrieved from http://emuprssa.com/2012/05/26/a-forefather-of-social-media-andrew-weinreich-and-sixdegrees-com/.

Protalinski, E. (2011, November 30). Facebook increases status update character limit to 63,206. *ZDNet*. Retrieved from http://www.zdnet.com/blog/facebook/facebook-increases-status-update-character-limit-to-63206/5754.

Risen, T. (2014, January 27). Don't predict Facebook's decline yet. *US News*. Retrieved from http://www.usnews.com/news/articles/2014/01/27/dont-predict-facebooks-decline-yet.

Rusli, E. M., & MacMillan, D. (2013, November 13). Snapchat spurned $3 billion acquisition offer from Facebook. *The Wall Street Journal*. Retrieved from http://blogs.wsj.com/digits/2013/11/13/snapchat-spurned-3-billion-acquisition-offer-from-facebook/?mg=blogs-wsj&url=http%253A%252F%252Fblogs.wsj.com%252Fdigits%252F2013%252F11%252F13%252Fsnapchat-spurned-3-billion-acquisition-offer-from-facebook.

Schulten, K. (2013, October 23). Why do we take selfies? *The New York Times*. Retrieved from http://learning.blogs.nytimes.com/2013/10/23/why-do-we-take-selfies/.

Shefrin, H. (2013, November 8). Why Twitter's IPO was really a failure. *Forbes*. Retrieved from http://www.forbes.com/sites/hershshefrin/2013/11/08/why-twitters-ipo-was-really-a-failure/.

Smith, C. (2013, October 6). By the numbers: 13 MySpace stats and facts then and now. *Digital Market Ramblings*. Retrieved from http://expandedramblings.com/index.php/myspace-stats-then-now/#.UxPX7NxblCo.

Smith, C. (2014, January 18). By the numbers: 50 amazing Pinterest stats. *Digital Market Ramblings*. Retrieved from http://expandedramblings.com/index.php/pinterest-qstats/#.UxPS3NxblCo.

Stinson, L. (2014, February 20). Fantastic infographics, drawn from a study of Instagram selfies. *Wired*. Retrieved from http://www.wired.com/design/2014/02/explore-world-selfies-new-data-visualization-tool/.

Strassburg, J., & Bunge, J. (2012, May 18). Social network's debut on Nasdaq interrupted by technical glitches, trader confusion. *The Wall Street Journal*. Retrieved from http://online.wsj.com/news/articles/SB10001424052702303448404577412251723815184?mod=googlenews_wsj&mg=reno64-wsj&url=http%3A%2F%2Fonline.wsj.com%2Farticle%2FSB10001424052702303448404577412251723815184.html%3Fmod%3Dgooglenews_wsj.

Stross, R. (2012, January 7). Sifting the professional from the personal. *The New York Times*. Retrieved from http://www.nytimes.com/2012/01/08/business/branchout-and-beknown-vie-for-linkedins-reach.html?_r=2.

Tsukayama, H. (2012, June 15). Facebook moves to consolidate cases. *The Washington Post*. Retrieved from http://www.washingtonpost.com/business/technology/facebook-moves-to-consolidate-cases/2012/06/15/gJQAry7SfV_story.html.

Twitter. (2014). About. *Twitter*. Retrieved from https://twitter.com/about.

Upbin, B. (2012, April 9). Facebook buys Instagram for $1 billion. Smart arbitrage. *Forbes*. Retrieved from http://www.forbes.com/sites/bruceupbin/2012/04/09/facebook-buys-instagram-for-1-billion-wheres-the-revenue/.

Zukerman, E. (2011, December 16). Why Google+ will overtake Facebook in 2 years or less (opinion). Make Use Of. Retrieved from http://www.makeuseof.com/tag/google-overtake-facebook-2-years-opinion/.

V

Conclusions

24

Other New Technologies

Jennifer H. Meadows, Ph.D.*

Introduction

Every two years, the editors of this book struggle to decide which chapters to include. There are always more options than there is space because new communication technologies develop almost daily. This chapter out-lines some of these technologies deemed important but not ready for a full chapter. The 15th edition of this book may cover these technologies in full chapters just as digital signage was once featured in an "Other New Technologies" chapter. Among the choices, two technologies stand out as exemplars of the newest technologies: wearable technologies and 3D printing

Wearable Technologies

A fast growing segment of communication technologies is wearable technologies. While wearable technologies such as headphones have been around for some time, the current definition of wearable technologies includes five major areas: fitness and wellness, healthcare and medical, infotainment, military, and industrial (IHS, 2013). IHS (2013) defines wearable technologies as "...products that must be worn on the user's body for an extended period of time, significantly enhancing the user's experience as a result of the product being worn. Furthermore, it must contain advanced circuitry, wireless connectivity and at least a minimal level of independent processing capability." n the consumer market these technologies are often found in three areas: fitness, glasses, and smartwatches.

* Professor and Chair, Department of Communication Arts, California State University, Chico (Chico, California)

Fitness

Wearable technologies for fitness are often wristbands that people wear to track activity, sleep, heart rate, distance traveled, calories and other metrics. Companies Fitbit, Jawbone and Garmin make popular fitness bands. Fitbit, for example, offers the Fitbit Flex, Force, One, and Zip. The Flex, Force and One are all bands that users wear that track activity and sleep. The Zip only tracks activity. Wearers of the Fitbit Flex can measure steps taken, calories burned, distance traveled, active minutes, hours slept and quality of sleep. This data can be synched with an online report where users can earn badges for reaching certain goals and set up competitions with other Fitbit users (Fitbit, n.d.). Fitbit took a hit in 2014 when it was found that their new Fitbit Force caused skin rashes on some users. The company recalled the product and plans to re-release it once the problem is solved (Rowgowsky, 2014).

Figure 24.1

Jawbone UP24 Fitness Band

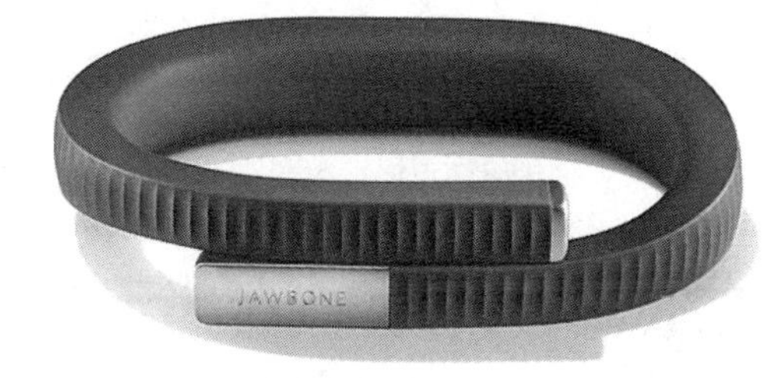

Source: Jawbone

Each band on the market offers slightly different features. The Jawbone UP24 and its associated smartphone app lets users track sleep, food and drink, workouts, patterns in daily activities, and moods. Users can set idle alerts to remind them to get up and move after a period of inactivity (Jawbone, 2014).

Garmin has been making satellite watches for runners, swimmers and cyclists for years. These watches track pace, distance, cadence, altitude, splits and more. Users can connect the devices to upload their data and keep track of workouts, share routes, and even complete with other runners. People usually wear these watches only for workouts, not all the time. Garmin introduced its own fitness band in 2014 called the Vivofit designed to be worn all the time. Similar to the other bands mentioned above, the Vivofit tracks steps, distance, calories, and time. It also connects to goals and reminds the user to move during time of inactivity (Vivofit, n.d.).

All of these bands come in a variety of colors and price points, although the full-featured bands tend to cost in the $99 to $130 range. There is even now fashion advice on what clothing goes with a particular band and which bands go with different personal styles (Rossman, 2014). The market for fitness bands is expected to grow. A forecast by Canalys reports that 2014 shipments will be 17 million, growing to 23 million in 2015 and 43 million in 2017 (Lomas, 2014). The 2013-2014 market leader in the United States is Fitbit with over 50% of the market. Jawbone is second with about 20% and Nike is a distant third with 13%. The Nike FuelBand measures activity with users earning Nike Fuel points. Nike has indicated that will be abandoning the Fuelband—speculation is that they are working with Apple on an iWatch (Shaughnessy, 2014).

Smartwatches

Ever since Dick Tracy communicated through his wristwatch, the idea of a smartwatch has been a technological dream for many. That dream has become reality with companies including Pebble, Samsung, and Sony selling smartwatches that communicate with properly-equipped smartphones, take pictures, access the Internet and of course, tell time. Smartwatchnews (n.d.) defines a smartwatch as "a wearable computing device in the form of a computerized wristwatch. Early smartwatches had limited functionality and were able to perform elementary calculations and basic data storage and retrieval. Modern devices typically possess enhanced data processing functionality similar to a smartphone or tablet and are able to run mobile apps. Modern smartwatches also include specialty features such as biometric monitoring, GPS and mapping capability, and can even function independently as a smartphone (smartwatchnews, n.d.)

Pebble has been making smartwatches since 2013 and is considered a market leader. One of the hottest smartwatches in mid 2014, the Pebble Steel, was launched at the Consumer Electronics Show in 2014 and comes with either a stainless steel or leather band. Costing $249, this smartwatch notifies users of incoming calls, texts, and email. The Steel allows users to control music on a smartphone and integrates with popular apps such as Yelp, Runkeeper, Twitter, and Pandora (Discover Pebble, n.d.).

Figure 24.2

Pebble Steel Smart Watch

Source: Pebble

Samsung smartwatches are probably the best known because of their ubiquitous television commercials. The Galaxy Gear, Gear 2, and Gear Neo are three smart watches that work with Samsung smartphones. The original smartwatch from Samsung, the Galaxy Gear, only worked a few Samsung phones and used an Android operating system. The Gear 2 and Neo, released in 2014 switch to a Tizen operating system. The new watches also work with a wider variety of devices including Samsung's smartphones and tablets. The Gear 2 does everything the original Gear did and now offers music player, 2 mega-pixel camera and a heart rate sensor (Dolcourt, 2014).

Other notable smartwatches include the Sony Smartwatch 2 SW2, I'm Watch Smartwatch, and the Martian Smartwatch. According to Next Market Research, smartwatch shipments are expected to reach 15 million in 2014 while BI Intelligence reports that they expect sales of 91.8 million smart watches worldwide by 2018 (Danova, 2014). As of mid-2014 Samsung has the greatest market share with about 30% of the market. This percentage can be misleading because these reports also include fitness watches like the Garmin 410 and the Nike SportWatch.

Smart Glasses

Smart glasses are still fairly rare as of mid 2014. The biggest name in smart glasses is Google Glass but other smart glasses are offered by companies such as Epsom and Vuzix. Smart glasses can be defined as glasses with lenses that allow for adjustable high definition display of data either stored in the glasses or accessed over the Internet via Bluetooth or Wi-Fi. Smart glasses often have other features such as cameras and voice recognition.

Google initially limited availability of its Google Glass product, limiting sales to Glass Explorers who applied to purchase the device. But Google made Google Glass available to the public for one day on April 15, 2014. People could go online and buy a pair for $1,500. Google Glass has a small adjustable lens that displays transparent and semi-transparent data. Glass has a camera and voice operated controls. Uses for Glass have been wide ranging from displaying building plans for firefighters, anatomy information for doctors, and even recipes for chefs.

The introduction of smart glasses has necessitated a need to revisit regulations and policies pertaining to technology use. Already states and municipalities are looking to ban smart glasses use while driving. Driving with smart glasses is illegal in Canada. Google has also found that many people are uncomfortable with Glass wearers and has even issued some etiquette guidelines for wearers. Others have offered these guides too such as Kevin Simtunguang of the *Wall Street Journal*. His rules include "always remember you have a camera on your head," "use voice commands only when you need to," "don't use Glass to make phone calls in public," and my favorite, "don't be creepy" (Simtunguang, 2013). Several businesses have banned the use of smart glasses to protect the privacy of other customers. This is because it is very easy to take a picture or video without anyone noticing. With Google Glass, all you need to do is touch the side of the frame to start or stop a recording.

Smart glasses, as creepy as they can be, are poised for market growth. Jupiter Research forecasts that sales will rise from 87,000 in 2013 to 10 million by 2018. Google Glass won't be the only choice, either. VentureBeat reports that at least 16 different smart glasses were expected to come onto the market in the between 2014 and 2015 (Takahashi, 2014).

3D Printing

Do you want a new pair of glasses? Soon enough there may be no need to order them because you can just print them from home using a 3D printer. Everyone is familiar with 2D printing—printing out pictures and papers and other two dimensional objects. 3D printing adds another dimension, allowing the creation of three-dimensional objects—from small machinery parts to jewelry, food, homes, guns, and even human body parts.

While it seems as if 3D printing just appeared on the scene it has been around since the mid-80's when it was developed by Charles Hull. 3D printing uses computer aided designs (CAD) to create three dimensional objects using an additive process. The model design is created using CAD software and then saved as a stereolithography (STL) file, which is then sent to a printer. The printer creates the object layer by layer using different materials such as plastic, metal, or even chocolate.

There are three different methods for 3D printing, SLS (selective laser sintering), FDM (fused deposition modeling) and SLA (stereolithography). SLS uses a laser to heat a fuse a powered material (Selective Lasering, n.d.). FDM printers heat a thermoplastic material into a semi-liquid state and then extrude the material layer by layer (FDM, n.d.). Finally, SLA uses a high powered light source that hardens a cross section of light-sensitive liquid plastic, building the object from the bottom up (Flaherty, 2012).

3D printing has a variety of uses across multiple industries. Printed objects range from tiny medical parts to entire buildings. Often 3D printed models are created as prototypes allowing a manufacturer to see if an object is going to function properly before it is manufactured. 3D printing itself is also used to create large quantities of objects. Companies including Nike use 3D printing for football cleats (Stinson, 2014) and Boeing uses 3D printing to make metal airplane parts (Hagerty & Linebaugh, 2014).

One of the most innovative uses for 3D printing is the creation of customized products for the medical and dental fields. This type of customized production makes for better fitting prosthetics for amputees and denture and crowns for dental patients. 3D printed jawbones and skulls have been successfully implanted in patients (Medical, n.d.). 3D printing is even being used to create living tissue. Oganovo and Wake Forest Institute for Regenerative Medicine use a type of 3D printing called bioprinting, to create bone, skeletal muscle, lung, and cardiac tissues (Craig, 2014).

One fun use for 3D printers is making food. Hershey is working with 3D Systems to create a 3D printer for chocolate (Hargreaves, 2014). 3D printer company Cubify offers designs for gummy candy, and the ChefJet and ChefJet Pro from 3D Systems are 3D printers designed to make edible objects in flavors ranging from chocolate to watermelon in a professional kitchen. The Pro model adds color to models (3D Systems sweetens, 2014). NASA has even funded a 3D printer that makes pizza for astronauts in space (Sandhana, 2014).

Figure 24.3

Cube 3D Printer

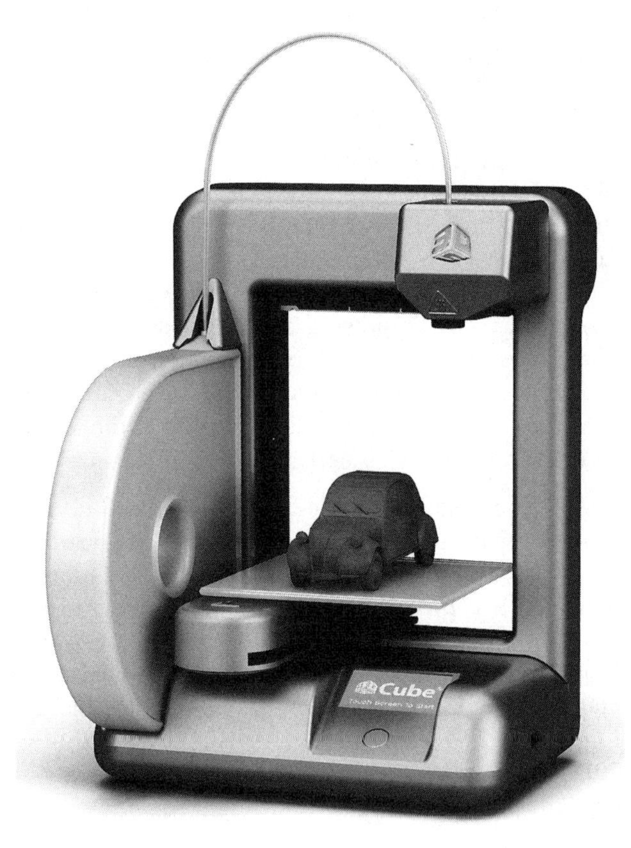

Source: Cubify

3D printing isn't just for big manufacturers, or even small offices now. The easy-to-use home 3D printer hit the market in 2014. Home 3D printers in the past were most likely owned by technology hobbyists who made their own printer or used kits that can be easily ordered online, but ready-made consumer 3D printers are now available. Companies that offer the printers include MakerBot and Cubify (part of 3D Systems). The MakerBot Replicator Mini costs $1,375 and is being marketed for the home and school. The printer offers one touch printing and comes with design software to allow users to design their own model or purchase designs from MakerBot. There are also a large number of free designs (MakerBot Replicator, 2014). The Cube 2 from Cubify costs $1,200 and comes in 5 different colors. The printer uses material cartridges that come in 16 different colors including glow in the dark, and the printer comes with 25 designs. Users can make objects from cooking utensils to toys and shoes (Buy a, 2014). As mentioned earlier, kits are a popular way to bring a 3D printer into the home, available on various websites including 3dstuffmaker.com and ultimaker.com.

There are some legal issues surrounding the growing use of 3D printing. One, of course, is copyright. With the right software and a 3D printer, any person or company can replicate any object. The intellectual property of physical design objects is going to become increasingly important.

Another legal area surrounds creating and using guns that have been made using 3D printing. Texan Cody Wilson designed a working 3D printed plastic gun called the Liberator and shared the CAD design. The Federal government, states and municipalities responded quickly, drafting and extending legislation regulating these guns. It is illegal to possess a 3D printed gun in Philadelphia. At the federal level, the Undetectable Firearms Act was extended in 2013 and the law clearly states that guns must be detectible in a metal detector or x-ray machine (Wagstaff, 2013).

The 3D printer market is expected to grow from $2 billion in 2104 to $6 billion in 2018 (Dorrier, 2014). This level of growth is slower than initially forecast. One reason is that consumers aren't ready for the technology. 3D printers are still in the early adopter stage. Also full scale adoption by industry hasn't taken off yet (Jakab, 2014).

Conclusion

This chapter briefly reviewed two of the communication technologies that we didn't get to fully cover in this edition. By the time the 15th edition of this book is published, there will be new communication technologies that haven't even been developed at the time this book is published in mid-2014.

Bibliography

3D Systems sweetens its offering with ChefJet 3D printer series. (2014). 3D Systems. Retrieved from http://www.3dsystems.com/press-releases/3d-systems-sweetens-its-offering-new-chefjettm-3d-printer-series.

Buy a Cube (2014) Cubify. Retrieved from http://cubify.com/en/Products/Cube2Order.

Craig, M. (2014). Could 3D printed organs be the future of medicine? *Forbes*. Retrieved from http://www.forbes.com/sites/ptc/2014/03/31/could-3-d-printed-organs-be-the-future-of-medicine/.

Davona, T. (2014). Smart Watch Forecast: Smart watches will bring wearable computing to the masses. *Business Insider*. Retrieved from http://www.businessinsider.com/global-smartwatch-sales-set-to-explode-2014-3#!HTT5I.

Discover Pebble, n.d. (2014). Pebble. Retrieved from https://getpebble.com/discover.

Dolcourt, J. (2014). Tizen based Samsung Gear 2 ditches Android adds music player (hands on). Cnet. Retrieved from http://www.cnet.com/products/samsung-gear-2/.

Dorrier, J. (2014). Beyond the hype and hope of 3D printers: what consumers should expect. Singularity Hub. Retrieved from http://singularityhub.com/2014/04/29/beyond-the-hype-and-hope-of-3d-printing-what-consumers-should-expect/.

FDM Technology (2014). Stratasys. Retrieved from http://www.stratasys.com/3d-printers/technology/fdm-technology.

Fitbit (n.d.) Fitbit. Retrieved from http://www.fitbit.com/.

Flaherty, J. (2012). 3D printers that don't suck. *Wired*. Retrieved from http://www.wired.com/2012/07/3-d-printers-that-dont-suck/.

Hagerty, J. and Linebaugh, K. (2014). Next 3-D frontier: printed plane parts. *The Wall Street Journal*. Retrieved from http://online.wsj.com/news/articles/SB10001424052702303933404577505080296858896.

Hargreaves, S. (2014). Hershey's to make 3D chocolate printer. CNN. Retrieved from http://money.cnn.com/2014/01/16/technology/3d-printer-chocolate/.

IHS Electronics and Media (2013). Wearable Technology Market Assessment. Retrieved from http://www.ihs.com/pdfs/Wearable-Technology-sep-2013.pdf.

Jakab, S. (2014). 3D Systems prints everything but money. *The Wall Street Journal*. Retrieved from http://online.wsj.com/news/articles/SB10001424052702304071004579409351529211822?KEYWORDS=3D+printing&mg=reno64-wsj.

Jawbone (n.d.) Up by Jawbone. Retrieved from https://jawbone.com/up.

Lomas, N. (2014). Wearables market heating up, with more than 17m bands forecast to ship this year says Canalys. Techcrunch. Retrieved from http://techcrunch.com/2014/02/12/wearables-market-heating-up/.

MakerBot Replicator Mini (2014). MakerBot. Retrieved from http://store.makerbot.com/replicator-mini.

Medical (n.d.) Stratasys. Retrieved from http://www.stratasys.com/industries/medical.

RepRap Options (n.d.) RepRap. Retrieved from http://reprap.org/wiki/RepRap_Options.

Rossman, K. (2014, March 21). Fitness trackers get stylish. *The Wall Street Journal*. Retrieved from http://online.wsj.com/news/articles/SB10001424052702304747404579445872844666550.

Rowgowsky, M. (2014). Fitbit Force recall is bad news for the company and wearable tech, but is it necessary? *Forbes*. Retrieved from http://www.forbes.com/sites/markrogowsky/2014/03/13/fitbit-force-recall-is-bad-for-the-company-and-wearable-tech-but-is-it-necessary/.

Sandhana, L. (2014). 3D printed pizza—a quick and easy meal for astronauts? Gizmag. Retrieved from http://www.gizmag.com/3d-printed-pizza-astronauts/30685/.

Selective Laser Sintering (n.d.). Solid Concepts. Retrieved from http://www.solidconcepts.com/technologies/selective-laser-sintering-sls/.

Shaughnessy, H. (2014). Nike FuelBand reveals truth of competitive advantage. *Forbes*. Retrieved from http://www.forbes.com/sites/haydnshaughnessy/2014/04/26/nike-fuelband-reveals-new-source-of-competitive-advantage/.

Simtunguang, Kevin (2013, May 3) Google Glass: An Etiquette Guide. *The Wall Street Journal*. Retrieved from http://online.wsj.com/news/articles/SB10001424127887323982704578453031054200120.

Stinson, L. (2014). For Super Bowl Nike uses 3D printing to create a faster football cleat. *Wired*. Retrieved from http://www.wired.com/2014/01/nike-designed-fastest-cleat-history/.

Takahashi, D. (2014). You'll soon have 16+ different smart glasses to choose from. Here's how to pick the right one. VentureBeat. Retrieved from http://venturebeat.com/2014/03/30/which-smart-glasses-will-be-right-for-you/.

Vivofit (n.d.) Garmin. Retrieved from http://sites.garmin.com/vivo/.

Wagstaff, K. (2013). Despite plastic gun ban, 3D printed firearms still have a future. NBC News. Retrieved from http://www.nbcnews.com/tech/tech-news/despite-plastic-gun-ban-3-d-printed-firearms-still-have-f2D11718212.

25

Your Future & Communication Technologies

August E. Grant, Ph.D.*

This book has introduced you to a range of ideas on how to study communication technologies, given you the history of communication technologies, and detailed the latest developments in about two dozen technologies. Along the way, the authors have told stories about successes and failures, legal battles and regulatory limitations, and changes in lifestyle for the end user.

So what can you do with this information? If you're entrepreneurial, you can use it to figure out how to get rich. If you're academically inclined, you can use it to inform research and analysis of the next generation of communication technology. If you're planning a career in the media industries, you can use it to help choose the organizations where you will work, or to find new opportunities for your employer or for yourself.

More importantly, whether you are in any of those groups or not, you are going to be surrounded by new media for the rest of your life. The cycle of innovation, introduction, and maturity of media almost always includes a cycle of decline as well. As new communication technologies are introduced and older ones disappear, your media use habits will change. What you've learned from this book should help you make decisions on when to adopt a new technology or drop an old one. Of course, those decisions depend upon your personal goals—which might be to be an innovator, to make the most efficient use of your personal resources (time and money), or to have the most relaxing life style.

This chapter explores a few ways you can apply this information to improve your ability to use and understand these technologies—or simply to profit from them.

Making Money from New Technologies

You have the potential to get rich from the next generation of technologies. Just conduct an analysis of a few emerging technologies using the tools in the "Fundamentals" section of the book; choose the one that has the best potential to meet an unmet demand (one that people will pay for); then create a business plan that demonstrates how your revenues will exceed your expenses from creating, producing, or distributing the technology.

Regardless of the industry you want to work in, there are five simple steps to creating a business plan:

1) **Idea:** Every new business starts with an idea for a new product or service, or a new way to provide or enhance an existing business model. Many people assume that having a good idea is the most important part of a business plan, but nothing could be further from the truth. The idea is just the first step in the process.

2) **Team:** Bringing together the right group of people to execute the idea is the next step in the process. The most important consideration in

* J. Rion McKissick Professor of Journalism, School of Journalism and Mass Communications, University of South Carolina (Columbia, South Carolina).

putting together a team is creating a list of competencies needed in your organization, and then making sure the management team includes one expert in each area. Make sure that you include experts in marketing and sales as well as production and engineering. The biggest misstep in creating a team is bringing together a group of people who have the same background and similar experiences. Once you've identified the key members of the team, you can create the organizational structure, identifying the departments and divisions within each department (if applicable) and providing descriptions for each department and responsibilities for each department head. It is also critical to estimate the number of employees in each position.

3) **Competitive Analysis:** The next step is describing your new company's primary competition, detailing the current strengths and weaknesses of competing companies. In the process, this analysis needs to identify the challenges and opportunities they present for your company.

4) **Finance:** The most challenging part of most business plans is the creation of a spreadsheet that details how and where revenues will be generated, including projections for revenues during the first few years of operation. These revenues are balanced in the spreadsheet with estimates of the expected expenses (e.g., personnel, technology, facilities, and marketing). This part of your business plan also has to detail how the initial "start-up" costs will be financed. Some entrepreneurs choose to start with personal loans and assets, growing a business slowly and using initial revenues to finance growth, a process known as "bootstrapping." Others choose to attract financing from venture capitalists and angel investors who receive a substantial ownership stake in the company in return for their investment. The bootstrapping model allows an entrepreneur to keep control—and a greater share of profits—but the cost is much slower growth and more limited financial resources

5) **Marketing Plan:** The final step is creating a marketing plan for the new company that discusses the target market and how you will let those in the target know about your product or service. The marketing plan must identify the marketing and promotional strategies that will be utilized to attract consumers/audiences/users to your product or service. Don't forget to discuss how both traditional and new media will be employed to enhance the visibility and use of your product, then follow through with a plan for how the product or service will be distributed and priced.

Conceptually, this five-step process is deceptively easy. The difficult part is putting in the hours needed to plan for every contingency, solve problems as they crop up (or before they do), make the contacts you need in order to bring in all of the pieces to make your plan work, and then distribute the product or service to the end users. If the lessons in this book are any indication, two factors will be more important than all the others: the interpersonal relationships that lead to organizational connections—and a lot of luck!

Here are a few guidelines distilled from 30-plus years of working, studying, and consulting in the communication technology industries that might help you become an entrepreneur:

- **Ideas are not as important as execution.** If you have a good idea, chances are others will have the same idea. The ones who succeed are the ones who have the tools and vision to put the ideas into action.
- **Protect your ideas**. The time and effort needed to get a patent, copyright, or even a simple nondisclosure agreement will pay off handsomely if your ideas succeed.
- **There is no substitute for hard work.** Entrepreneurs don't work 40-hour weeks, and they always have a tool nearby to record ideas or keep track of contacts.
- **There is no substitute for time away from work.** Taking one day a week away from the job gives you perspective, letting you step back and see the big picture. Plus, some of the best ideas come from bringing in completely unrelated content, so make sure you are always scanning the world around you for developments in the arts, technology, business, regulation, and culture.
- **Who you know is more important than what you know**. You can't succeed as a solo act in the communication technology field. You have to a) find and partner with or hire people who are better than you in the skill sets you don't have, and

b) make contacts with people in organizations that can help your business succeed.

- **Keep learning**. Study your field, but also study the world. The technologies that you will be working with have the potential to provide you access to more information than any entrepreneur in the past has had. Use the tools to continue growing.
- **Create a set of realistic goals**. Don't limit yourself to just one goal, but don't have too many. As you achieve your goals, take time to celebrate your success.
- **Give back**. You can't be a success without relying upon the efforts of those who came before you and those who helped you out along the way. The best way to pay it back is to pay it forward.

This list was created to help entrepreneurs, but it may be equally relevant to any type of career. Just as the communication technologies explored in this book have applications that permeate industries and institutions throughout society, the tools and techniques explored in this book can be equally useful regardless of where you are or where you are going.

The one constant you will encounter is change, and that change will be found most often at the organizational level and in the application of system-level factors that impact your business. As discussed in the first chapter of this book, new, basic technologies that have the potential to change the field come along infrequently, perhaps once every ten or twenty years. But new opportunities to apply and refine these technologies come along every day. The key is training yourself to spot those opportunities and committing yourself to the effort needed to make your ideas a success.

Index

To continue your search for insight into the technologies explored in this book, Communication Technology Update and Fundamentals, please visit the companion website:
www.tfi.com/ctu